KNOW
FORMAT
FOR
FINAL

**ADDISON-WESLEY PUBLISHING COMPANY**

Reading, Massachusetts • Menlo Park, California

London • Amsterdam • Don Mills, Ontario • Sydney

# PROBLEM
# SOLVING
# AND STRUCTURED
# PROGRAMMING
# IN
# FORTRAN

## Second Edition

**FRANK L. FRIEDMAN**
**ELLIOT B. KOFFMAN**
Temple University

This book is in the

**Addison-Wesley Series in**

**Computer Science and Information Processing**

**Library of Congress Cataloging in Publication Data**

Friedman, Frank L.
 Problem solving and structured programming in FORTRAN.

 Includes index.
 1. FORTRAN (Computer program language)
2. Structured programming. 3. Electronic digital computers—
Programming. I. Koffman, Elliot B., joint author. II. Title.
QA76.73.F25F74   1981      001.64'24      80-20943
ISBN 0-201-02461-6

Copyright © 1981 by Addison-Wesley Publishing Company, Inc. Philippines
      copyright 1981 by Addison-Wesley Publishing Company, Inc.

All rights reserved. No part of this publication may be reproduced, stored
in a retrieval system, or transmitted, in any form or by any means, electronic,
mechanical, photocopying, recording, or otherwise, without the prior written
permission of the publisher. Printed in the United States of America. Published
simultaneously in Canada. Library of Congress Catalog Card No. 80-20943

ISBN 0-201-02461-6
  CDEFGHIJK-DO-898765432

To our families
Martha, Shelley, and Dara Friedman
and
Caryn, Richard, Deborah, and Robin Koffman
who graciously accepted us as part-time
family members during the writing of this book

# PREFACE

This book is designed as a text for a one semester, introductory course in computer programming. No background other than high school algebra is assumed. The material presented reflects the authors' view that good problem solving and programming habits should be introduced at a very early stage in the development of a student's programming skills, and that they are best instilled by examples, by frequent practice, and through instructor-student interaction. Therefore we have concentrated on demonstrating problem solving and programming techniques through the use of more than 30 completely solved problems.

Discipline and planning in both problem solving and programming are illustrated in the text from the beginning. We have attempted to integrate a number of relatively new pedagogic ideas into a unique, well-structured format that is uniformly repeated for each problem discussed. Three basic phases of problem solving are emphasized: the analysis of the problem; the stepwise specification of the algorithm (using the flow diagrams); and finally, the language implementation of the program.

Our goal is to bridge the gap between textbooks that stress problem solving approaches divorced from implementation and language considerations and programming manuals that provide the opposite emphasis. Language independent problem analysis and algorithms are described in the same text as the language features required to implement the problem solution on the computer. For each new problem introduced in the text, the problem analysis and algorithm description are presented along with the complete syntactic and semantic definitions of the new language features convenient for the implementation of the algorithm.

The top-down or stepwise approach to problem solving is illustrated repeatedly in the solution of each of the problems solved in the text. Three pedagogic tools—a data definition table, a flow diagram, and a program system chart—are used to provide a framework through which students may practice the definition and documentation of program variables in parallel with the stepwise development of algorithms.

The data definition table provides a description of the attributes (initial values, types, sizes, etc.) of each variable or parameter appearing in the problem solution. The flow diagrams that are used to represent the algorithms are similar to the D-Charts of Dijkstra. Each diagram is a short sequence of individual flow diagram patterns representing the subtasks of an algorithm; refinements of the subtasks are diagrammed separately.

The main goal of this revision of *Problem Solving and Structured Programming in FORTRAN* is to incorporate the new features of FORTRAN 77 into the text. This process has been greatly simplified by the fact that the original text already reflected the new pedagogy of the "structured programming era"; it also described many of the features of FORTRAN 77 (e.g. list-directed input/output, the character data type, and the IF-THEN-ELSE) that were generally available as extensions to FORTRAN IV compilers.

The text now corresponds to the FORTRAN 77 standard with one important exception: we have retained the WHILE loop because we feel it is a very useful pedogogical tool. It provides a natural way to express general loop structures (not involving a counter), and is available in at least three compilers: M77 (University of Minnesota), WATFIV (University of Waterloo, Ontario, Canada) and the Digital Equipment Corporation VAX-11 compiler (DO-WHILE). Students using a compiler that does not support the WHILE loop may easily implement it using GO TO's, as is demonstrated in the text.

We have tried to emphasize programming style and the stepwise approach to algorithm development a little more rigorously than in the first edition. Program Style displays have been inserted throughout the text to explain and discuss techniques of good programming style. Separate diagrams are used for the refinements of algorithm steps rather than providing all refinements in one diagram. We have also been careful to add program variables to a data table only as the need for them arises during algorithm development.

The text is organized so that students may begin working with the computer as soon as possible. Chapter 1 contains a short discussion of the basic computer hardware components and an introduction to machine and high level languages and the role of a compiler. This is followed by a description of some basic FORTRAN statements that perform simple input and output operations, the arithmetic operations of addition, subtraction, multiplication, and division, and the STOP. A sample problem is introduced and its solution is presented in terms of the basic FORTRAN statements. Students are provided with a glimpse of the FORTRAN language and with sufficient information about batch and timesharing systems to enable them to write short programs and run them on the computer. We suggest that this latter material be covered early in the chapter so that students can begin running programs and getting firsthand experience that will complement the discussions in the text.

The stepwise approach to programming is introduced in Chapter 2. The data definition table and flow diagram are described, and problems involving decisions and loops are examined and solved. The algorithms are represented using special flow diagram patterns for decision and loop steps.

The standard IF-THEN-ELSE and nonstandard WHILE structures are informally introduced in this chapter, and several of the solved problems are written in FORTRAN to provide illustrations of these structures. We believe it is best to introduce these structures as soon as possible, so that students may continue using the computer throughout Chapter 2. Students should be motivated early to select the best control structures to implement each flow diagram pattern.

The PARAMETER statement is also described in Chapter 2, and some discussion concerning the differences between program constants and variables is provided. Guidelines are given to aid the student in deciding whether to treat an item of information as a program variable, parameter or an in-line constant.

The IF-THEN-ELSE (limited form of the general block IF) and WHILE loop structures are formally introduced in Chapter 3, and several completely solved problems are presented to illustrate their use. The WHILE loop implementation in standard FORTRAN is also illustrated. The FORTRAN conditional and unconditional transfers (using the GO TO) are introduced for this purpose, but students are strongly encouraged to use them only to implement the WHILE loop if it is not supported. Both the nonstandard and standard WHILE loop implementations are illustrated throughout the text in each program that involves a WHILE loop.

A limited form of the standard FORTRAN DO loop is also introduced in Chapter 3 for convenient implementation of counting loops. A special section on program testing and debugging is included toward the end of the chapter.

In Chapter 4, the fundamental data types—real, integer, character and logical—are formally introduced. Since students have been working with both real and integer data since Chapter 1, much of this material will be review and they should concentrate on the study of logical and character data. The rules of evaluation of multiple-operator arithmetic statements are given, and some common library functions are described. The logical operations AND, OR and NOT are introduced.

The use of format descriptors for output is described at the end of Chapter 4. Here we have taken advantage of the new FORTRAN 77 feature that allows the specification of a format as a character string inserted directly into a READ or PRINT statement. It is possible to postpone studying this material until Chapter 8 if desired.

Arrays are introduced in Chapter 5, and several different techniques for using arrays are demonstrated in the problems solved in this chapter and in Chapter 6. Two types of array declarations are described: a restricted form (lower subscript bound of one, as per the old standard), and a general form (as described in the new standard) which is introduced toward the end of the chapter.

In Chapter 6, the block IF and DO loop structures are defined in their full generality, and numerous examples of their use are provided. Several somewhat more complicated examples requiring the use of nests of structures are illustrated. The use of the GO TO statement for loop exit and starting the next loop iteration is also discussed.

Chapter 7 concerns subprograms; the definition and use of functions and subroutines, argument lists and common blocks are presented. Module independence is stressed in this chapter and the program system chart is introduced as a means of representing the interrelationships and data flow among the modules of a program system. The development of a small program system is described in detail in order to illustrate the techniques of planning for the use of subprograms and for documenting and implementing them.

List-directed READ and PRINT statements are used throughout the first seven chapters in order to simplify input and output. This allows students to concentrate on mastering the concepts of algorithm development, control structure and data representation, and coding and debugging with a minimum of interference from the details of format specification. However, the rudiments of formats can be introduced conveniently as early as Chapter 4 if the instructor so desires. This may be especially important for handling output if your compiler does not provide fixed-width fields for list-directed output.

Details concerning the use of the basic format descriptors for input and output are provided in Chapter 8. The material usually appears less threatening to students who have had a liberal sprinkling of format usage prior to this chapter.

Chapter 8 also includes new material on the manipulation of sequential and direct access files. The new FORTRAN 77 OPEN and CLOSE statements are discussed, and the use of the input/output specifiers FMT, UNIT, REC, ACCESS, and END are illustrated.

Chapters 9 and 10 contain a number of moderately difficult problems illustrating the use of logical and character data (Chapter 9), and multidimensional arrays (Chapter 10). In Chapter 9, the new FORTRAN 77 character string concatenation operator is described along with several new character string functions.

It is the authors' intention to provide in this text sufficient material to accommodate the needs of a wide variety of students in a first semester computer programming course. The depth of understanding of the basic methods, as well as the proficiency demonstrated in the application of these methods, will vary according to the skill of the student and the expectations of the instructor. For the prospective computer science major, the introductory material on computers and the discussion of the role of the compiler are highly relevant and will provide a good foundation for future study.

In some cases, descriptions of computers and language processors that are more detailed than those provided in this text may be warranted. However, much of this detail is often provided in subsequent computer science courses and should be postponed in deference to the immediate goal: the presentation of methods of problem solving, algorithm specification, and computer program implementation. This is at least a one-semester task. The material concerning the computer and the language processor is important insofar as it serves to elucidate this presentation—not to complicate or add to it. At first, some of this material may not be fully appreciated by the student. However, we believe that it provides a necessary foundation for numerous subsequent discussions that will inevitably occur, if only in response to questions from students eager to understand how and why things work as they do.

## Acknowledgments

There are many people whose talents and influence are reflected in this text. We wish to thank Charles Hughes, Brian Kernighan and Charles Pfleeger, who carefully read and commented on the first edition of the textbook; and Thomas Byther, Enrique Gonzales, Jane Huerta, Bruce Martin, Tony O'Hare, Mary Beth Ruskai, and Harvey Shapiro who provided extensive reviews of the revised text. We are also indebted to Loren Meissner who unselfishly gave advice and helpful suggestions during all phases of manuscript preparation.

The errors that remain in the text are, of course, our own responsibility. The fact that they are not more numerous is due to the efforts of those mentioned above and to the efforts of our colleagues and students at Temple University who have used the manuscript.

To Anita Girton, Nancy Klein, Robert Solomon and Judy and Steve Stebulis we owe thanks for the numerous supporting tasks required in putting together the text. To Mary McCutcheon, a special note of thanks for her incredible diligence in converting our often unintelligible notes into a readable, typed manuscript.

Fran Palmer Fulton did an excellent job as production manager. She performed the multiple tasks of text design and copy-editing, and coordinated the artwork and typesetting for the final textbook. We are indebted to her for her effort in pulling the book together. We would also like to thank Bob Lambiase and Lisa Antonucci for their meticulous drawing of all art and diagrams in the text.

As always, it was a great pleasure working with Bill Gruener and Peter Gordon, the Addison-Wesley Computer Science Editors. Bill's persistance and dedication have been an inspiration to both of us, and his encouragement and support over the years have been invaluable. His confidence and guidance were instrumental in the initial development of this text.

Philadelphia                                        F.L.F.
January 1981                                      E.B.K.

# CONTENTS

# 3

## BASIC CONTROL STRUCTURES
### 81

# 4

## DATA TYPES
### 127

# 5

## ARRAYS
### 181

# 6

## ADVANCED CONTROL STRUCTURES
### 233

# 7
## SUBPROGRAMS
### 277

# 8
## FORMATS AND FILES
### 343

# 9
## LOGICAL AND CHARACTER STRING DATA
### 389

# 10
## MULTIDIMENSIONAL ARRAYS
### 435

# INTRODUCTION TO COMPUTERS AND PROGRAMMING

## 1.1    INTRODUCTION

In this chapter we describe the general organization of computers and discuss languages used for communicating with computers (programming languages). We will see that all computers consist of four basic components: memory, central processor, input devices and output devices; we will learn how information is represented in computer memory and how it is manipulated.

In the discussion of programming languages, two kinds of languages are described: machine language and high-level language. One high-level language, FORTRAN, is introduced, and examples of how to specify some basic computer operations in FORTRAN are given.

## 1.2    COMPUTER ORGANIZATION

### 1.2.1    Components of a Computer

A computer is a tool for representing and manipulating information. There are many different kinds of computers, ranging in size from hand-held calculators to large and complex computing systems filling several rooms or entire buildings. In the recent past, computers were so large and expensive that they could be used only for large-scale business or scientific computations; now there are personal computers available for use in the home (see Fig. 1.1).

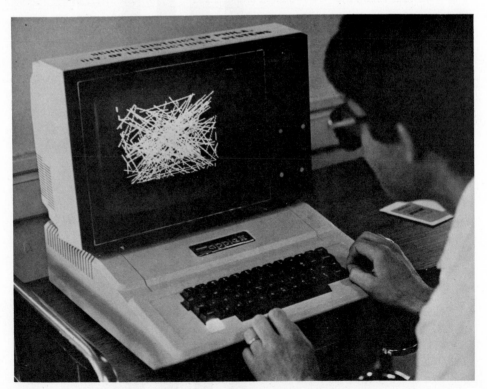

**Fig. 1.1**    Apple computer (Photo courtesy of Caryn Koffman).

The size and cost of a computer is generally dependent upon the amount of work it can turn out in a given time unit. Larger, more expensive computers have the capability of carrying out many operations simultaneously, thus increasing their work capacity. They also have more devices attached to them for performing special functions, all of which increase their capability and cost.

Despite the large variety in the cost, size and capabilities of modern computers, they are remarkably similar in a number of ways. Basically, a *computer* consists of four components as shown in Fig. 1.2. The lines connecting the components represent possible paths of information flow. The arrows show the direction of information flow.

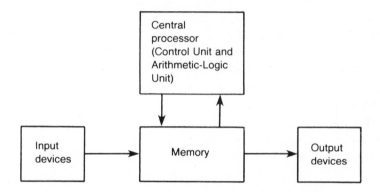

**Fig. 1.2**   Diagram of the basic components of a computer.

All information that is to be processed by the computer must first be entered into the computer memory via an input device. The information in memory is manipulated by the arithmetic and logic unit, and the results of this manipulation are also stored in memory. Information in memory can be displayed through the use of appropriate output devices. All of these operations are coordinated by the control unit. These components and their interaction are described in more detail in the following sections.

## 1.2.2 | The Computer Memory

The memory of a computer may be pictured as an ordered sequence of storage locations called *memory cells*. Each cell has associated with it a distinct *address*, which indicates its relative position in the sequence. Figure 1.3 depicts a computer memory consisting of 1,000 cells numbered consecutively from 0 to 999. Some large-scale computers have memories consisting of millions of cells.

The memory cells of a computer are used to store information. All types of information—numbers, names, lists and even pictures—may be represented in the memory of the computer. The information that is contained in a memory cell is called the *contents* of the memory cell. Every memory cell contains some information—a cell is never empty. Furthermore, a cell can only contain one data item. Whenever a data item is placed into a memory cell, any information already there is destroyed and cannot be retrieved.

It is important to distinguish the address of a memory cell from its contents.

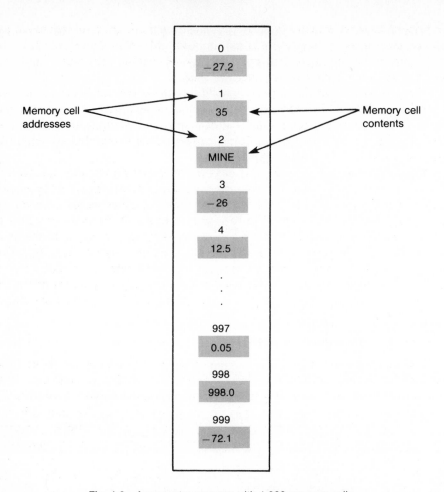

**Fig. 1.3**  A computer memory with 1,000 memory cells.

The address indicates the relative position of the cell in memory; the contents is the information stored in the cell. For example, in Fig. 1.3, the memory cell with address 1 contains the value 35; the value −72.1 is stored in the cell with address 999.

**Exercise 1.1:**   What are the contents of memory cells 0, 2 and 997 shown in Fig. 1.3?

**Exercise 1.2:**   What are the addresses of the cells containing 12.5, −26 and 998.0, respectively?

### 1.2.3   The Central Processor Unit

The information representation capability of the computer would be of little use to us by itself. Indeed, it is the *manipulative* capability of the computer that enables us to study problems that would otherwise be impossible because of the computational requirements. With appropriate directions, modern computers can generate large quantities of new information from old, solving many of these otherwise impossible problems, and providing useful insights into others; and they can do so in exceptionally short periods of time.

The heart of the manipulation capability of the computer is the *central processor unit* (CPU), which consists of a *control unit* and an *arithmetic-logic unit* (ALU). The control unit of the CPU coordinates and controls the activities of the various computer components. It determines which manipulations should be carried out and in what order.

The arithmetic part of the ALU consists of electronic circuitry designed to perform a variety of operations, including addition, subtraction, multiplication and division. The speed with which it can perform these operations is on the order of a millionth of a second. The logic part consists of electronic circuitry to compare information and to make decisions based upon the results of the comparison. It is this feature, together with its storage capability (the memory), that distinguishes the computer from the simple, hand-held calculators that many of us have used. Most of these calculators can be used only to perform arithmetic operations on numbers; they cannot compare these numbers, make decisions or store large quantities of numbers.

### 1.2.4 Input and Output Devices

The manipulative skills of the computer would be of little use to us if we were unable to communicate with the computer. Specifically, we must be able to enter information into the computer memory, and display information (usually the results of a manipulation) that is in the computer memory. The input devices are used to *enter* into the computer memory data to be manipulated by the computer. The output devices are used to *display* the results of this manipulation (program output) in a readable form.

There are many types of input and output devices. Examples of input devices include card-readers, paper-tape readers and computer terminals. A common input device is the *card reader* (see Fig. 1.4). This device reads pieces of lightweight cardboard called *punch cards*. Some card readers can process up to 1,000 punch cards per minute.

An example of a typical punch card is shown in Fig. 1.5. The punch card may contain up to 80 columns of information, and each column may contain one character. In the card shown in Fig. 1.5, columns 1 to 26 contain the letter characters A to Z, respectively, and columns 36 to 45 contain the decimal digit characters 0 to 9. Additional characters shown here include = ' ( * ) . $ , − / + and the blank. Each character is represented by its own unique configuration of holes in a card column. The blank is represented by the absence of any holes.

Cards to be read into the computer are punched on a standard keypunch, such as the one shown in Fig. 1.6. Most keypunches will also print the punched character at the top of the card. This printed information is solely for our benefit and cannot be read by the card reader. Only the holes in the card are read by the card reader.

**Fig. 1.4** A punch card reader (photo courtesy of Sperry Univac, a division of Sperry Corporation).

**Fig. 1.5** A typical punch card.

A high-speed *line printer* (Fig. 1.7) is a very common output device. It normally prints 120 to 132 characters of information across a line at speeds of up to 2,000 or more lines per minute.

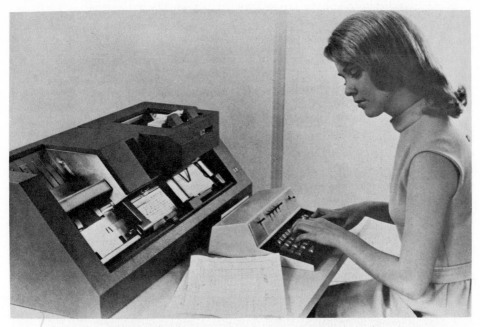

**Fig. 1.6**   A keypunch machine.

**Fig. 1.7**   A high-speed line printer (photo courtesy of Control Data Corporation).

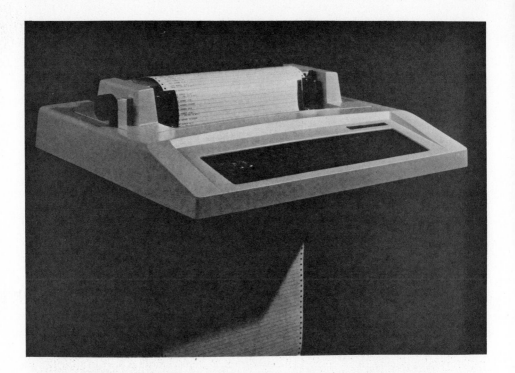

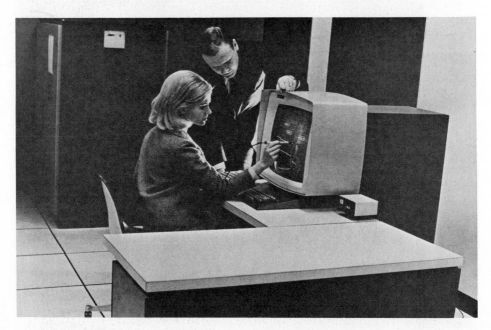

**Fig. 1.8**  Computer terminals: a standard "hard copy" terminal (top, photo courtesy Digital Equipment Corp.); a graphics terminal (bottom, photo courtesy IBM Corp.).

Another type of device that provides both input and output capability is a computer terminal. Terminals contain a typewriter-like keyboard on which information required by the computer is typed (see Fig. 1.8, top). The results of a computation may be printed on a roll of paper fed through the terminal carriage or displayed on a video screen as *alphanumeric characters* (letters and numbers). Some terminals are equipped with *graphics capabilities* (see Fig. 1.8, bottom), which enable the output to be displayed as a two-dimensional graph or picture, and not just as rows of letters and numbers. With some graphics devices, the user can communicate with the computer by pointing at information displayed on the screen with an electronic pointer called a *light pen,* as shown in Fig. 1.8.

Personal computers normally have a built-in keyboard for program and data entry. Sometimes a cathode-ray tube is part of the computer as well; other personal computers use a separate television monitor for display of results.

In many computer systems, another type of input/output device (a *secondary storage device*) is used to provide additional capability for information storage and retrieval. These devices, which include *magnetic tapes, disks,* and *drums* (see Fig. 1.9), can be used to store huge quantities of information. During a computer session, information saved previously may be retrieved from a secondary storage device, and new information may be saved for future retrieval and use.

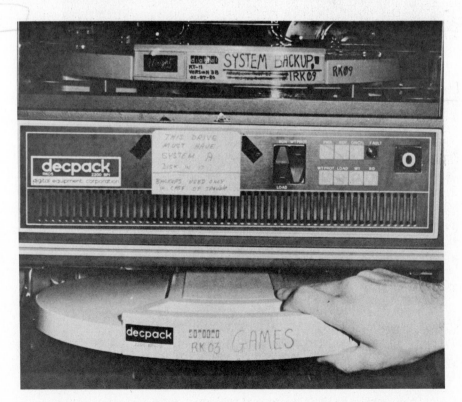

**Fig. 1.9**  Disk drive and magnetic disk (photo courtesy of Caryn Koffman).

**Fig. 1.10** Typical computer system: a Digital Equipment Corporation VAX-11 (photo courtesy of Digital Equipment Corporation).

The reason for having two types of memory is that the fast, main memory is very expensive. Secondary storage is slower but less expensive, making it perfectly suited for storing large amounts of data that are used infrequently.

A typical computer system, including memory, a central processor unit, magnetic tape and disk drives, and terminals is pictured in Fig. 1.10. In the remainder of this text, we will learn how to use this *computer hardware* by writing appropriate *software* (computer programs) for specifying data manipulation.

**Exercise 1.3:** Describe the purpose of the control unit; the arithmetic-logic unit; the memory.

## 1.3 PROGRAMS AND PROGRAMMING LANGUAGES

### 1.3.1 Introduction

The computer is quite a powerful tool. Information (*input data*) may be stored in its memory and manipulated at exceptionally high speeds to produce a result (*program output*). The problem is that the computer has no will of its own. It cannot do anything without first being told what to do. We can describe a data manipulation task to the computer by presenting it with a *list of instructions* (called a *program*) to be carried out. Once this list has been provided, the computer can then assume responsibility and carry out (*execute*) the instructions.

The process of making up a list of instructions (writing a program) is called *programming.* Writing a computer program is very similar to describing the rules of a complicated game to people who have never played the game. In both cases, a language of description understood by all parties involved in the communication is required. For example, the rules of the game must be described in some language, and then read and carried out. Both the inventor of the game and those who wish to play must be familiar with the language of description used.

Languages used for communication between man and the computer are called *programming languages.* All instructions presented to a computer must be represented and combined (to form a program) according to the *syntactic rules* (grammar) of the programming language. There is, however, one significant difference between a programming language and a language such as French, English or Russian: the rules of a programming language are very precise, and have no exceptions or ambiguities. The reason for this is that a computer cannot think! It can only follow instructions exactly as given. It cannot interpret these instructions to figure out, for example, what the program writer (*programmer*) meant it to do. An error in writing an instruction will change the meaning of a program, and cause the computer to perform the wrong action. Even the omission of a single comma can be disastrous.

### 1.3.2 Machine Language

*Machine language* is the "native tongue" of the computer. It is the only programming language that the computer understands. Unfortunately, it is a language of numbers. If we wish to communicate directly with the computer, we must do so in machine language. Each machine-language instruction contains a numeric code for the operation to be performed. The information to be manipulated must also have a numeric representation, which is usually the address of the memory cell containing the information. The exact form of these numeric encodings differs from computer to computer.

It is rather cumbersome to write a program in machine language since it is far removed from our own natural language: the programmer must remember the numeric code for each operation and the address in memory of each data item. The slightest error in the use of these codes or addresses can be extremely difficult to find and will usually produce meaningless program results.

To make matters even worse, each machine language is unique to a particular family of computers, so that machine-language programs written for one computer are not likely to execute on another. It is for these reasons that high-level languages were developed.

### 1.3.3  High-level Languages and Program Translation

When writing programs in machine language, the major problem is in translating the initial formulation of the program from a language that we understand to the machine language of the computer. It clearly would be advantageous to be able to provide instructions to the computer directly, in a form that is meaningful to us, rather than in machine language. While this objective has not been entirely satisfied, considerable progress has been made. Computer scientists have designed *high-level programming languages* that enable a programmer to write programs that utilize many familiar symbols and terminology. It is much easier to write programs using high-level languages, and high-level language programs are usually easier to read and to change. They also may be more easily moved from one computer to another.

There is, however, a price to be paid for all of this luxury: the symbols and terminology used in a high-level language cannot be understood by the computer. Therefore, high-level language programs must first be translated into machine language before they can be executed by the computer. This translation is performed by a large program called a *compiler* (see Fig. 1.11).

**Fig. 1.11**  Translating a high-level language program.

There are many high-level languages available for use today. In this text, we will study the high-level programming language FORTRAN. However, many of the programming and problem-solving concepts you learn will be applicable to other programming languages as well.

FORTRAN was developed in 1957 and was the first high-level programming language. It was originally used primarily for scientific computation but has evolved over the years into a general-purpose programming language. The version of FORTRAN used in this text, FORTRAN 77, was approved as the latest, *standard* form of FORTRAN in 1978. In the next section, we will discuss a few of the basic features of FORTRAN; other features will be introduced throughout the text. We will not present all of the FORTRAN 77 language features; however, we will provide enough of the major features to enable you to solve a wide variety of problems using the computer.

### 1.3.4   Loading and Executing a Program

Once a high-level language program has been translated successfully into machine language, it must then be stored in the memory of the computer in order to be executed. On most computers this is accomplished by a special program called a *loader*. The function of the loader is to take the machine language program produced by the compiler, store it in the memory of the computer, and tell the computer where the first instruction is located. The computer can then execute the program beginning at the specified location. The loading process is depicted in Fig. 1.12.

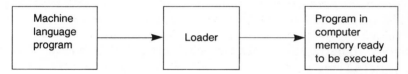

**Fig. 1.12**   Loading a program for execution by the computer.

Computers that execute programs stored in memory are called *stored program computers.* To execute a stored program, the computer control unit examines each program instruction in memory starting with the first, and sends out the command signals appropriate for carrying out the instruction. Normally, the instructions are executed in sequence; however, as we shall see later, it is possible to have the control unit skip over some instructions or execute some instructions more than once.

During execution, data may be entered into the memory of the computer, and the results of the manipulations performed on this data may be displayed. Of course, these things will happen only if the program contains instructions telling the computer to enter or display the appropriate information.

Figure 1.13 shows the relationship between a program for computing a payroll and its input and output, and indicates the *flow of information* through the computer during execution of the program. The data to be manipulated by the program (employee time cards) must first be entered into the computer memory (Step 1 in Fig. 1.13). As directed by the program instructions, the central processor unit manipulates the data in memory, and places the results of these computations back into memory (Step 2). When the computation process is complete, the results can be output from the memory of the computer (Step 3) in the desired forms (as employee checks and payroll reports).

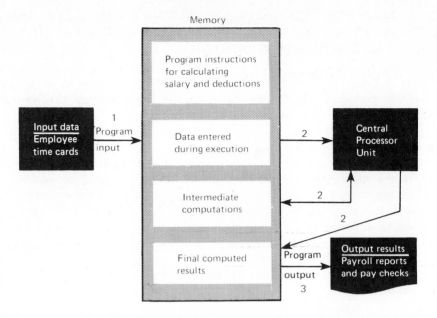

**Fig. 1.13**   The flow of information through the computer.

## 1.4   INTRODUCTION TO FORTRAN AND COMPUTER OPERATIONS

### 1.4.1   Some Basic Computer Operations

There are many different computers available today, and each has a unique set of *basic operations* that it can perform. These operations generally fall into three categories:

Input and output operations
Data manipulation and comparison
Control operations

Despite the large variety of operations in these categories, there are a few operations in each that are common to most computers. These operations are summarized next.

### Input/Output Operations

Read
Print

### Data Manipulation and Comparison

| | | | |
|---|---|---|---|
| Add | Subtract | Multiply | Divide |
| Negate | Assign | Compare | |

### Control Operations

Transfer
Conditional execution
Stop

In the remainder of this chapter, we will describe some of these operations by showing how they are written in FORTRAN. We will do this by way of illustration, using a payroll processing problem.

### 1.4.2 A FORTRAN Payroll Program

Fig. 1.14 contains the FORTRAN statements for a program that solves the following problem.

**Problem 1A:** Compute the gross salary and net pay for an employee of a company, given the employee's hourly rate, the number of hours worked, and a fixed tax deduction amount of $25.

```
REAL HOURS, RATE, GROSS, NET
READ, HOURS, RATE
PRINT, HOURS, RATE
GROSS = HOURS * RATE
NET = GROSS - 25.00
PRINT, GROSS, NET
STOP
END
```

**Fig. 1.14** The FORTRAN program for a simple payroll problem (Problem 1A).

One of the nicest features of FORTRAN is that it enables us to write program statements that resemble English. Even at this point, we can read and understand the program in Fig. 1.14, although we have no idea how to write our own FORTRAN statements. We will briefly describe this program next; an explanation of each statement is given in the sections that follow.

The first line in Fig. 1.14 tells the compiler the names that will be used to reference data in the program. For example, HOURS is used to represent the hours worked, and RATE represents the hourly pay rate; GROSS represents the gross salary and NET represents the employee net pay (after taxes). The next two lines read and print the *input data* for this problem (the hours worked (HOURS) and the hourly rate (RATE)). Then the gross salary (GROSS) and net pay (NET) are computed as functions of the input data, and, finally, these results are printed as the *program output*. The STOP statement terminates the execution of the program and the END statement marks the physical end of the list of FORTRAN statements.

### 1.4.3 Use of Symbolic Names in FORTRAN

An important feature of FORTRAN is that it permits the use of descriptive symbolic names (called *variable names,* or simply *variables*), rather than numeric addresses, to designate memory cells. This is accomplished through the compiler, which *associates* one memory cell for each variable name used in our program. We need not be concerned with the address of the cell assigned. We simply reference by name each variable that we wish to manipulate, and let the compiler determine the address of the cell assigned to that variable.

There are some rules that must be followed in the formation of FORTRAN variable names. These rules are given next.

---

**Variable Names**

- May contain only combinations of letters (A–Z) and the numbers (0–9)
- Must always begin with a letter (A–Z)
- May be from 1 to 6 characters in length

---

For the payroll problem, we used the variables HOURS (for hours worked), RATE (for hourly rate), GROSS (for gross salary), and NET (for net pay). These are pictured in Fig. 1.15. The question mark in each box indicates that we have no idea of the current values of these variables (although variables always have values).

HOURS          RATE          GROSS          NET

**Fig. 1.15**   Using meaningful variable names to designate memory cells.

The FORTRAN statement used to inform the compiler that these variable names will designate the memory cells used in a program is

```
REAL HOURS, RATE, GROSS, NET
```

This statement is called a *declaration statement.* Note that each variable name is separated from the next one in the list by a comma.

This statement also informs the compiler that these variables will be used for storing *real numbers.* In FORTRAN, only those numbers that contain a decimal point are real numbers (examples: 3.14, −35., .0005, 0.0). Real numbers are used to represent quantities that are likely to have a fractional part (e.g., hourly salary, average speed).

In the payroll program, we manipulated only real numbers. However, FORTRAN also allows the manipulation of numbers, called *integers,* which have no decimal point or fractional part (examples: −99, 35, 0, 1). Integer variables are often used to represent counts of items. A declaration statement similar in form to the REAL statement, but beginning with the word INTEGER is used to indicate the names of variables to be used for storing integers.

The general form of the declaration statement is described in the next display.

---

**Variable Declaration (REAL, INTEGER)**

REAL *list of variables*
or   INTEGER *list of variables*

**Interpretation:** A memory cell is allocated for each variable in the *list of variables.* The word REAL or INTEGER indicates the type of data to be stored. Commas are used to separate the variable names in the *list of variables.* Variable declarations should come first in a FORTRAN program.

---

**Program Style**

*Using meaningful variable names*

Throughout this text we will use boxes such as this to discuss certain points that contribute to good program style. You may think that it is more important to get a program working correctly without being concerned with its style. However, computer scientists have found that programs that are written initially to conform to widely accepted style conventions are much easier to read and understand. They are also less likely to contain errors; moreover, any errors that might exist are easier to locate and correct.

As an example of good programming style, it is very important to choose variable names that can be readily associated with the information stored in the variable. Well chosen names can make programs considerably easier to read and understand. Avoid the temptation to use single letter variable names, such as R and H, instead of more meaningful names such as RATE and HOURS.

**Exercise 1.4:**   Which of the following "strings" of characters can be used as legal variable names in FORTRAN? Indicate the errors in the strings that are illegal.

| | | | |
|---|---|---|---|
| 1. ARK | 2. MICHAEL | 3. ZIP12 | 4. 12ZIP |
| 5. ITCH | 6. P3$ | 7. GROSS | 8. X123459 |
| 9. NINE+T | | | |

## 1.4.4  Simple Data Manipulation—Assignment Statements

As shown earlier, we will assume that the variables HOURS and RATE represent the number of hours worked and the hourly wage rate, respectively. GROSS and NET will be used to represent the computed gross and net salary, respectively. Our problem is to perform the two computations:

- Compute gross salary as the product of hours worked and hourly wage rate.
- Find net pay by deducting the tax amount from the gross salary.

We need to write FORTRAN instructions to tell the computer to perform these computations. This is done in Fig. 1.14 using the FORTRAN *assignment statements*

```
GROSS = HOURS * RATE
NET = GROSS - 25.00
```

These data manipulation statements are called assignment statements because they specify the assignment of a value to a given variable. Figure 1.16 illustrates the effect of these statements.

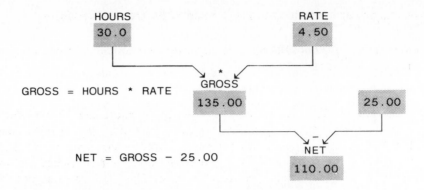

**Fig. 1.16**  Effect of assignment statements.

The first statement causes the value of the variable GROSS to be replaced by the *product* (indicated by *) of the values of the variables HOURS and RATE, or 135.00. The second statement causes the value of the variable NET to be replaced by the *difference* between the values of the variable GROSS and the tax amount, 25.00. (We are assuming, of course, that meaningful data items are already present in the variables HOURS and RATE and that the value of 25.00 is also present in a memory cell.) Only the contents of GROSS and NET are changed by this sequence of *arithmetic operations;* the variables HOURS and RATE retain their original values.

The general form of the assignment statement is shown in the following display.

---

**Assignment Statement**

*result = expression*

**Interpretation:** The variable specified by *result* is assigned the value of the *expression*. The *expression* can be a single variable or constant, or a computation involving constants and variables and the arithmetic operators listed in Table 1.1. The previous value of *result* is destroyed when the expression value is stored.

---

| Arithmetic operator | Meaning |
| --- | --- |
| + | Addition |
| − | Subtraction |
| * | Multiplication |
| / | Division |

**Table 1.1 FORTRAN arithmetic operators.**

As we shall see, any arithmetic formula can be specified as an expression. For the time being, we will focus on simplified forms involving, at most, one ,ja arithmetic operator, as illustrated above and in the following examples.

**Example 1.1:**   In FORTRAN, it is perfectly permissible to write assignment statements of the form

SUM = SUM + ITEM

where the variable SUM is used on both sides of the equal sign. This is obviously not a mathematical equation, but it illustrates something that is often done in FORTRAN. As shown next, this statement instructs the computer to add the current value of the variable SUM to the value of the variable ITEM and assign the result as the new value of the variable SUM. The previous value of SUM is destroyed in the process.

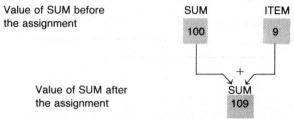

The above statement is discussed further in Chapter 2, where it is used to *accumulate* the sum of a large number of data items.

**Example 1.2:**   Assignment statements can also be written with a single variable or constant as the expression. The statement

ABSX = X

instructs the computer to *copy* the value of the variable X into ABSX. The statement

ABSX = −X

instructs the computer to *negate* the value of the variable X and store the result in ABSX. Neither of these statements affects the contents of the variable X. Negating a number is equivalent to multiplying it by −1. Thus, if the variable X contains −3.5, then the statement

ABSX = −X

will cause 3.5 to be stored in the variable ABSX.

### 1.4.5   Storing Information in Memory—Program Constants

Information cannot be manipulated by the computer unless it is first stored in memory. In this section and the next, we discuss two ways of initially placing information to be manipulated into computer memory: by writing constant values directly in an assignment statement, or by reading data into memory during the execution of the program. Normally, the first approach is taken for information whose value is known *a priori*. Such values are usually given ahead of time, often in the statement of the problem to be solved. The second approach is taken for data that are likely to vary each time the program is used.

For example, in the payroll problem, the withholding tax amount is given as $25 regardless of which employee's net pay is to be computed. Since this value is given ahead of time as a constant, we may write it directly in the assignment statement

$$NET = GROSS - 25.00$$

When a constant is written *in-line* in this manner, it is the responsibility of the compiler to ensure that the constant is placed in some memory cell prior to the execution of the program.

### 1.4.6  The READ Statement

Unlike the tax amount, which is given as part of the problem statement, the number of hours worked per week and the employee hourly rate may vary with each employee. Hence, these values should be read into memory during program execution. This operation must be performed prior to carrying out any calculations involving these values.

In the program of Fig. 1.14, the statement

$$READ*, HOURS, RATE$$

causes the computer to enter a data item into each of the variables listed (HOURS and RATE in this case). The prior values of these variables are destroyed by the data entry process. Depending on the computer system, the data items may be entered at a terminal during program execution, or the program may read them from data cards. In either case, the program and its data must be kept separate because the program must be translated before it can be executed.

The READ statement is illustrated below and described in the next display.

$$READ*, HOURS, RATE$$

---

### READ Statement

$$READ*, \textit{input list}$$

**Interpretation:** Data are entered into each variable specified in the *input list* (a list of variables). Commas are used to separate the variable names in the *input list*. (The data items will be provided separately with spaces between items.)

To verify that the correct values have been entered, it is advisable to display or *echo print* the value of each variable used for storage of input data. Such a printout also provides a record of the data manipulated by the program. This record is often quite helpful to the programmer and to those who must read and interpret the program output. The statement used to display or print out the value of a variable is described in the next section.

### 1.4.7  The PRINT Statement

Thus far, we have discussed the FORTRAN instructions required for the entry of employee hours and wage rate, and the computation of gross salary and net pay. The computational results have been stored in the variables GROSS and NET, respectively. Yet all of this work done by the computer is of little use to us since we cannot physically look into a memory cell to see what is there. Therefore, we must have a way to instruct the computer to display or print out the value of a variable, especially those variables that represent computational results.

The instruction

```
                    PRINT*, GROSS, NET
```

causes the values of the variables GROSS and NET to be printed on a line as part of the program output (see Fig. 1.17). The values of GROSS and NET are unchanged. The PRINT statement is described in the next display.

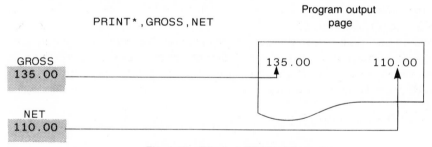

**Fig. 1.17**  Effect of PRINT statement.

---

**PRINT Statement**

```
                 PRINT*, output list
```

**Interpretation:** Each **PRINT** statement initiates a new line of output. The value of each item in the *output list* is printed in sequence across the output line. Commas are used to separate items in the *output list*. These items may be constants, variables, or expressions.

---

There are two **PRINT** statements in Fig. 1.14

```
              PRINT*, HOURS, RATE
              PRINT*, GROSS, NET
```

The first PRINT statement follows the READ statement and echo prints the data values for HOURS (30.0) and RATE (4.50). The second PRINT statement displays the computed results, GROSS and NET.

The output for a sample run of this program is shown below.

```
30.00000000000        4.500000000000
135.0000000000        110.0000000000
```

From the first line we see that the employee worked thirty hours (value of HOURS), and earned $4.50 per hour (value of RATE). The gross salary is $135.00 and the net pay is $110.00.

### 1.4.8  STOP and END

Once all desired calculations have been performed and the results displayed, the computer must be instructed to stop execution of the program. The instruction that does this consists of a single word:

STOP

The last statement in every program is

END

This statement marks the physical end of our FORTRAN program; it signals the compiler that there are no more statements left to be translated. The declaration statement (beginning with REAL) and END statement are examples of *non-executable statements*. These statements are not executed by the computer. Rather, they provide information to be used by the compiler in translating a program.

We have now completed the discussion of the payroll program. It is repeated in Fig. 1.18; make sure that you understand the purpose and form of each statement shown.

```
REAL HOURS, RATE, GROSS, NET
READ*, HOURS, RATE
PRINT*, HOURS, RATE
GROSS = HOURS * RATE
NET = GROSS - 25.00
PRINT*, GROSS, NET
STOP
END
```

**Fig. 1.18**  The payroll program.

**Program Style**

*Use of blank space*

Program style is a very important consideration in programming. A program that "looks good" is easier to read and understand than a sloppy program. Most programs, at some time or another, will be examined or studied by someone else. It is certainly to everyone's advantage if a program is neat and its meaning is clear.

The consistent and careful use of blanks can significantly enhance the style of a program. Blanks in FORTRAN programs are ignored by the compiler and may be inserted as desired to improve the style and appearance of a program. As shown in Fig. 1.18, we shall always leave a blank space after a comma and before and after operators such as *, —, and =.

All of these measures are taken for the sole purpose of improving the style and, hence, the clarity of the program. They have no effect whatsoever on the meaning of the program.

**Exercise 1.5:**   Can any of the statements in the program in Fig. 1.18 be moved without altering the results of the program? Which statements can be moved? Which cannot be moved? Why?

**Exercise 1.6:**   What values will be printed by the payroll program for the alternate pair of data items 35.0 and 3.80?

**Exercise 1.7:**   Let H, R, and T be symbolic names of memory cells containing the information shown below:

| H | R | T |
|---|---|---|
| 40.0 | 16.25 | 0.18 |

What values will be printed following the execution of this sequence of instructions?

```
G = H * R
T = G * T
N = G — T
PRINT*, H, R, G, T, N
```

## 1.5   LOADING AND EXECUTION OF A PROGRAM

Once translated by the compiler, a program in its machine language form is loaded into the memory of the computer and executed. In this section, we illustrate this process using the payroll program shown in Fig. 1.18.

Once the loading of the payroll program has been completed, but before execution starts, the program and the memory cells assigned to the variable names of the program would appear as shown on the left side of Fig. 1.19.

Note the following important points concerning this figure.

- Only the executable portion of the original program is loaded into memory. The REAL and END statements are used by the compiler during the translation of the program and do not generate machine language instructions.
- Specific memory cells have been assigned to each of the four variable

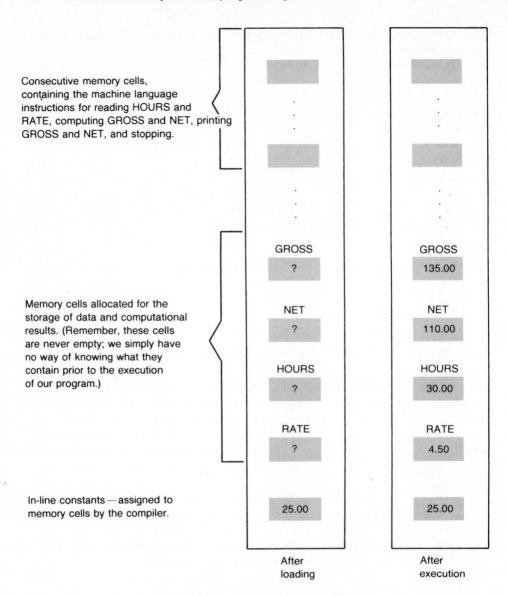

Consecutive memory cells, containing the machine language instructions for reading HOURS and RATE, computing GROSS and NET, printing GROSS and NET, and stopping.

Memory cells allocated for the storage of data and computational results. (Remember, these cells are never empty; we simply have no way of knowing what they contain prior to the execution of our program.)

In-line constants—assigned to memory cells by the compiler.

GROSS ? / NET ? / HOURS ? / RATE ? / 25.00 — After loading

GROSS 135.00 / NET 110.00 / HOURS 30.00 / RATE 4.50 / 25.00 — After execution

**Fig. 1.19**   The payroll program in memory.

names used in the program. However, the values of these variables are unknown prior to execution of the program.

- All in-line constants, such as 25.00, used in the program are stored in memory prior to program execution.

After the execution of the program is complete, the configuration of memory is as shown on the right side of Fig. 1.19. The final values of the variables HOURS, RATE, GROSS and NET will depend upon the input values of

HOURS and RATE for a given employee. The program instructions are not changed by execution.

**Exercise 1.8:**  Illustrate the appearance of memory allocated for the storage of data (as in the lower part of Fig. 1.19) before and after the execution of the following program with a data value of 1.0 for RADIUS.

```
REAL AREA, RADIUS, RADSQ
READ*, RADIUS
PRINT*, RADIUS
RADSQ = RADIUS * RADIUS
AREA = 3.14159 * RADSQ
PRINT*, AREA
STOP
END
```

What value of AREA will be printed if RADIUS is 2.0?

## 1.6  USING THE COMPUTER

### 1.6.1  Batch and Timesharing Operating Systems

Once your program has been written, it must be entered into the computer and translated by the FORTRAN compiler before it can be executed. The actual process of entering a program and its data differs on each computer system.

On computers that serve many users at one time, a *user program* is processed by a supervisory program called the *operating system*. The operating system schedules the resources of the computer and controls the order in which programs are processed.

There are two basic types of operating systems: batch and timesharing. If you are using a batch operating system, you will normally keypunch your program and data on punch cards. If you are using a timesharing operating system, your program will be typed at a computer terminal. Your data values may be entered all at once, or as needed, while the program is executed.

The operating system must have some information about your job in order to process it. This information may include:

- Your account number and password
- The compiler or other processor needed (e.g., FORTRAN)
- The time and memory requirements of the job
- The location of any data
- The physical end of the job

If you are using a personal computer or microcomputer, then you will normally interact with a simpler operating system called a *monitor* when entering your programs and data through a terminal. In this case, it is not necessary to provide all of the information listed above. However, it is still necessary to learn how to direct the system monitor to process your program. You will need a list of the available *monitor commands* and a description of how to use them.

The next three sections are oriented towards batch use of FORTRAN with punch cards. Section 1.6.2 describes the placement of a FORTRAN statement on a punch card and is also relevant for FORTRAN statements typed at a terminal. Consequently, we recommend that everyone read Section 1.6.2. If you are using a timesharing system at a terminal or if you are using a personal computer, then you should read Sections 1.6.5 and 1.6.6 instead of Sections 1.6.3 and 1.6.4.

### 1.6.2   Keypunching a FORTRAN Statement

The punch card contains space for a maximum of 80 characters of information. When keypunching FORTRAN statements, this space may be considered to be divided into four sections called fields (see Fig. 1.20).

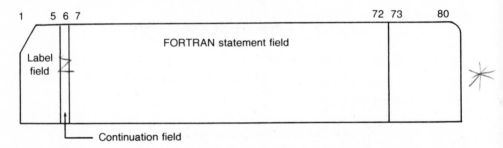

**Fig. 1.20**   The four fields of a FORTRAN card.

Certain rules must be followed in keypunching a FORTRAN statement on a punch card. These rules are summarized next.

---
**Rules for Keypunching FORTRAN Statements**

1. The FORTRAN statement must be punched in the statement field (card columns 7–72 inclusive), and may contain blanks anywhere the programmer desires to improve program readability. Blanks in the statement field have no meaning and are ignored by the compiler. (Columns 1–5 of the card are used for statement labels, which are introduced in Chapter 3.)
2. If a FORTRAN statement is too long to fit on one card, it may be continued in columns 7–72 of subsequent cards (called *continuation cards*) simply by punching any character except zero in column 6 of each continuation card.
3. Columns 73–80 are generally unused and are always ignored by the compiler. Some programmers use these columns for numbering their FORTRAN statements.

---

An example of a keypunched FORTRAN statement is shown in Fig. 1.21. Figure 1.22 shows the payroll program (listed in Fig. 1.18) keypunched on cards.

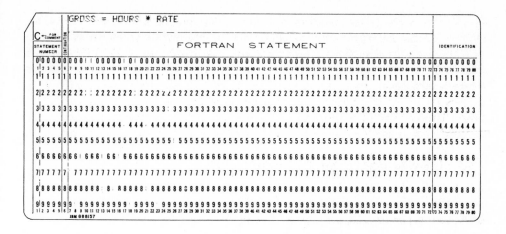

**Fig. 1.21** A keypunched FORTRAN statement.

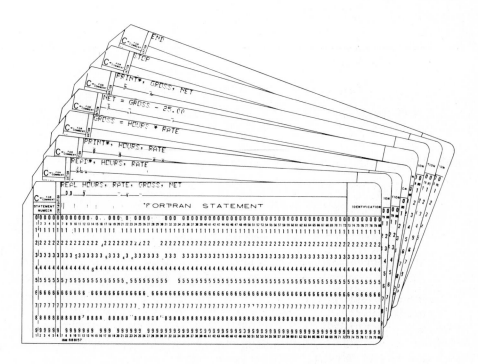

**Fig. 1.22** The payroll program on punch cards.

Remember to be very careful in keypunching your program cards. The computer cannot guess what you mean. If you misspell a variable name or keypunch an extra character somewhere, your program may not run at all. If it does run, it may produce incorrect results. Also, be sure that all FORTRAN statements are keypunched in columns 7 through 72.

### 1.6.3  Keypunching Data*

When you are using a batch system, you will have to keypunch on *data cards* all data values to be read by your program. Normally, you will have the same number of data cards as there are READ statements in the program. The first card should contain the data values to be entered by the first READ statement, the second card should contain the values to be entered by the second READ, and so on. All data cards should be placed behind your FORTRAN program. The data cards will usually be separated from the program cards by a special separator card, as illustrated in Fig. 1.23, for the payroll problem.

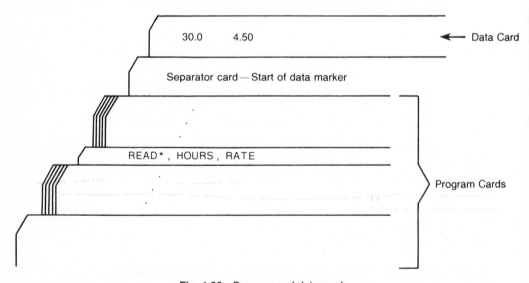

**Fig. 1.23**  Program and data cards.

When keypunching data, all 80 columns of a data card may be used. If there is more than one item on a card, one or more blank spaces must be left between each pair of values. The number and order of the data values punched on a card will be determined by the READ statement that processes the card. If the input list contains two variables, then two data values should be keypunched on the corresponding data card. The first data value will always be stored in the first variable; the second value in the second variable, etc.

---

* For batch users only.

**Example 1.3:** Given the two data cards

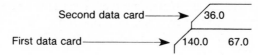

the statements

```
READ*, WEIGHT, HEIGHT
READ*, AGE
```

would cause the real data on the cards to be read into the variables listed as shown in Fig. 1.24.

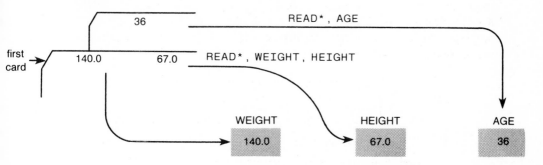

**Fig. 1.24**  Correspondence between data cards and READ statements.

If the two data cards shown in Fig. 1.24 were somehow interchanged, the values would not be stored as desired. This is a frequent source of error in programming. To minimize the chance of this or other similar input errors going undetected, you should always echo print the value of each input variable immediately following a READ statement involving that variable.

### 1.6.4  Control Cards*

As mentioned in Section 1.5.1, the program user and the resource requirements of the program must be specified before a job can be processed. Special cards, called *control cards,* are used to convey this information to the operating system. The actual form of these cards is very much dependent on your particular computer and installation. Your instructor will provide you with the format for the control cards needed. Figure 1.25 shows the relationship among the control cards, program statements, and data cards of your *job deck.*

**Exercise 1.9:** Keypunch the payroll program shown in Fig. 1.18, and run this program (with the necessary control cards and some sample data) on your computer.

---

* For batch users only.

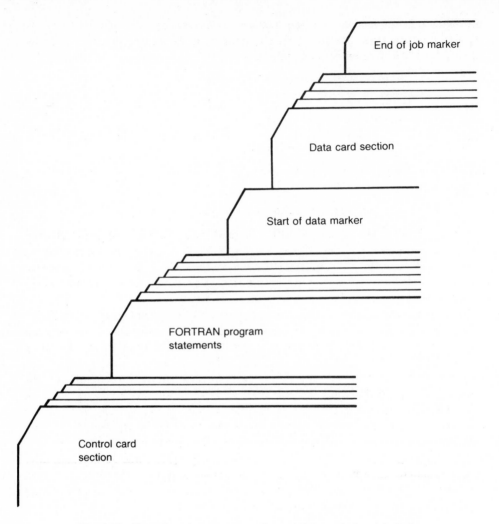

**Fig. 1.25**  Control cards, program cards, and data cards in a job deck.

### 1.6.5  Entering FORTRAN Programs at a Terminal

The procedures used in entering a FORTRAN program at a terminal are different on each computer system. We shall try to present some of the fundamental concepts here; your instructor will provide you with explicit details for your computer.

Every program statement must be numbered. The *line number* indicates the relative position of each statement in the FORTRAN program. It is not necessary to enter the statements in order (although we generally do), and the same statement (line number) may be entered more than once. In this case, the last version typed is the one retained.

As discussed in Section 1.6.2, the FORTRAN statement itself must start in the seventh column after the line number. Consequently, you should first type the line number, press the space bar six times, type the statement, and then press the RETURN key to enter each statement.

```
100        REAL HOURS, RATE, GROSS, NET
110        READ*, HOURS, RATE
120        PRINT*, HOURS, RATE
130        GROSS = HOURS * RATE
140        NET = GROSS - 5.00
150        PRINT*, GROSS, NET
160        STOP
170        END
140        NET = GROSS - 25.00
```

**Fig. 1.26**  The payroll program typed at a terminal.

As shown in Fig. 1.26, line 140 is entered twice because the initial attempt contained an error (5.00 was typed instead of 25.00). If you make a mistake in typing a line, simply retype it—line number and all. On many systems, a special *rubout key* may be used to "erase" an erroneous character (or characters) as the line is being entered.

Some systems provide facilities for automatically numbering FORTRAN statements. On some systems, it is also not necessary to enter six blanks between the line number and the FORTRAN statement, as this is done by the operating system.

After your program is typed in, you will normally want the computer to display it for you so that you can see whether it contains any typing errors. On many systems, this is done by typing the *system command LIST*. After you have checked that your program is entered correctly, it should be translated and executed. Normally, this is accomplished by typing the *system command RUN*.

If the program statements are all correct, the program will be executed after it is translated. However, if any lines contain invalid FORTRAN statements, they will be listed and the program will not be executed. You should correct each invalid statement, reenter it, and try again.

### 1.6.6   Entering Data at a Terminal

Whenever a READ statement is reached during program execution, your program will pause and wait for data. It will indicate this by typing a ? at the terminal. You must then enter one data value for each variable in the input list of the READ statement (see Section 1.4.6). The order of the data values must correspond to the order of the variables. The data values should be separated from each other by one or more blanks. After the data required are typed, press the RETURN key to resume program execution.

For the payroll program in Fig. 1.18, the data entry line would appear as

```
?        30.0        4.50
```

after the data items for HOURS and RATE were typed. In Section 1.7.2, we shall describe a technique for specifying what data are required just before the program pauses.

**Exercise 1.10:**    Enter, list, and run the payroll program on your computer.

## 1.7    ADDITIONAL INPUT AND OUTPUT FEATURES

### 1.7.1    Annotated Output

The printout for our payroll program consists of four numbers only, with no indication of what these numbers mean. In this section, we shall learn how output values may be annotated, using *strings* of characters, to make it easier for us to identify the variable values they represent. We will learn more about the use of strings in Chapter 4.

A string is a sequence of symbols enclosed in apostrophes. We can insert strings directly in PRINT statements in order to provide descriptive messages in the program output. The string will be displayed exactly as it is typed (with the apostrophes removed).

**Example 1.4:**    Given the input values 30.0 and 4.5, the modified payroll program below

```
REAL HOURS, RATE, GROSS, NET
READ*, HOURS, RATE
PRINT*, 'HOURS = ', HOURS, '    RATE = ', RATE
GROSS = HOURS * RATE
NET = GROSS - 25.00
PRINT*, 'GROSS = ', GROSS, '    NET = ', NET
STOP
END
```

would generate the output lines

```
HOURS =    30.00000000000    RATE =    4.500000000000
GROSS =    135.0000000000    NET =    110.0000000000
```

Clearly, the lines printed above convey more information than the original output. We suggest that you adopt the convention of always using descriptive strings to help identify information that you are printing.

Once again, commas are used to separate items in the output list, whether they are variables or strings. The placement of the commas is very important; a computer error will result if a comma or apostrophe is missing, or if they are in the wrong place. (For example, if the order of a comma and apostrophe were reversed, the comma would be considered part of the string instead of a separator.)

**Example 1.5:**    An additional example of the use of strings to annotate output is provided in the program shown in Fig. 1.27. This program computes the estimated time and cost of an automobile trip using the formulas below:

1) time = distance / speed
2) gallons used = distance / miles per gallon
3) cost of trip = gallons used × cost per gallon

There are four input data items for this program: trip distance (DSTNCE), average speed (SPEED), number of miles travelled on a gallon of gas (MPG), and cost of a gallon (GALCOS). The program computes the estimated time of the trip (TIME) and the total cost of gasoline (TOTCOS).

```
REAL DSTNCE, SPEED, MPG, GALCOS, TIME, TOTCOS, GALUSE
PRINT*, 'COMPUTE TRIP TIME'
READ*, DSTNCE, SPEED
PRINT*, 'DISTANCE = ', DSTNCE, ' MILES'
PRINT*, 'AVERAGE SPEED = ', SPEED, ' MILES PER HOUR'
TIME = DSTNCE / SPEED
PRINT*, 'TIME OF TRIP = ', TIME, ' HOURS'
PRINT*, ' '
PRINT*, 'COMPUTE TRIP COST'
READ*, MPG, GALCOS
PRINT*, 'MILEAGE RATE = ', MPG, ' MILES PER GALLON'
PRINT*, 'COST PER GALLON = ', GALCOS, ' DOLLARS'
GALUSE = DSTNCE / MPG
TOTCOS = GALUSE * GALCOS
PRINT*, 'TOTAL GASOLINE COST = ', TOTCOS, ' DOLLARS'
STOP
END
```

**Fig. 1.27** Use of strings to annotate output.

The computations performed by this program are quite simple. There is an assignment statement corresponding to each of the equations (1), (2) and (3) above. The remaining statements in the program body are used for data entry and display.

The output generated for the data lines

```
600.00              50.0
20.0                1.25
```

is shown in Fig. 1.28. The first data line contains the values of DSTNCE and SPEED; the last data line contains the values of MPG and GALCOS.

```
COMPUTE TRIP TIME
DISTANCE =    600.0000000000     MILES
AVERAGE SPEED =    50.00000000000     MILES PER HOUR
TIME OF TRIP =    12.00000000000     HOURS

COMPUTE TRIP COST
MILEAGE RATE =    20.00000000000     MILES PER GALLON
COST PER GALLON =   1.250000000000     DOLLARS
TOTAL GASOLINE COST =    37.50000000000     DOLLARS
```

**Fig. 1.28** Output for the trip cost program.

The first output line in Fig. 1.28 is generated by the initial PRINT statement which prints a descriptive message only. The blank output line is generated by the PRINT statement with the output list ' ' (a blank string).

**Exercise 1.11:** Explain the purpose of the blank spaces in the string '    RATE = ' in Example 1.4.

**Exercise 1.12:** Give the seven PRINT statements needed to print the TIC-TAC-TOE board configuration shown below:

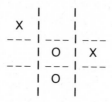

### 1.7.2  Strings as Prompting Messages*

If you are using a computer terminal, you will normally enter your data interactively during the execution of your program. It is always most helpful to print a string as a *prompting message* just before each READ statement is executed. For example, the statements

```
PRINT*, 'ENTER HOURS AND RATE'
READ*, HOURS, RATE
```

would cause the lines

```
ENTER HOURS AND RATE
?
```

to be printed. The ? indicates that the computer is waiting for data. The program user must type the two data values requested following the question mark.

In most cases, it would be redundant to also echo-print the data values after they are entered interactively. These values and their use in the program would be clear from the prompting message and the numbers typed right after it.

We shall continue the practice of listing the batch versions of programs. If you are writing programs for interactive use, you should remove the echos and insert appropriate prompts before each READ statement.

**Exercise 1.13:** Provide prompting messages for the program in Fig. 1.27.

## 1.8  POSSIBLE ERRORS

One of the first things you will discover in writing programs is that a program very rarely runs through to completion the first time it is submitted. Mur-

---

\* For users of terminals only.

phy's Law, "If something can go wrong, it will," seems to be written with the computer programmer, or programming student, in mind. We will discuss four common errors next.

1. *Control card and deck arrangement errors* are most frustrating because they usually result in a situation where the operating system will not know what it is to do with your program. It may not be able to determine that the FORTRAN compiler will be needed to translate your program. Still worse, it may not even be able to tell that you are the owner of the deck of cards it is processing. Control-card and deck-arrangement errors are easy to avoid. Your instructor will describe to you the control cards and deck arrangement to be used for your computer. If you follow this description to the letter, you should have little difficulty.

2. *FORTRAN syntax errors* are caused by FORTRAN statements that do not follow the precise rules of formation (*syntax rules*) of FORTRAN. Such statements cannot be translated by the compiler. Often, careless mistakes made in typing will cause syntax errors. The compiler will identify these syntax errors by printing error diagnostics for each illegal instruction in your program. It will be up to you to read these diagnostics and review the rules of FORTRAN in order to correct your errors.

At first, the error diagnostics may be difficult to understand. It is worthwhile to try to decipher them yourself rather than to seek help immediately. If you persist, as the course progresses, you will become more proficient at recognizing and correcting your own errors.

An example of a FORTRAN syntax error would be keypunching RESL instead of REAL. Another common syntax error is caused by the failure to type the symbols *, after the words READ and PRINT in input and output statements. When a syntax error occurs, the compiler is not likely to recognize the statement that has been punched and will tell you so with an appropriate diagnostic. Usually this diagnostic will inform you only that you have punched an *unrecognizable statement.* Some compilers will, however, try to indicate to you what might be wrong with the statement. Whenever you are told of an unrecognizable statement, you should carefully check your punctuation. Remember, spelling errors and the misuse of commas are most likely to produce unrecognizable statement diagnostics.

Recall that FORTRAN statements must begin in column 7 or beyond. If you accidentally start a statement to the left of column 7, you will cause the compiler much confusion. The compiler might think that you have a continuation card, and it will not be able to translate your program correctly.

The words REAL, INTEGER, READ, PRINT, STOP, and END are examples of FORTRAN *keywords.* You will encounter additional keywords in subsequent chapters of the text. Although the rules of FORTRAN syntax do not prohibit it, we suggest that you avoid using these keywords as variable names, since this may cause syntax errors on some compilers.

3. *Inconsistent spelling of variable names* is a common error but is not normally detected by the compiler. For example, the statement sequence in Fig. 1.29

will be translated by most compilers with no error indication given.

```
REAL COUNT, X
READ*, COUNT
X = COUNTR / 2.0
PRINT*, X
STOP
END
```

**Fig. 1.29**  A program segment containing a spelling error.

The statement

```
X = COUNTR / 2.0
```

in Fig. 1.29 will be translated using a new memory cell associated with the variable COUNTR as the dividend, even though the intention was to use the variable COUNT. The error may not become apparent until your program executes and prints an incorrect value of X.

However, some compilers will try to ensure that all variables have been *defined*—that is, have had data placed in them by the program—before they are to be manipulated or printed. There are two ways of defining a variable:

- Using a READ statement to enter variable information during execution
- Using an assignment statement to store the value of an expression

Obviously, it makes little sense to try to manipulate or display the value of a variable that has never been defined in one of these ways. Unfortunately, most compilers will not inform a programmer when such a mistake is made. If your compiler does, the diagnostic "attempt to use an undefined variable" would help you find most misspelled variable names.

4. *Program logic errors* are caused by mistakes that have been made in the logical organization of the instructions in your program. Many of these errors can be avoided if a careful, reasoned approach is taken to problem solving and program development. Logic errors that do occur can often be more easily diagnosed if some care and discipline have been applied in the design and coding of the program. It is our intention to provide, through numerous examples, some useful guidelines for problem solving and program writing.

**Exercise 1.14:**  What will be the value of X printed by the program segment shown in Fig. 1.29 once the spelling error has been corrected?

**Exercise 1.15:**  Find the syntax errors in the following FORTRAN statements:

```
REAL NET GROSS HOURS X Y Z
READ*, HOURS * RATE
NETPAY = GROSS - TAX
READ * HOURS
X + Y = Y + Z
PRINT*, 'HOURS = ' , HOURS
RATE
PRINT*, 'GROSS = , GROSS
PRINT*, 'NET = ' NET
```

## 1.9   SUMMARY

You have been introduced to the basic components of a computer: the memory, the central processor unit, and the input and output devices. The next display contains a summary of important facts about computers that you should remember.

---

**Summary of Important Facts about Computers**

1. A memory cell is never empty, but its initial contents may be meaningless to your program.
2. The current contents of a memory cell are destroyed whenever new information is placed in that cell (via an assignment or READ statement).
3. The computer cannot think for itself, and must be instructed to perform a task in a precise and unambiguous manner, using a programming language.
4. Programs must first be placed in the memory of the computer before they can be executed.
5. Data must first be stored in memory before it can be manipulated.
6. Programming a computer can be fun—if you are patient, organized, and careful.

---

You have also seen how to use the FORTRAN programming language to perform some very fundamental operations. You have learned how to instruct the computer to read information into memory, perform some simple computations, and print from memory the results of the computation. All of this has been done using symbols (punctuation marks, variable names, and special operators such as * and −) that are familiar, easy to remember and easy to use. You needed to know virtually nothing about the computer you are using in order to understand and use FORTRAN.

In Table 1.2 we have provided a summary of all of the FORTRAN statements introduced in this chapter. An example of the use of each statement is also given. These examples can serve as guides to ensure that you use the correct syntax in the program statements that you write.

In the last part of this chapter, we showed two ways of using the computer:

- By creating a job deck of computer punch cards containing control cards, program cards, and data cards
- By using a timesharing system to run a program interactively

Although the control cards and statements for each computer system may differ, the FORTRAN programs will be very similar. The primary difference between a program keypunched on cards and a program entered at a terminal involves the use of prompting messages before each READ statement in the latter. There is also no need to echo print data values when working at a terminal since the values are normally displayed as they are entered.

The small amount of FORTRAN that you have seen in this chapter should enable you to solve a large variety of complex problems using the computer. The

| Statement type and use | Examples |
|---|---|
| DECLARATION: Informs the compiler of the list of variable names to be used in a program and the type of data to be stored in each variable | REAL GROSS, RATE, NET, TAX<br>INTEGER COUNT |
| ASSIGNMENT: Assigns a new value to a variable | TAX = 25.00<br>GROSS = HOURS * RATE<br>GROSS = NET |
| INPUT: Enters input data into a variable | READ*, HOURS, RATE |
| OUTPUT: Displays the value of a variable or constant | PRINT*, 'NET PAY = ', NET |
| STOP: Terminates execution | STOP |
| END: Informs the compiler that there are no more program statements to be translated | END |

**Table 1.2 Summary of FORTRAN Statements in Chapter 1.**

more you learn about FORTRAN, the easier it will be for you to write programs to solve more complicated problems on the computer. However, you will only learn by doing! Programming is not a spectator activity; it requires practice on a variety of different types of problems.

In the remainder of this text we will introduce more features of the FORTRAN language, and provide descriptions of the rules for using these features. You must remember throughout that, unlike the rules of English, the rules of FORTRAN are quite precise, and allow no exceptions. FORTRAN statements formed in violation of these rules will cause syntax errors in your programs.

You should not find the mastery of the rules of FORTRAN particularly difficult. The rules are precise and relatively few in number, especially compared to English. By far the most challenging aspect of your work will be the formulation of the logic and organization of your programs. For this reason, we will introduce you to a methodology for problem solving with a computer in the next chapter, and continue to emphasize this methodology throughout the remainder of the book.

## PROGRAMMING PROBLEMS*

**1B**    Write a program to read in the weight (in pounds) of an object, and compute and print its weight in kilograms and grams. *Hint:* one pound is equal to 0.453592 kilograms or 453.59237 grams.

**1C**    A cyclist coasting on a level road slows from a speed of 10 miles/hr. to 2.5 miles/hr. in one minute. Write a computer program that calculates the cyclist's constant rate of acceleration and determines how long it will take the cyclist to come to rest,

---

* For all text problems, output should be annotated. If you are using an interactive system, provide a prompt for each input item. If your are using a batch system, echo-print each item.

given an initial speed of 10 miles/hr.
*Hint:* Use the equation

$$a = \frac{v_f - v_i}{t}$$

where a is acceleration, t is time, $v_i$ is initial velocity, and $v_f$ is the final velocity.

**1D** Write a program to read three data items into variables X, Y, and Z, and find and print their product and sum.

**1E** Eight track stars entered the mile race at the Penn Relays. Write a program that will read in the race time in minutes, (MIN), and seconds, (SEC), for any one of these runners, and compute and print the speed in feet per second and in meters per second. *Hints:* There are 5280 feet in one mile and one meter equals 3.282 feet. Test your program on one of the times (minutes and seconds) given below.

| | | |
|---|---|---|
| 3.0 minutes 52.83 seconds | 3.0 minutes 56.22 seconds | 3.0 minutes 59.83 seconds |
| 4.0 minutes 00.03 seconds | 4.0 minutes 16.22 seconds | 4.0 minutes 19.00 seconds |
| 4.0 minutes 19.89 seconds | 4.0 minutes 21.21 seconds | |

**1F** You are planning to rent a car to drive from Boston to Philadelphia. Cost is no consideration, but you want to be certain that you can make the trip on one tankful of gas. Write a program to read in the miles-per-gallon and tank size in gallons for a particular rent-a-car, and print out the distance that can be travelled on one tank. Test your program for the following data:

| miles-per-gallon | tank size (*gallons*) |
|---|---|
| 10.0 | 15.0 |
| 40.5 | 20.0 |
| 22.5 | 12.0 |
| 10.0 | 9.0 |

**1G** Write a program that prints your initials in large block letters. *Hint:* Use a 6 × 6 grid for each letter and print six messages. Each message should consist of a row of *'s interspersed with blanks.

**1H** Every Sunday afternoon you take an afternoon drive in your car. Before leaving, you fill your gas tank with lead-free gasoline, which costs $1.25 a gallon. Write a program that you could use to read in the distance traveled (DIST) each week, the time (TIME) required to make the trip, and the miles-per-gallon estimate (MPG) for your car. Compute and print the average speed traveled, and the estimated cost of your trip.

**1I**    Write a program that reads a pair of numbers and prints out the product of these numbers.

**1J**    Write a program that reads in a single value RADIUS, and then computes and prints:

- the area of a circle having the given radius
- the volume of a sphere having the same radius

*Hint:* The area of a circle is given by:

$$\text{area} = \pi r^2$$

and the volume of a sphere is

$$\text{volume} = \frac{4}{3}\pi r^3$$

where $\pi = 3.14159$ and r is the radius.

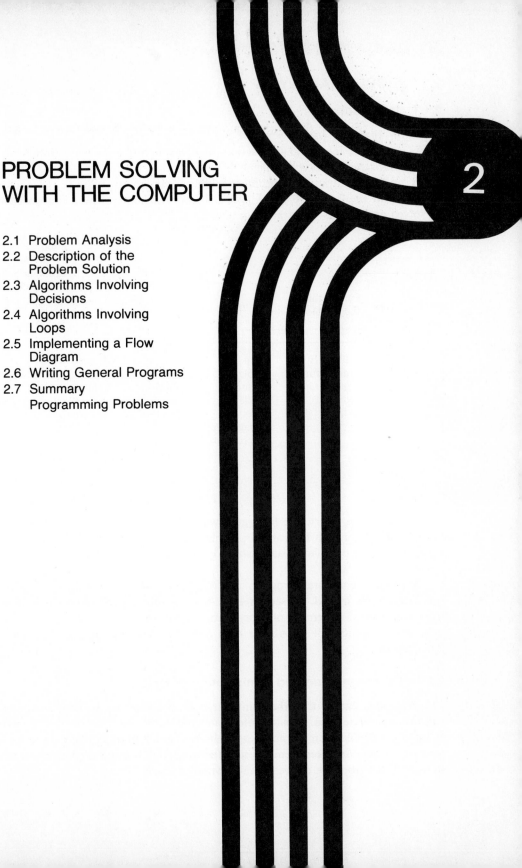

# PROBLEM SOLVING WITH THE COMPUTER

2

## 2.1    PROBLEM ANALYSIS

### 2.1.1    Introduction

Now that you have been introduced to the computer—what it is, how it works and what it can do—it is time to turn our attention to learning how to use the computer to solve problems.

Using the computer for problem solving is similar to trying to put a man on the moon in the late 1950's and 1960's. In both instances, there is a problem to be solved and a final "program" for solving it.

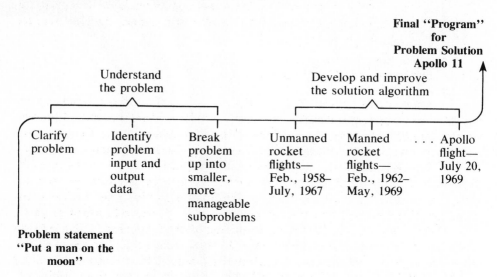

In the moon effort, the final goal was not achieved directly. Rather, it was brought about through the careful planning and organization of subtasks, each of which had to be completed successfully before the Apollo 11 flight could even be attempted.

Writing a computer program also requires careful planning and organization. It is rare, indeed, to see an error-free computer program written directly from the original statement of a problem. Usually, the final program is achieved only after a number of steps have been followed. These steps are the subject of this chapter.

### 2.1.2    Representation and Manipulation of Data

We stated earlier that the computer is a tool that can be used to represent and to manipulate data. It is, therefore, not too surprising that the first two steps in solving a problem on the computer require the definition of the data to be represented in the computer memory, and the formulation of an *algorithm*—a list of steps that describes the desired manipulation of these data.

These two steps are not entirely unrelated. Decisions that we make in defining the data may be subject to numerous changes throughout the algorithm formulation. Nevertheless, it is essential that we perform the data definition in as complete and precise a fashion as possible before constructing the algorithm. Careless errors, or errors in judgement in deciding what information is to be represented and what form this information is to take, can result in numerous difficulties in the later stages of solving a problem on the computer. Such mistakes can make the algorithm formulation extremely difficult, and sometimes even impossible.

Once the definition of the information to be represented in the computer has been made and a precise statement of the problem is available, the algorithm for solving the problem can be formulated.

### 2.1.3  Understanding the Problem

The definition of the data to be represented in the computer memory requires a clear understanding of the stated problem. First, we must determine what information is to be computed and printed by the computer. Then it is necessary to identify the information needed as input in order to produce the desired output. Once the input and output data have been identified, we must ask if sufficient information is available to compute the required output from the given input. If the answer to this question is no, we must determine what additional information is needed and how this information can be provided to the program.

When identifying the data items associated with the problem, it is helpful to assign to each item a descriptive variable name that can be used to represent the computer memory cell containing the data item. Recall from Chapter 1 that we do not have to be concerned with the actual memory cell associated with each variable name. The compiler will assign a unique memory cell to each variable name and it will handle all bookkeeping details necessary to retain this correspondence.

We will apply this data definition process to a specific problem.

**Problem 2A:**  Write a program to compute and print the sum and average of two numbers.

**Discussion:**  The first step is to make certain that we understand the problem and to identify the input and output data for the problem. Then we can obtain a more precise formulation of the problem in terms of these input and output items.

All items of information to be used to solve a given problem should be listed in a *data table*, along with a description of the variable used to represent each data item. The data table for Problem 2A is given next. The entries shown describe the input and output data for the problem.

**Data Table for Problem 2A**

| *Input variables* | *Output variables* |
|---|---|
| NUM1: First number to be used in computation | SUM: Sum of two numbers |
| NUM2: Second number to be used in computation | AVE: Average of two numbers |

There are clearly two items of information required as output for this problem. They are the sum and the average of two numbers. In order to compute these values, we must be able to store the data items to be summed and averaged into the memory of the computer. In this example, we will use the variables NUM1 and NUM2 to represent these two data items.

The table form just illustrated will be used for all data tables in the text. Variables whose values are entered through READ statements are listed as input variables; variables whose values represent final computational results required by the problem statement are listed as output variables. In all cases, it is important to include in the data table a short, concise description of how each variable is to be used in the program.

The data table is valuable not only during algorithm development but also as a piece of *program documentation*. It is a convenient reference document for associating variable names and their uses in the program. You should always prepare a data table, pay close attention to it during the algorithm development process, and save it along with your program listing. The data table may subsequently turn out to be your only reminder of how the variables in your program are being used.

Now that the information required for Problem 2A has been identified, a more precise formulation of the problem is possible: we must read two data items into the variables NUM1 and NUM2, find the sum and the average of these two items, and print the values of the sum and the average.

## 2.2   DESCRIPTION OF THE PROBLEM SOLUTION

### 2.2.1   Developing an Algorithm

At this point we should have a clearer understanding of what is required for the solution of Problem 2A. We can now proceed to organize the problem formulation into a carefully constructed list of steps—the algorithm—that will describe the sequence of manipulations to be performed in carrying out the problem solution.

**Algorithm for Problem 2A (Level One)**

Step 1     Read the data items into the variables NUM1 and NUM2 (echo print these values if using a batch system).

Step 2    Compute the sum of the data items in NUM1 and NUM2 and store the result in the variable SUM.

Step 3    Compute the average of the data items in NUM1 and NUM2 and store the result in the variable AVE.

Step 4    Print the values of the variables SUM and AVE.

Step 5    Stop.

## 2.2.2    Algorithm Refinement

Note that this sequence of events closely mirrors the problem formulation given earlier. This is as it should be! If the problem formulation is complete, it should provide us with a general outline of what must be done to solve the problem. The purpose of the algorithm is to provide a detailed and precise description of the individual steps to be carried out by the computer. The algorithm is essentially a *refinement* of the general outline provided by the original problem formulation. It is often the case that several *levels of refinement* of the general outline are required before the algorithm formulation is complete.

The key question in deciding whether or not further refinement of an algorithm step is required is this:

Is it clear what FORTRAN instructions are necessary in order to tell the computer how to carry out the step?

If it is not obvious what the FORTRAN instructions are, then the algorithm should be further refined.

What is obvious to some programmers may not be at all clear to others. The refinement of an algorithm is, therefore, a personal matter to some extent. As you gain experience in developing algorithms and converting them to programs, you may discover that you are doing less and less algorithm refinement. This may also happen as you become more familiar with the FORTRAN language.

If we examine the level one algorithm for Problem 2A, we see that only Step 3 may require further refinement. We already know how to write instructions for reading, printing, adding and stopping. However, we may not know how to tell the computer to find the average of two numbers.

### Refinement of Step 3

Step 3.1    Divide the sum (stored in SUM) by the number of data items (2).

We now have a complete description of the algorithm and the data table, and can proceed to write the FORTRAN representation for the algorithm (Fig. 2.1). We do this on a step-by-step basis, starting with the necessary declarations as indicated by the data table.

All variables used to store program data should already be listed in the data table; each of these variables should appear in a declaration statement (REAL or INTEGER) at the beginning of the program. Following the declarations, each step in the algorithm is translated into FORTRAN and listed in the program. If a step has been refined, the refinement is expressed in FORTRAN instead.

```
C COMPUTE THE SUM AND AVERAGE OF TWO NUMBERS
C
      REAL NUM1, NUM2, SUM, AVE
C
C READ AND PRINT DATA
      READ*, NUM1
      PRINT*, 'THE FIRST NUMBER IS ', NUM1
      READ*, NUM2
      PRINT*, 'THE SECOND NUMBER IS ', NUM2
C
C COMPUTE THE SUM AND AVERAGE OF THE DATA
      SUM = NUM1 + NUM2
      AVE = SUM / 2
C
C PRINT RESULTS
      PRINT*, 'THE SUM IS ', SUM
      PRINT*, 'THE AVERAGE IS ', AVE
C
      STOP
      END


*** PROGRAM EXECUTION OUTPUT ***

THE FIRST NUMBER IS   13.50000000000
THE SECOND NUMBER IS  -6.000000000000
THE SUM IS   7.500000000000
THE AVERAGE IS   3.750000000000
```

**Fig. 2.1**   FORTRAN program and sample output for Problem 2A.

The output printed during execution of this program is shown at the bottom of Fig. 2.1. The message

```
    *** PROGRAM EXECUTION OUTPUT ***
```

is printed by our computer system above the program output.

The lines in Fig. 2.1 that start with C (in column one) are descriptive comments. They are ignored by the compiler during translation and are listed with the program statements to aid the programmer in identifying or documenting the purpose of each section of the program. Guidelines for the use of comments are provided next.

---

**Program Style**

*Use of comments*

All programs should start with a comment describing the purpose of the program. Some programmers also insert a comment that provides their name and the date the program was created (or last modified). Each subsequent comment should describe the purpose of the program statements that follow it. There should be enough comments to clarify the intent of each section of your program; however, too many comments can clutter the program, make it difficult to read and waste time and space. A good rule of thumb is to use a comment to identify the FORTRAN implementation of each step in the level one algorithm as well as any other steps requiring further refinement. In this way, the correspondence between the level one algorithm and its FORTRAN implementation becomes obvious.

Comments should be carefully worded. One suggestion is to use an abbreviated form of the corresponding algorithm step description. For example, the comment in the program segment

```
C COMPUTE THE SUM AND AVERAGE OF THE DATA
      SUM = NUM1 + NUM2
      AVE = SUM / 2
```

conveys more information and, hence, is better than the comment

```
C ADD NUM1 TO NUM2 AND DIVIDE BY 2
```

which is simply an English description of the FORTRAN statements that follow the comment.

---

**Program Style**

*Echo prints and prompts*

In Fig. 2.1, the implementation of Step 1 of the algorithm includes a statement to echo print the input data values, NUM1 and NUM2. We strongly recommend that you do this whether or not you are requested to in the problem specification. If you are writing an interactive program at a terminal, it is better to print prompting messages to the user immediately prior to each read. These messages should specify what data is to be entered when the program pauses.

---

**Exercise 2.1:** Rewrite the "READ AND PRINT DATA" section of the sum and average program (Fig. 2.1) to provide prompting messages for interactive excution.

**Exercise 2.2:** Write a data table and an algorithm for computing the sum and average of four real numbers.

### 2.2.3  Flow Diagram Representations of Algorithms

As problems become more complicated, precise English descriptions of algorithms for solving these problems become more complex and difficult to follow. It is, therefore, helpful if some kind of notation can be used to specify an algorithm. We will use one such descriptive notation, called a *flow diagram*, throughout this text.

Not everyone in the computer field believes that flow diagrams are useful and many experienced programmers do not always use them. However, we believe that flow diagrams are helpful because they provide a graphical, two-dimensional representation of an algorithm. Consistent use of the special flow diagram symbols and forms shown in the text will make algorithms easy to write, easy to refine and still easier to follow.

Flow diagram representations of two levels of the algorithm for Problem 2A are shown in Fig. 2.2. They contain a number of symbols that should be noted.

1. Ovals are used to indicate the starting and stopping points of an algorithm.
2. Rectangular boxes are used to indicate the manipulation of information in the memory of the computer.
3. A box in the shape of a computer card (with one corner cut off) is used to indicate the reading of information into the computer.
4. A box with a wavy bottom is used to indicate the printing of information stored in the computer memory.
5. Arrows are used to indicate the "flow of control" of an algorithm from one step to another.

You will find it convenient to represent all levels of algorithms with a flow diagram. The first level will often be quite general. It will contain a summary, usually written in English, of the basic steps of an algorithm, as shown on the left side of Fig. 2.2. In some cases, usually when the step is very simple, these summaries may be precise and detailed. However, in other cases, one or more levels of refinement will be necessary before a sufficiently detailed diagram is completed. Such refinements are illustrated in the right side of Fig. 2.2. The dotted arrows point to the refinements of Steps 1, 2, 3 and 4; the solid arrows indicate flow of control from one step to the next. Remember that refinements are for your benefit; they should be used when additional detail is needed to clarify what must be done to complete an algorithm step.

### 2.2.4  Problem Solving Principles

Up to now we have presented a few suggestions for solving problems on the computer. These suggestions are summarized below.

1. Understand what you are being asked to do.
2. Identify all problem input and output data. Assign a variable name to each input or output item and list it in the data table.
3. Formulate a precise statement of the problem in terms of the input and output data and make certain there are sufficient input items provided to complete the solution.
4. State clearly the sequence of steps necessary to produce the desired problem output; i.e., develop the algorithm and represent it as a flow diagram.
5. Refine this flow diagram until it can be easily implemented in the programming language to be used. List any additional variables required in the data table.
6. Transform the flow diagram to a program.

Steps 4 and 5 are really the most difficult of the steps listed; they are the only truly creative part of this process. People differ in their degree of capability to formulate solutions to problems. Some find it easy to develop algorithms for the most complex problem, while others must work diligently to produce an al-

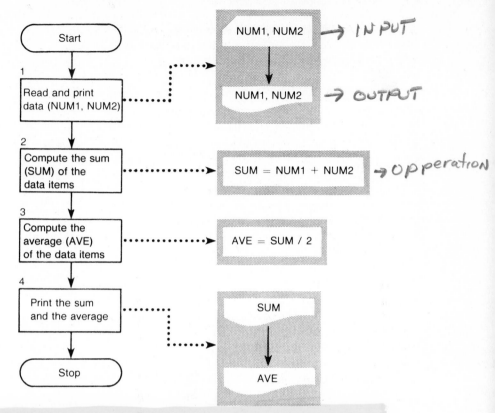

**Fig. 2.2**  Level one flow diagram and refinements for the sum and average problem (2A).

gorithm for solving a simple problem.

The ability to solve problems is fundamental to computer programming. The transformation of the refined algorithm to a working program (Step 6) is a highly skilled clerical task that requires a thorough knowledge of the programming language available. This detailed knowledge can normally be acquired by anyone willing to devote the necessary effort. However, a flow diagram that correctly represents the necessary problem-solving operations and their relationship must first be developed.

In this book, we will provide detailed solutions to many sample problems. Examining the text solutions carefully should enable you to become more adept at formulating your own solutions, because the techniques used for one problem may frequently be applied in a slightly different way to solve another. Often, new problems are simply expansions or modifications of old ones.

The process of outlining and refining problem solutions can be used to break a complex problem up into more manageable subproblems that can be solved individually. This technique will be illustrated in all of the problems solved in the text. We suggest you practice it in developing your own solutions to the programming problems.

## 2.3   ALGORITHMS INVOLVING DECISIONS

### 2.3.1   Decision Steps and Conditions

Normally, the steps of an algorithm are performed in the order in which they are listed. In many algorithms, however, the sequence of steps to be performed is determined by the input data. In such cases, decisions must be made, based upon the values of certain variables, as to which sequence of steps is to be performed. Such decisions require the evaluation of a condition that is expressed in terms of the relevant variables. The result of the evaluation determines which algorithm steps will be executed next. An example of this is illustrated in the following problem, which is a modification of the payroll problem discussed in Chapter 1.

**Problem 2B:** *Modified payroll problem.* Compute the gross salary and net pay for an employee of a company, given the number of hours worked and the employee's hourly wage rate. Deduct a tax amount of $25 but only if the employee's gross salary exceeds $100.

The data table for this problem is shown below. The flow diagrams are drawn in Fig. 2.3.

**Data Table for Problem 2B**

| *Input variables* | *Output variables* |
|---|---|
| HOURS: Number of hours worked | GROSS: Gross salary |
| | NET: Net pay |
| RATE: Hourly wage rate | |

In numbering flow diagrams and their refinements, we will use a scheme that is analogous to the numbering of sections in this text. For example, the refinements of Step 3 are numbered 3.1, 3.2, 3.3. If Step 3.1 were to be refined further, its refinements would be numbered 3.1.1, 3.1.2, etc. All steps in a level one flow diagram will be numbered. Normally, only those refinement steps that are referred to in the text narrative will be numbered.

As shown in the flow diagram refinement of Step 3 (see Fig. 2.3), the *decision step* (3.1) describes a *logical condition* ("GROSS greater than 100.00") that is evaluated in order to determine which algorithm step should be executed next. If the condition is true, Step 3.2 (deduct tax) is performed next, as indicated by the arrow labelled T. Otherwise, Step 3.3 (assign GROSS to NET) is performed, as indicated by the arrow labelled F. In either case, Step 4 will be carried out following the completion of the chosen step.

As illustrated in Fig. 2.3, a logical condition describes a particular relationship between a pair of variables, or a variable and a constant. The value of the

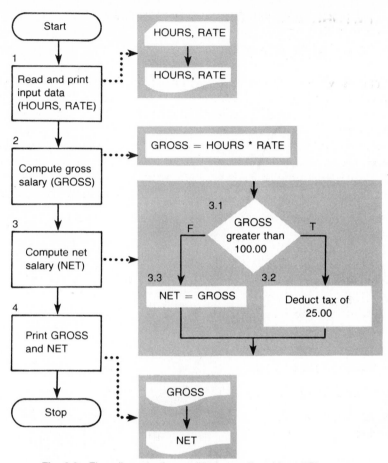

**Fig. 2.3**  Flow diagrams for modified payroll problem (2B).

condition is true if the specified relation holds for the current variable values; otherwise, the condition value is false. Examples of conditions are shown in Table 2.1. The *relational operators* that may be used in writing logical conditions are summarized in Table 2.2.

| Condition | FORTRAN form |
|-----------|--------------|
| G greater than MAX | G .GT. MAX |
| X equal to SUM | X .EQ. SUM |
| X not equal to 0 | X .NE. 0 |
| G less than or equal to 10 | G .LE. 10 |

**Table 2.1 Examples of FORTRAN conditions.**

| Relational operator | Meaning |
| --- | --- |
| .EQ. | equal to |
| .NE. | not equal to |
| .LT. | less than |
| .GT. | greater than |
| .LE. | less than or equal to |
| .GE. | greater than or equal to |

**Table 2.2 FORTRAN relational operators.**

The decision step (Step 3.1) used in Problem 2B involves a choice between two alternatives—a sequence of one or more steps to be executed if the condition is true (the True Task) and a sequence to be executed if the condition is false (the False Task). Such a decision step is called a *double-alternative decision step*. The general flow diagram pattern for this step is shown in Fig. 2.4.

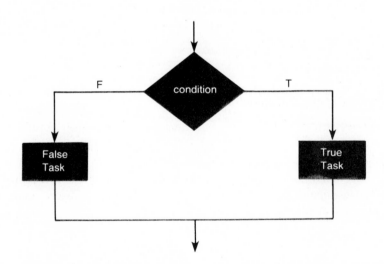

**Fig. 2.4**  Flow diagram pattern for the double-alternative decision step.

Quite often, a decision step in an algorithm will involve only one alternative: a sequence of one or more steps that will be carried out if the given condition is true, but skipped if the condition is false. The flow diagram pattern for this *single-alternative decision step* is shown in Fig. 2.5.

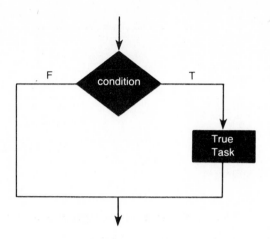

**Fig. 2.5**   Flow diagram pattern for the single-alternative decision step.

**Exercise 2.3:**   Let A and B have the values $-3.5$ and $6.2$, respectively. Indicate the values (true or false) for each of the following conditions.

        a) A .NE. B                c) B .LE. A

        b) A .GT. B              d) B .EQ. 6.2

## 2.3.2   The Largest Value Problem

In the next problem, we illustrate the use of both the single- and double-alternative decision steps.

**Problem 2C:**   Read and print three numbers and find and print the largest of these numbers.

The data table for Problem 2C follows; the flow diagrams are drawn in Fig. 2.6.

**Data Table for Problem 2C**

| *Input variables* | *Output variables* |
|---|---|
| N1: First number | LARGE: The largest |
| N2: Second number | number found at any |
| N3: Third number | point in the program |

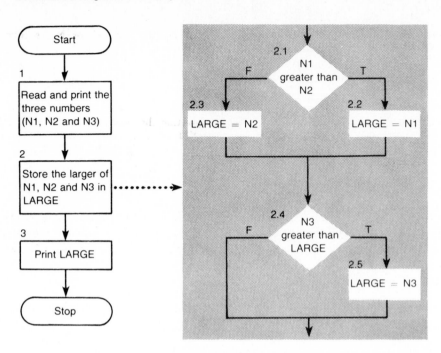

**Fig. 2.6** Flow diagrams for largest of three numbers problem (2C).

As shown in Fig. 2.6, once the data are read into N1, N2 and N3 and printed (Step 1), the double-alternative decision step (2.1–2.3) is used to store the larger of N1 and N2 in LARGE. The single-alternative decision Step (2.4–2.5) is used to compare N3 to this value, and copy N3 into LARGE if N3 is larger. If LARGE is already greater than or equal to N3, no action is taken following Step 2.4.

In the next section, we will see how to express decision steps in FORTRAN. We will see that it is relatively easy to go from a flow diagram to a program, even when complicated decision steps are required.

**Exercise 2.4:** Write the flow diagram pattern to represent the following English descriptions:

    a)   If ITEM is not equal to ZERO, then multiply PRODCT by ITEM. Otherwise, skip this step. In either case, then print the value of PRODCT.

    b)   If ITEM exceeds LARGER, store the value of ITEM in LARGER. Otherwise skip this step. In either case, then print the value of ITEM.

    c)   If X is larger than 0.0, add X to SUMPOS. Otherwise, if X is smaller than 0.0, add X to SUMNEG. Otherwise, if X = 0.0, add one to CNTZRO.

**Exercise 2.5:** What values would be printed by the algorithm in Fig. 2.3 if HOURS is 37.5 and RATE is 3.75? If HOURS is 20.0 and RATE is 4.0? "Execute" the program yourself to determine the results.

**Exercise 2.6:** What happens in Fig. 2.6 if N1 is equal to N2 or N3 is equal to LARGE? Does the algorithm work for these cases?

**Exercise 2.7:** Modify the flow diagram in Problem 2C to find the largest of four numbers.

**Exercise 2.8:** Draw a flow diagram for an algorithm that computes the absolute difference between two numbers. If X is greater than Y, the absolute difference is $X-Y$; if Y is greater than X, the absolute difference is $Y-X$.

### 2.3.3  Implementing Decision Structures in FORTRAN

We can implement the algorithms in the previous section quite easily in FORTRAN through the use of the block-IF structure. The block-IF will be discussed more formally in Chapter 3; for the time being, however, we will use the simplified form

```
IF (condition) THEN
     True  Task
ELSE
      False  Task
ENDIF
```

to implement a double-alternative decision. The form

```
IF (condition) THEN
     True  Task
ENDIF
```

will be used to implement a single alternative decision. These forms directly mirror the flow diagram patterns described earlier.

The FORTRAN implementation of the modified payroll problem (Fig. 2.3) is shown in Fig. 2.7; and the largest of three numbers problem (Fig. 2.6) is shown in Fig. 2.8. You should compare each FORTRAN program with its corresponding flow diagram representation.

### 2.3.4  Computing Social Security Tax

The following problem provides an illustration of the points discussed thus far in this chapter.

**Problem 2D:** Write a program to read an employee's salary and compute the Social Security tax due. The Social Security tax is 5.6 percent of an employee's gross salary. Only the first $25,000 earned is taxable. (This means that all employees earning $25,000 or more pay the same tax.)

The data table for Problem 2D follows; the level one algorithm for Problem 2D is given in Fig. 2.9a.

```
C MODIFIED PAYROLL PROBLEM - DEDUCTS TAX CONDITIONALLY
C
      REAL HOURS, RATE, GROSS, NET
C
C READ AND PRINT HOURS AND RATE
      READ*, HOURS, RATE
      PRINT*, 'HOURS = ', HOURS
      PRINT*, 'RATE = ', RATE
C
C COMPUTE GROSS SALARY
      GROSS = HOURS * RATE
C
C COMPUTE NET PAY
      IF (GROSS .GT. 100.00) THEN
         NET = GROSS - 25.00
      ELSE
         NET = GROSS
      ENDIF
C
C PRINT GROSS AND NET
      PRINT*, 'GROSS = ', GROSS
      PRINT*, 'NET = ', NET
C
      STOP
      END


*** PROGRAM EXECUTION OUTPUT ***

HOURS =   30.00000000000
RATE =    4.500000000000
GROSS =   135.0000000000
NET =   110.0000000000
```

**Fig. 2.7**  FORTRAN program and sample output for Problem 2B.

**Data Table for Problem 2D**

| Input variables | Output variables |
|---|---|
| SALARY: Employee's salary | SOCTAX: Social Security tax |

In order to refine Step 2, a new variable, TAXSAL, is introduced to represent the portion of SALARY on which the Social Security tax will be computed. Since TAXSAL is neither given as input data nor requested as output, but is instead the result of an intermediate computation, we should list it as a *program variable* in the data table.

```
C FIND THE LARGEST OF THREE NUMBERS
C
      REAL N1, N2, N3, LARGE
C
C READ AND PRINT THE THREE NUMBERS
      READ*, N1, N2, N3
      PRINT*, 'THE THREE NUMBERS ARE '
      PRINT*, N1, N2, N3
C
C STORE THE LARGEST OF N1, N2 AND N3 IN LARGE
      IF (N1 .GT. N2) THEN
         LARGE = N1
      ELSE
         LARGE = N2
      ENDIF
      IF (N3 .GT. LARGE) THEN
         LARGE = N3
      ENDIF
C
C PRINT THE LARGEST VALUE
      PRINT*, 'THE LARGEST VALUE IS ', LARGE
C
      STOP
      END

*** PROGRAM EXECUTION OUTPUT ***

THE THREE NUMBERS ARE
   20.00000000000      30.00000000000      10.00000000000
THE LARGEST VALUE IS   30.00000000000
```

**Fig. 2.8**  FORTRAN program and sample output for PROBLEM 2C.

## Additional Data Table Entry for Problem 2D

*Program variable*

TAXSAL: The portion of
SALARY for which
SOCTAX is computed

The level two and three refinements of Step 2 are shown in Fig. 2.9b. The program for Problem 2D is shown in Fig. 2.10.

---

**Program Style**

*Indenting true and false tasks*

In Figs. 2.7, 2.8 and 2.10, we have indented the statements representing the True and False Tasks of a decision structure. This is not required; however, it is a good practice to follow as it clarifies the meaning of the block-IF and improves the readability of the program.

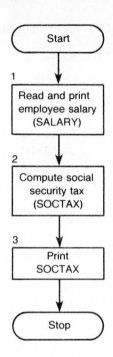

**Fig. 2.9a** Level one algorithm for social security tax problem (2D).

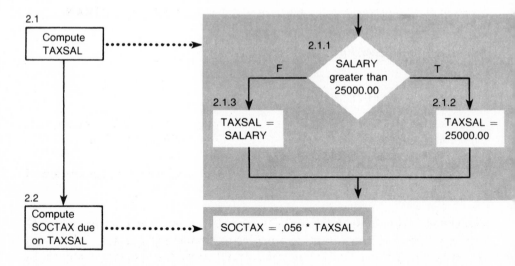

**Fig. 2.9b** Level two and three refinements of Step 2.

```
C MODIFIED PAYROLL PROBLEM - WITH SOCIAL SECURITY TAX DEDUCTION
C
      REAL SALARY, SOCTAX, TAXSAL
C
C READ AND PRINT SALARY
      READ*, SALARY
      PRINT*, 'SALARY IS ', SALARY
C
C COMPUTE SOCIAL SECURITY TAX (AFTER DETERMINING TAXABLE SALARY)
      IF (SALARY .GT. 25000.00) THEN
         TAXSAL = 25000.00
      ELSE
         TAXSAL = SALARY
      ENDIF
      SOCTAX = 0.056 * TAXSAL
C
C PRINT SOCIAL SECURITY TAX
      PRINT*, 'SOCIAL SECURITY TAX IS ', SOCTAX
C
      STOP
      END


*** PROGRAM EXECUTION OUTPUT ***

SALARY IS    23000.00000000
SOCIAL SECURITY TAX IS    1288.000000000
```

**Fig. 2.10**   FORTRAN program and sample output for Problem 2D.

**Exercise 2.9:**   Implement your solutions to Exercises 2.7 and 2.8 in FORTRAN.


## 2.4   ALGORITHMS INVOLVING LOOPS

### 2.4.1   The Motivation for Loops

The algorithm for finding the sum and average of two numbers works quite well. Suppose, however, that we are asked to solve a different problem.

**Problem 2E:**   Write a program to compute and print the sum and average of 2000 data items.

**Discussion:**   The first question to be answered now concerns whether or not the approach previously taken will be satisfactory for this problem too. The answer is clearly *no*! It is not that the approach won't work, but rather that no reasonable person is likely to have the patience to carry out this solution for 2000 numbers. Our difficulties would begin in attempting to produce a data table listing differently named variables for each of the 2000 items involved.

**Data Table for Problem 2E**

| *Input variables* | *Output variables* |
|---|---|
| NR1: First data item | SUM: Sum of 2000 data items |
| NR2: Second data item | |
| | AVE: Average of the 2000 data items |
| NR3: Third data item | |
| ⋮ | |
| NR2000: 2000th data item | |

This in itself is a horrendous task. Then, assuming we could finally name all 2000 variables, we would have quite a boring task in describing the algorithm for solving the problem. Not even little children enjoy drawing pictures that much!

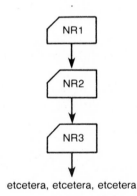

etcetera, etcetera, etcetera

A new approach is needed in order to solve this problem. Regardless of what this new approach involves, it will still be necessary to tell the computer to read in and add together 2000 numbers. The essence of the problem is to find a way to do this without writing separate instructions for the reading and the addition of each of the 2000 data items needed to compute the sum. It would be ideal if we could write one step for reading, one step for accumulating the sum, and then repeat these two steps for each of the 2000 items.

It happens that we can actually achieve this goal quite easily. All that is necessary is to (a) solve the problem of naming each data item, (b) learn how to describe a repeated sequence of steps in a flow diagram, and (c) learn how to specify the repetition of a sequence of steps in FORTRAN.

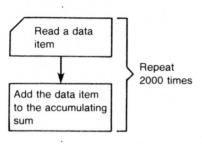

The solution to the naming problem rests upon the following realization:

*Once a data item has been read into the computer memory and added to the sum, it is no longer needed in the computer memory.*

Thus, each input data item can be read into the same variable. After each item is entered, the value of this variable can be added to SUM, and the next data item can be read into the same variable. This, of course, destroys the previous data item, but it is no longer needed for the computation.

To see how this works, consider what happens if we try to carry out an algorithm consisting solely of the repetition of the steps

1. Read a data item into the variable named ITEM.
2. Add the value of ITEM to the value of SUM and store the result in SUM.

To begin, the memory cells ITEM and SUM appear as shown below. The initial value of SUM must be zero; otherwise, the final result will be incorrect by an amount equal to whatever was initially stored in SUM.

ITEM        SUM
?           0.0

Let us assume that the first three data items are the numbers 10.0, −11.0, and 6.0. After steps (1) and (2) are performed the first time, the variables ITEM and SUM will be defined as follows:

Note that the number 10.0 has now been incorporated into the sum that we are computing, and is no longer required for this problem. We may therefore read the next data item into the variable ITEM. After the second execution of (1) and (2), we have:

and upon completion of the third execution of (1) and (2), we obtain:

This process continues for all 2000 data items. With each execution of Steps 1 and 2, the data item just read in is used as required by the problem and can be replaced in memory by the next data item.

With this solution to the naming problem, the data table for Problem 2E can be rewritten relatively easily.

### Revised Data Table for Problem 2E

| *Input variables* | *Output variables* |
|---|---|
| ITEM: Contains each data item as it is being processed | SUM: Used to accumulate the sum of the data items |
| | AVE: Average of all data items |

We can also write a level one flow diagram for our algorithm (Fig. 2.11).

From this diagram, it is clear what is required in Step 2 and in the output phase (Step 3) of the algorithm. However, part of the computation phase (Step 1) requires further refinement before the program can be written.

In order to refine algorithm Step 1, we need to have a flow diagram representation for a sequence of repeated steps. This representation, shown in Fig. 2.12, is called a *loop*.

The loop is always entered through a box containing the *loop repetition test*. The *loop body* is the sequence of steps that is to be repeated if the loop repetition test is true. It is connected to the rest of the flow diagram by an arrow drawn to the right of the diagram (labelled T). The *exit* arrow (labelled F) always points to the first step in the algorithm that is to be carried out upon completion of the loop; i.e., when the loop repetition test is false. The dashed line in Fig. 2.12 serves as a reminder that control returns to the loop repetition test each time the loop body is executed. It is not part of the flow diagram and we will omit it in later chapters.

How do we know when the loop is complete? More importantly, how can we tell the computer when it has completed the execution of the loop? A person might do it ten or 100 times and then ask, "Am I done yet?" However, we are

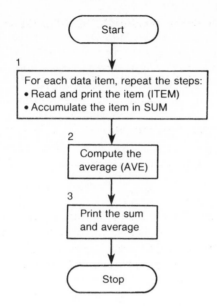

Fig. 2.11  Level one flow diagram for the sum and average of 2000 numbers problem (2E).

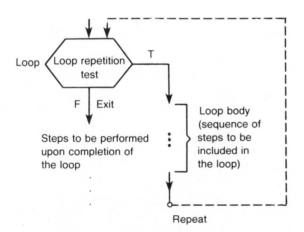

Fig. 2.12  Flow diagram pattern for a loop.

developing an algorithm that will eventually take the form of a sequence of steps to be performed by a computer—and, the computer cannot think!! Therefore, if we want to tell it to repeat a sequence of steps, it is not enough to tell it what those steps are. We must also tell the computer when to stop performing these steps. This requires additional *loop control steps* in the algorithm.

Let us examine what is required in defining the loop control steps in the algorithm for Problem 2E. Clearly, we want the steps in the loop (the *loop body*) to be executed once for each data item to be input. This means that we must perform the input and summation steps of the loop exactly 2000 times. One way to do this is to use a variable in the program as a *counter* to count the number of data items processed at any point during the execution of the algorithm. This variable will serve as the *loop control variable* for the summation loop. The value of the counter will initially be zero, because at the start of the algorithm no data items have been read in. After each data item is read into the computer and processed, the value of the counter will be *incremented* (increased) by one. When this value reaches 2000, we can tell the computer to *exit* from the loop. At this point, 2000 data items will have been read into the computer memory and accumulated in the variable SUM.

To follow this sequence of steps, we should first complete the data table for Problem 2E. There is an additional program variable entry to be made as shown below.

**Additional Data Table Entry for Problem 2E**

*Program variables*

COUNTR: Loop control
variable representing the
number of data items
processed at any point.

Now we can complete the refinement of Step 1 in the level one flow diagram of Fig. 2.11. This refinement is shown in Fig. 2.13.

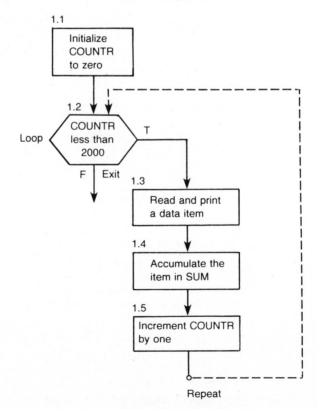

**Fig. 2.13**  Refinement of Step 1 (Fig. 2.11).

As indicated earlier, it is necessary that COUNTR initially be set to zero in order to indicate that the loop body has not yet been executed. Otherwise, the loop will not be repeated the correct number of times. Further, a careful analysis of the flow diagram in Fig. 2.13 would indicate a serious omission: namely, failure to initialize SUM to 0.0 prior to entry to the loop. This error can be corrected by inserting the initialization of SUM as the first step (Step 1.0) in the refinement of Step 1.

The level one flow diagram and revised refinement for Problem 2E are shown in Fig. 2.14. We have numbered all steps for reference purposes.

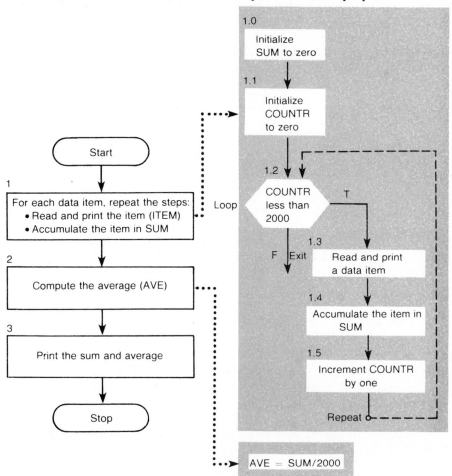

**Fig. 2.14**   Complete flow diagrams for Problem 2E.

Every loop that we write will follow the general pattern shown in Fig. 2.12. The loop control will always consist of three steps that manipulate the loop control variable (COUNTR in this case). These steps involve

- A loop repetition test (Step 1.2 in Fig. 2.14)
- A loop control initialization step (Step 1.1 in Fig. 2.14)
- A loop control update step (Step 1.5 in Fig. 2.14)

The diagram in Fig. 2.15 shows the relationship among these steps. The loop control variable (LCV) must be set to an initial value when the loop is first reached. The loop repetition condition is always tested prior to each execution of the loop body. As long as the condition is true, the loop body will be repeated and the LCV will be updated. The LCV update is usually the last step in the loop body. When the condition becomes false, the loop will be exited and the next algorithm step will be executed.

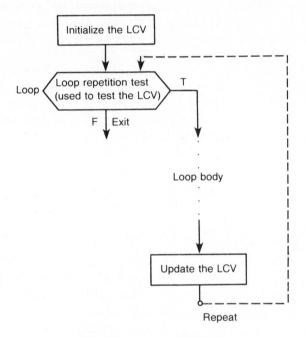

**Fig. 2.15**   Loop control steps.

Exactly how the loop repetition test is constructed depends upon the problem to be solved. For each problem, we must decide how to define the repeat condition to ensure precisely the number of loop executions that we require: no more, and no less. For some loops, this is an extremely difficult task; for others it is easy.

**Exercise 2.10:**   Follow the steps shown in the algorithm of Fig. 2.14 to determine the sum of the four items 8.2, $-6.1$, 0.0, 2.4.

## 2.5   IMPLEMENTING A FLOW DIAGRAM

### 2.5.1   Tracing a Flow Diagram

Once the algorithm and data table for a problem are complete, it is important to verify that the algorithm specifies the correct steps before you begin to write the program. This algorithm verification can be carried out by manually following or *tracing* the sequence of steps indicated by the algorithm. Such traces can often lead to the discovery of a number of logical errors in the flow diagram. The correction of these errors prior to writing the FORTRAN instructions can save considerable effort during the final checkout, or *debugging,* of the FORTRAN program.

It is clear that we cannot trace the program for Problem 2E for 2000 data items. However, we will perform the trace for just three data items. If the algorithm works properly for this simpler case, it should work for 2000 data items as well.

The trace is shown in Table 2.3. The algorithm step numbers are from the flow diagram in Fig. 2.14. Only the new value of the variable affected by an algorithm step is shown to the right of the step. All other variable values are unchanged. The values of all variables at the start of execution are undefined as shown in the first line. The data items being tested are 12.5, 15.0, and $-3.5$.

| Algorithm step | ITEM | Variable changed COUNTR | SUM | AVE | Effect |
|---|---|---|---|---|---|
| | ? | ? | ? | ? | |
| 1.0 | | | 0.0 | | |
| 1.1 | | 0 | | | |
| 1.2 | | | | | $0 < 3$ is true — execute loop |
| 1.3 | 12.5 | | | | Read & print 12.5 |
| 1.4 | | | 12.5 | | |
| 1.5 | | 1 | | | |
| 1.2 | | | | | $1 < 3$ is true — execute loop |
| 1.3 | 15.0 | | | | Read & print 15.0 |
| 1.4 | | | 27.5 | | |
| 1.5 | | 2 | | | |
| 1.2 | | | | | $2 < 3$ is true — execute loop |
| 1.3 | $-3.5$ | | | | Read & print $-3.5$ |
| 1.4 | | | 24.0 | | |
| 1.5 | | 3 | | | |
| 1.2 | | | | | $3 < 3$ is false — exit loop |
| 2 | | | | 8.0 | |
| 3 | | | | | Print 24.0 and 8.0 |

(1st loop repetition: steps 1.2, 1.3, 1.4, 1.5)
(2nd loop repetition: steps 1.2, 1.3, 1.4, 1.5)
(3rd loop repetition: steps 1.2, 1.3, 1.4, 1.5)

**Table 2.3   Trace of algorithm in Fig. 2.14.**

If a program step does something other than change a variable value, then its effect is indicated in the rightmost column of Table 2.3. As an example, for the first execution of Step 1.2, the value of COUNTR is zero, which is less than 3. (The symbols $0 < 3$ mean "0 is less than 3") Therefore, the loop body is executed. We have used brackets to indicate each repetition of the loop. The loop is executed exactly three times, as desired. The loop exit occurs when the value of COUNTR becomes 3. Note that the value of AVE remains undefined until after the loop is exited.

Program traces must be done diligently, or they are of little use. The flow diagram must be traced carefully, on a step-by-step basis, without making any assumptions about what is happening. Changes in variable values must be noted at each step and compared to the expected results of the program. This should be done for at least one carefully chosen set of test data for which the intermediate and final results are easily determined. If an algorithm contains decision steps, then extra sets of test data should be used to ensure that all paths through the algorithm are traced.

**Example 2.1:** The flow diagram for Problem 2C (the largest value problem) is redrawn in Fig. 2.16. Since there is a true path (T) and false path (F) out of each condition, there are four distinct paths through the algorithm:

Path 1:    Step 2.1 - T, Step 2.4 - T
Path 2:    Step 2.1 - T, Step 2.4 - F
Path 3:    Step 2.1 - F, Step 2.4 - T
Path 4:    Step 2.1 - F, Step 2.4 - F

The actual path traced depends on the test data. Table 2.4 illustrates the trace for the three data items: 3, 6 and 10.

| | Variable changed | | | | |
|---|---|---|---|---|---|
| Algorithm step | N1 | N2 | N3 | LARGE | Effect |
| | ? | ? | ? | ? | |
| 1 | 3 | 6 | 10 | | |
| 2.1 | | | | | $3 > 6$ is false — execute Step 2.3 |
| 2.3 | | | | 6 | |
| 2.4 | | | | | $10 > 6$ is true — execute Step 2.5 |
| 2.5 | | | | 10 | |
| 3 | | | | | Print 10 |

**Table 2.4   Trace of Fig. 2.16 flow diagrams.**

The trace shown in Table 2.4 follows path 3 (Step 2.1-F, Step 2.4-T). To verify that the algorithm is correct, you should provide additional test data that causes the three remaining paths to be traced.

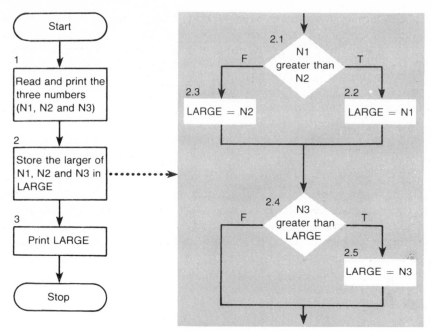

**Fig. 2.16**   Flow diagram for Problem 2C, redrawn.

---

**Program Style**

*Importance of algorithm traces*

The algorithm trace may seem like an unnecessary step and there is often a temptation to implement the algorithm without performing a trace. However, if a trace is done carefully, it may reveal errors in logic that are more easily corrected in the algorithm design stage, before the program is written. After the program is written, correcting logic errors is a very tedious and time-consuming process.

---

**Exercise 2.11:**   Carry out a complete trace of the flow diagram shown in Fig. 2.14 for the data items

$$-12.5, \ 8.25, \ 0.0, \ -16.5, \ .25$$

**Exercise 2.12:**   a) Trace the algorithm for the largest of three numbers problem for N1, N2, N3 equal to 5, 20, 15, respectively. Which of the four paths described in Example 2.1 did you follow?

b) Assign three values to N1, N2 and N3 to take you through path 1 as defined in Example 2.1. Does the algorithm produce the correct answer for these data?

c) If N1 = 16, N2 = −20 and N3 = 0, which of the four paths of Example 2.1 are followed? Does this test, combined with those of Example 2.1 and parts a) and b) of this exercise adequately test the algorithm? Explain.

**Exercise 2.13:**   Will the algorithm for the largest of three numbers problem (Fig. 2.16) work if two of the numbers have the same value? If all three numbers have the same value? Justify your answer.

### 2.5.2  Implementing a Loop in FORTRAN

In this section, we shall informally introduce one way to implement a loop conforming to the flow diagram pattern in Fig. 2.12. Such a loop is often called a *WHILE loop* and we will implement it using the WHILE loop control structure. The WHILE loop shown is not included as part of FORTRAN 77. However, it is implemented as an additional feature on some compilers. We shall illustrate other ways to implement loops in the next chapter.

The WHILE loop control structure has the form

> WHILE (*condition*) DO
> *statements to be repeated*
> ENDWHILE

where the *condition* corresponds to the loop repetition test. If the condition is true, the *statements to be repeated* are executed, and the *condition* is then retested. If the *condition* is false, the loop is exited and the first statement following the ENDWHILE statement is executed instead. The WHILE loop implementation of the algorithm for adding two thousand numbers is shown in Fig. 2.17.

```
C FIND THE SUM AND AVERAGE OF TWO THOUSAND NUMBERS
C
      INTEGER COUNTR
      REAL ITEM, SUM, AVE
C
C COMPUTE THE SUM
C READ AND PRINT EACH DATA ITEM AND ACCUMULATE IN SUM
      PRINT*, 'LIST OF ITEMS PROCESSED'
      SUM = 0.0
      COUNTR = 0
      WHILE (COUNTR .LT. 2000) DO
         READ*, ITEM
         PRINT*, ITEM
         SUM = SUM + ITEM
         COUNTR = COUNTR + 1
      ENDWHILE
C
C COMPUTE THE AVERAGE
      AVE = SUM / 2000
C
C PRINT THE SUM AND AVERAGE
      PRINT*, ' '
      PRINT*, 'SUM = ', SUM
      PRINT*, 'AVERAGE = ', AVE
C
      STOP
      END
```

**Fig. 2.17**  FORTRAN program for Problem 2E.

---

**Program Style**

*Printing column headings*

The statement

```
PRINT*, 'LIST OF ITEMS PROCESSED'
```

is used in Fig. 2.17 to print a column heading before any data values are printed. This heading is printed before the WHILE loop is entered. The statement

```
PRINT*, ITEM
```

inside the loop prints one data value per line for each repetition of the loop, thus forming a column of data underneath the heading.

---

**Program Style**

*Indenting the loop body*

In Fig. 2.17, the statements that are repeated (the loop body) are indented to clarify the structure of the program. This practice will be followed throughout the text.

---

### 2.5.3  Using Integers for Counting

In the sum and average program shown in Fig. 2.17, we have declared the variable COUNTR as type INTEGER and used integer constants (2000, 0 and 1) in the statements associated with this name. In general, variables and constants used in counting operations should be type INTEGER because their values can be easily represented as whole numbers. A useful guide when writing constants in a program is to use integer constants in statements associated with integer variables; real constants should be used in statements associated with real variables.

**Exercise 2.14:**  Write the flow diagram for a loop to read in a collection of seven data items, compute the product of all nonzero items in the collection, and print the final product. Trace your flow diagram to verify that it is correct. Implement it in FORTRAN.

## 2.6  WRITING GENERAL PROGRAMS

### 2.6.1  Data Collections of Arbitrary Size

Suppose that you are asked to solve Problem 2E for 20,000 data items instead of 2000; or for 1966 items; or for 10 data items. Will the approach just taken work here too? The answer, of course, is yes; however, the loop repetition condition must be changed for each case before the program can be run. Consequently, the algorithm given is not very general.

Algorithms and programs should be written so that any reasonable variation of a problem can be solved without altering the program. A general FORTRAN program for Problem 2E would compute the sum for an arbitrary, but prespecified, number of data items. To accomplish this, the number of data items should be treated as an input variable (rather than a constant) to be read in by the program at the beginning of execution. In this way, any collection of data may be processed by the same program, as long as the first item entered is the number of items in this data collection. The additional input variable required, NRITMS, should be added to the data table as shown next.

**Additional Data Table Entry for Problem 2E:**

*Input variable*

NRITMS: The number
of data items to
be processed

When preparing the data for this program, we must insert the number of data items to be processed (an integer) as the first data value. This value should be read into NRITMS as the first step in the algorithm. The revised level one algorithm for Problem 2E is shown in Fig. 2.18.

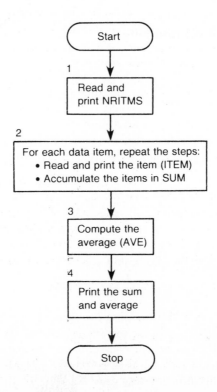

**Fig. 2.18**  Revised level one algorithm for Problem 2E.

The general version of the program for finding the sum and average of a collection of data items is shown in Fig. 2.19, along with the output for a sample run. The data values provided for this test case were the numbers:

$$5 \longleftarrow \text{number of items to be processed}$$

$$\left.\begin{array}{c} 37.5 \\ 50.2 \\ 14.5 \\ 20.5 \\ 19.5 \end{array}\right\} \quad \text{five data items to be summed}$$

Remember that the order of execution of the READ statements determines where each data value will be stored. The first data value above is an integer constant read into the type INTEGER variable NRITMS; all remaining data (the five numbers to be summed) are read into the type REAL variable ITEM.

**Exercise 2.15:**   What do you think would happen if the first data value for the sample run above was 3 instead of 5? 7 instead of 5?

**Exercise 2.16:**   What data would be required if you wished to use the program in Fig. 2.19 to find the sum and average of all multiples of ten in the range of 10 to 100?

### 2.6.2   Program Parameters

FORTRAN has another feature, the PARAMETER statement, which can be helpful in writing general programs and programs that are more readable and are easier to maintain (keep up-to-date). For example, in the modified payroll program (Fig. 2.7), the Block-IF structure

```
IF (GROSS .GT. 100.00) THEN
   NET = GROSS - 25.00
ELSE
   NET = GROSS
ENDIF
```

is used to deduct a tax amount of $25.00 for each employee earning over $100. If either the tax amount or minimum salary for a tax deduction were changed, the block-IF structure would have to be modified. Any other program statements that happened to use these values would have to be changed as well.

Alternatively, we could introduce two *program parameters* named TAX and MINSAL to represent the constants 25.00 and 100.00, respectively. These parameters would be declared at the beginning of the program as shown below.

```
C MODIFIED PAYROLL PROGRAM - DEDUCTS TAX CONDITIONALLY
C
C PROGRAM PARAMETERS
      REAL TAX, MINSAL
      PARAMETER (TAX = 25.00, MINSAL = 100.00)
```

The PARAMETER statement above associates the parameter TAX with the constant value 25.00 and the parameter MINSAL with the value 100.00. If the tax amount or minimum salary were to change, we could keep our program up-to-date simply by modifying the PARAMETER statement, rather than changing

```
C FIND THE SUM AND AVERAGE OF A COLLECTION OF REAL DATA ITEMS
C
      INTEGER COUNTR, NRITMS
      REAL ITEM, SUM, AVE
C
C READ AND PRINT THE NUMBER OF ITEMS TO BE PROCESSED
      READ*, NRITMS
      PRINT*, 'THE NUMBER OF ITEMS TO BE PROCESSED IS ', NRITMS
C
C COMPUTE THE SUM
C READ AND PRINT EACH DATA ITEM AND ACCUMULATE IN SUM
      PRINT*, 'LIST OF ITEMS PROCESSED'
      SUM = 0.0
      COUNTR = 0
      WHILE (COUNTR .LT. NRITMS) DO
          READ*, ITEM
          PRINT*, ITEM
          SUM = SUM + ITEM
          COUNTR = COUNTR + 1
      ENDWHILE
C
C COMPUTE THE AVERAGE
      AVE = SUM / NRITMS
C
C PRINT THE SUM AND AVERAGE
      PRINT*, ' '
      PRINT*, 'SUM = ', SUM
      PRINT*, 'AVERAGE = ', AVE
C
      STOP
      END


*** PROGRAM EXECUTION OUTPUT ***

THE NUMBER OF ITEMS TO BE PROCESSED IS                      5
LIST OF ITEMS PROCESSED
   37.50000000000
   50.20000000000
   14.50000000000
   20.50000000000
   19.50000000000

SUM =   142.2000000000
AVERAGE =   28.44000000000
```

**Fig. 2.19**  General program and sample output for Problem 2E.

all of the program statements that manipulate these amounts. The PARAME-
TER statement is described in the next display.

Once the parameter name is defined, it may be used anywhere in the pro-

**PARAMETER Statement**

PARAMETER $(pname_1 = const_1, pname_2 = const_2, \ldots)$

**Interpretation:** The PARAMETER statement is a nonexecutable statement that associates each symbolic name, $pname_1$, $pname_2$, . . . with a constant value $const_1$, $const_2$, . . . The symbolic name may be used to represent the constant value throughout the program.

*Notes:* A symbolic name used as a parameter may not be redefined via an assignment, READ, or another PARAMETER statement.

The PARAMETER statements must precede the executable statements of a program. The type of each parameter should be declared before its associated value is specified in a PARAMETER statement.

---

**Program Style**

*Use of in-line constants, program parameters, and input variables*

Constants such as the tax amount (25.00) and minimum taxable salary (100.00), which have a special meaning in a problem and change infrequently, if at all, should be treated as *program parameters*. There are two advantages to treating such constants as parameters rather than writing them as in-line constants in a program. First, as illustrated by the decision step at the top of page 76, the parameters convey what is happening more clearly than in-line constants (see Fig. 2.7). Instead of using two numbers, 100.00 and 25.00, which have no intrinsic meaning, we use names such as MINSAL and TAX, which are more descriptive. Second, if we wish to modify a constant associated with a parameter, we need only modify the parameter declaration. If the constant value was used in-line, then we would have to re-do all statements that manipulate it. This can become a tedious task in a large program with multiple references to a constant value.

Even if the value of a special constant is not likely to change, we may associate it with a parameter (for example, using the name PI for the value 3.14159). This use of the parameter feature makes the resulting program easier to read and understand.

On the other hand, constants such as 0, 1 and 0.0 are written in-line in Fig. 2.19 because they have no special meaning to the problem being solved, and because they represent data that will never change. (For example, in Problem 2E we will always want to initialize SUM and COUNTR to zero, and increment COUNTR by one regardless of the data collection being processed.

Finally, data values such as NRITMS and ITEM (Problem 2E) and HOURS and RATE (Problem 2B) that are likely to change with each use of a program should be treated as input variables.

gram, as shown below for the decision structure of Problem 2B.

```
IF (GROSS .GT. MINSAL) THEN
    NET = GROSS - TAX
ELSE
    NET = GROSS
ENDIF
```

We will continue the practice of treating special constants as parameters, and will list them in a separate section in the data table, as well as in the program. Whenever the PARAMETER statement is used, care should be taken to ensure that all parameter names are declared before they are used in the PARAMETER statement.

**Exercise 2.17:**   Redo Problem 2D using the PARAMETER statement.

**Exercise 2.18:**   We could also read in the values of MINSAL and TAX in Problem 2B. What is the reason for using these symbolic names as parameters instead of input variables?

## 2.7  SUMMARY

In the first part of this chapter we outlined a method for solving problems on the computer. This method stresses six points:

1. Understand the problem given.
2. Identify the input and output data for the problem as well as other relevant data.
3. Formulate a precise statement of the problem.
4. Develop an algorithm.
5. Refine the algorithm.
6. Implement the algorithm in FORTRAN.

In the remainder of the chapter, we introduced the flow diagram representation of the various steps in an algorithm. Flow diagrams provide a graphical representation of an algorithm consisting of a number of specially shaped boxes and arrows as well as several *patterns* of boxes and arrows used to describe decision steps and loops. These boxes and patterns are summarized in Fig. 2.20.

There was much discussion in this chapter of *program documentation* and good *programming style*. Guidelines for using program comments were presented. Well placed and carefully worded comments, combined with a complete and concise data table, will provide much of the documentation necessary for a program. The flow diagrams are also part of the program documentation as well as an integral step in the design process.

The major points of programming style that were emphasized involved echo printing (or prompting) all input data values, indenting the body of a loop and the True and False Tasks in a decision structure, the use of comments, and the use of the PARAMETER statement to associate symbolic names with the values of important program constants.

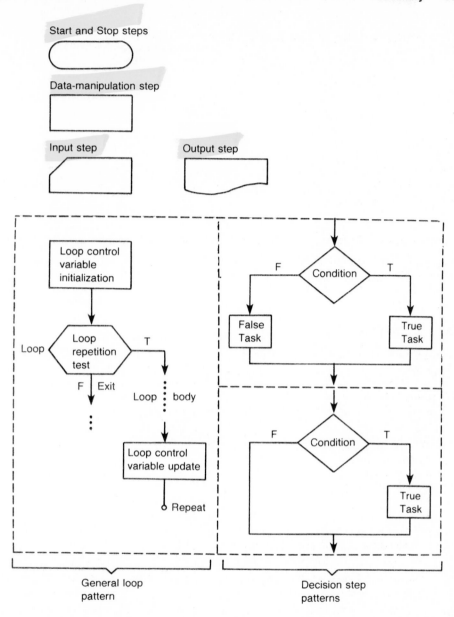

**Fig. 2.20**  Flow diagram symbols and patterns.

We believe that the issues of documentation and style are extremely important to the entire field of computing, for they contribute to making programs and
programming systems easier to read, to debug (locate and correct errors), and to
maintain (keep up-to-date). We urge you to pay particular attention to the documentation and style that you develop; they are as important as the FORTRAN

structures and programs that you will learn to write. You will find that complete documentation and consistent style will save you considerable time in the long run.

The algorithms that were developed in this chapter consisted of three major parts: data entry, data manipulation (computation), and the output of results. Of these, the data manipulation phase usually is the most complicated and requires the most attention. This phase can be started once the input data and desired problem output have been clearly defined, and a precise understanding of the problem has been achieved.

Often, additional entries are made to the data table as the data manipulation phase progresses. For example, in Problem 2E, the need for the program variable COUNTR was not readily apparent until the algorithm for manipulating the data was chosen.

Three new features of FORTRAN were introduced in this chapter: the PA-RAMETER statement, used to associate symbolic names with important program constants; and two control statements, the block-IF decision and WHILE loop (See Table 2.5). We will further illustrate how to use these features to facilitate implementing our flow diagrams as FORTRAN programs. This will enable us to solve some relatively complex problems on the computer, using programs that clearly reflect the careful planning and organization used in our algorithm development.

| Statement type and use | Examples |
|---|---|
| Comments | `C THIS IS A COMMENT` |
| PARAMETER: Associates a symbolic name with a constant value. | `PARAMETER (TAX = 25.00)` |
| The Block-IF Structure<br>single alternative:<br>for implementing decision steps involving one alternative | `IF (N3 .GT. LARGE) THEN`<br>`   LARGE = N3`<br>`ENDIF` |
| double alternative:<br>for implementing decision steps involving a choice between two alternatives | `IF (GROSS .GT. MINSAL) THEN`<br>`   NET = GROSS - TAX`<br>`ELSE`<br>`   NET = GROSS`<br>`ENDIF` |
| The WHILE Loop Structure<br>for implementing loops | `SUM = 0.0`<br>`COUNTR = 0`<br>`WHILE (COUNTR .LT. NRITMS) DO`<br>`   READ*, ITEM`<br>`   PRINT*, ITEM`<br>`   SUM = SUM + ITEM`<br>`   COUNTR = COUNTR + 1`<br>`ENDWHILE` |

**Table 2.5   Summary of FORTRAN features in Chapter 2.**

# PROGRAMMING PROBLEMS*

**2F**    Given the bank balance in your checking account for the past month and all the transactions in your account for the current month, write a program to compute and print your checking account balance at the end of the current month. You may assume that the total number of transactions for the current month is known ahead of time. *Hint:* Your first data card should contain your checking account balance at the end of last month. The second card should contain the number of transactions for the current month. All subsequent cards should represent the amount of each transaction for the current month, with a positive amount for a deposit, a negative amount for a withdrawal.

**2G**    *Continuation of Problem 2F.* Modify your data table, algorithm and program for Problem 2F to compute and print the number of deposits and the number of withdrawals.

**2H**    Write a program to simulate a state police radar gun. The program should read a single automobile speed (as an integer) and print the message "OK" if the speed is less than or equal to 55 mph or "SPEEDING" if the speed exceeds 55 mph.

**2I**    *Continuation of Problem 2H.* Modify your data table algorithm and program for Problem 2H to do the following:
a)   Read in a collection of automobile speeds as long as (WHILE) the speeds do not exceed 100 mph;
b)   count the number of speeds less than or equal to 55 mph, the number in excess of 55 mph, and the total;
c)   stop reading speeds and print the counts when a speed in excess of 100 mph is read.

**2J**    Write a program to read in one real value X and print a message indicating if "X is positive", "X is negative" or "X is zero".

**2K**    *Continuation of Problem 2J.* Modify the data table, flow diagram and program for Problem 2J to read 100 real values and to count the number of positive values, the number of negative values, and the number of zeros.

**2L**    Write a program for an algorithm to compute the factorial, $N!$, of a single arbitrary integer $N$. ($N! = N \times (N - 1) \times \cdots 2 \times 1$). Your program should read and print the value of N and print N! when done.

**2M**    If N contains an integer, then we can compute $X^N$ for any $X$, simply by multiplying $X$ by itself $N$ times. Write a program to read in a value of $X$ and a value of $N$, and compute $X^N$ via repeated multiplications. Test your program for

$$X = 6.0, \qquad N = 4$$
$$X = 2.5, \qquad N = 6$$
$$X = -8.0, \qquad N = 5$$

**2N**    *Continuation of Problem 2M*
a)   How many multiplications are required in your program for Problem 2M in order to compute $X^9$? Can you figure out a way of computing $X^9$ in fewer multiplications?
b)   Can you generalize your algorithm for computing $X^9$ to compute $X^N$ for any positive $N$?
c)   Can you use your algorithm in part (b) to compute $X^{-N}$ for any positive $N$? How?

---

* Provide a data table and flow diagram for each problem.

**2O**  Green Thumb brand grass seed costs $3.20 per pound, and one pound will cover 500 square feet. Write a program that will compute and print the number of pounds of seed and total cost to cover areas between 0 and 10,000 square feet, in steps of 250. Your program should read in the cost per pound of the seed and the coverage rate per pound, and then print a table, as shown below.

| Area in square feet | Amount of seed (pounds) | Total cost |
|:---:|:---:|:---:|
| 0 | 0.00 | 0.00 |
| 250 | — | — |
| 500 | — | — |
| 1000 | — | — |
| ⋮ | ⋮ | ⋮ |
| 9500 | — | ⋮ |
| 9750 | — | — |
| 10000 | — | — |

(If you are working in the metric system, use the following figures:

      Seed cost: $7.05 per kilogram
      Area of coverage for one kilogram: 0.2 square meters

Print your table of kilograms of seed and total cost to cover areas of from 0 to 2000 square meters in steps of 50.)

**2P**  Your neighbor owns a large drugstore chain in a state which has just raised its sales tax to 5 percent, effective in 3 days. He needs to produce copies of the new tax tables for his sales people and checkers. Write a program to produce a new tax-rate table, as shown below.

| Purchase amount | Tax amount |
|:---:|:---:|
| 0.00–0.09 | 0.00 |
| 0.10–0.29 | 0.01 |
| 0.30–0.49 | 0.02 |
| 0.50–0.69 | 0.03 |
| 0.70–0.89 | 0.04 |
| 0.90–1.09 | 0.05 |
| 1.10–1.29 | 0.06 |
| ⋮ | ⋮ |
| 9.50–9.69 | 0.48 |
| 9.70–9.89 | 0.49 |
| 9.90–10.09 | 0.50 |

# BASIC CONTROL STRUCTURES

## 3.1    INTRODUCTION TO CONTROL STRUCTURES

In Chapter 2, we introduced flow diagram patterns for decision and loop structures. We also introduced the FORTRAN condition and informally described the implementation of decisions and loops. In this chapter, we shall formally discuss the FORTRAN implementation of these flow diagram patterns and provide additional examples of their use.

We will show that the development of correct, precise algorithms is an important part of using the computer to solve problems. Furthermore, the initial English description of an algorithm is most critical, for if this description is incorrect or imprecise, all refinements as well as the resulting program will reflect these errors. Therefore, as we introduce new features of FORTRAN, we will continue to emphasize algorithm development through the use of the flow diagram.

## 3.2    DECISION STRUCTURES

### 3.2.1    Decision Structures with One and Two Alternatives

In this section we will be concerned with a formal definition of the FORTRAN control structures used to represent the single and double-alternative decision steps described in Chapter 2. In the double-alternative decision pattern (Fig. 3.1), if the condition is true, the algorithm steps specified by the True Task are carried out; otherwise, the steps specified by the False Task are performed. Only one of the paths from the condition test will be followed. Execution will then continue at the point indicated by the arrow at the bottom of the diagram. The True and False Tasks may each consist of a number of different boxes and flow diagram patterns. In general, however, it is a good idea to keep these task descriptions simple, and refine them in a separate diagram if additional details are needed.

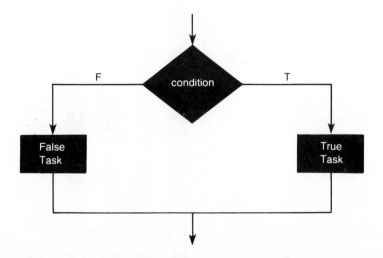

**Fig. 3.1**   Double alternative decision step pattern.

The FORTRAN control structure form for the double-alternative decision pattern is described in the next display.

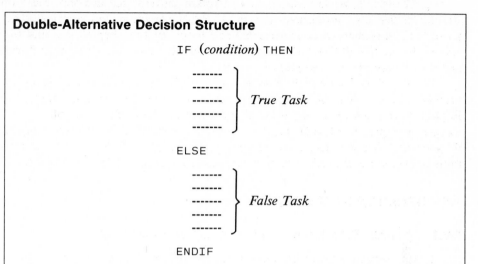

**Double-Alternative Decision Structure**

IF (*condition*) THEN

⎫
⎬  *True Task*
⎭

ELSE

⎫
⎬  *False Task*
⎭

ENDIF

**Interpretation:** The *condition* is evaluated. If the *condition* is true, then the *True Task* is executed, and the *False Task* is skipped. Otherwise, the *True Task* is skipped, and the *False Task* is executed. In either case, execution will continue with the first statement following the ENDIF.

For the single-alternative decision pattern (Fig. 3.2), if the *condition* is true, the *True Task* is executed. However, there is no task to be carried out if the *condition* is false. In either case, the algorithm continues at the point indicated by the arrow at the bottom of the diagram.

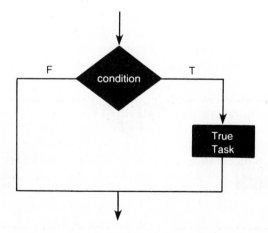

**Fig. 3.2**  Single-alternative decision step pattern.

The FORTRAN single-alternative decision structure is described in the next display.

---

**Single-Alternative Decision Structure**

$$\text{IF } (condition) \text{ THEN}$$

$$\left. \begin{array}{l} \text{-------} \\ \text{-------} \\ \text{-------} \\ \text{-------} \end{array} \right\} \quad \textit{True Task}$$

$$\text{ENDIF}$$

**Interpretation:** The *condition* is evaluated. If the *condition* is true, then the *True Task* will be executed. If the *condition* is false, then the *True Task* will be skipped, and execution will continue with the first statement following the ENDIF.

---

The single- and double-alternative decision structures are special cases of the general block-IF structure described in Chapter 6. In these structures, the statement

$$\text{IF } (condition) \text{ THEN}$$

is called the *header statement* of the structure. The statement ENDIF is called the *terminator statement.* These statements mark the beginning and the end of the control structure. For each structure header used in a program, a terminator statement must appear somewhere below the header statement. The terminator may also be written as END IF.

We have already seen examples of the use of the single- and double-alternative decision patterns in Chapter 2. Additional illustrations will be provided in the solved problems in this chapter.

**Problem 3A:**    Read two numbers into the variables X and Y and compute and print the quotient QUOT = X / Y.

**Discussion:**    This is a problem that looks quite straightforward, but it has the potential for disaster hidden between the lines of the problem statement. In this case, as in many others, the potential trouble spot is due to unanticipated values of input data—values for which one or more of the data manipulations required by the problem are not defined.

In this problem, the quotient X/Y is not defined mathematically if Y equals 0.0. If we instruct the computer to perform the calculation X/Y in this case, it will either produce an unpredictable, meaningless result, or it will not even be able to complete the operation and will prematurely terminate or *abort* our program. Most computers will provide the programmer with a diagnostic message if division by zero is attempted, but some will not. In order to avoid the problem entirely, we will have our program test for a divisor of zero and print a message of its own if this situation should occur. The data table is provided next; the flow diagram and program for this solution are shown in Fig. 3.3 and 3.4.

**Data Table for Problem 3A**

*Input variables*                                    *Output variables*

X: Dividend                                          QUOT: Quotient of X
                                                           and Y

Y: Divisor

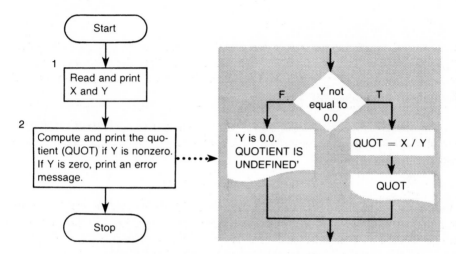

**Fig. 3.3**  Flow diagram for the quotient problem (3A).

```
C QUOTIENT PROBLEM
C
      REAL X, Y, QUOT
C
C READ AND PRINT X AND Y
      READ*, X, Y
      PRINT*, 'DIVIDEND = ', X
      PRINT*, 'DIVISOR = ', Y
C
C COMPUTE THE QUOTIENT IF Y IS NON-ZERO
      IF (Y .NE. 0.0) THEN
         QUOT = X / Y
         PRINT*, 'QUOTIENT = ', QUOT
      ELSE
         PRINT*, 'Y IS 0.0, QUOTIENT IS UNDEFINED.'
      ENDIF
C
      STOP
      END
```

*(Continued)*

```
           *** PROGRAM EXECUTION OUTPUT ***
           DIVIDEND =   14.00000000000
           DIVISOR =    3.500000000000
           QUOTIENT =   4.000000000000
```

**Fig. 3.4**  FORTRAN program with sample output for Problem 3A.

In the next problem, we illustrate the use of the single-alternative decision structure.

**Problem 3B:**   Read two numbers into variables X and Y and compare them. Place the larger in X and the smaller in Y.

### Data Table for Problem 3B

| *Input variables* | *Output variables* |
|---|---|
| X, Y: Items to be com-<br>    pared | X: Larger item<br>Y: Smaller item |

**Discussion:**   The flow diagram for this program is shown in Fig. 3.5. Note that the contents of variables X and Y are exchanged only if the condition "Y greater than X" is true. In the completed program for this problem (shown in Fig. 3.6), this exchange is implemented using an additional variable, TEMP, in which a copy of the initial value of X is saved.

### Additional Data Table Entry

*Program variables*

TEMP: Temporary vari-
able used in exchange

To verify the need for TEMP, we trace the program execution for the data values 3.5 and 7.2 below.

| *Program Trace* | *Variables Affected* | | |
|---|---|---|---|
| FORTRAN statements | X | Y | TEMP |
| READ*, X | 3.5 | | |
| READ*, Y | | 7.2 | |
| TEMP = X | | | 3.5 |
| X = Y | 7.2 | | |
| Y = TEMP | | 3.5 | |

As indicated in the trace, after the assignment statement

```
           X = Y
```

is executed, the value 3.5 is no longer available in X. Copying X into TEMP first, prevents this value from being lost.

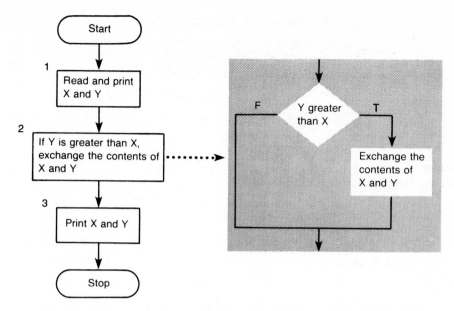

**Fig. 3.5**  Flow diagram for determining the larger of two numbers (Problem 3B).

```
C LARGER OF TWO NUMBERS PROBLEM
C
        REAL X, Y, TEMP
C
C READ AND PRINT DATA
        READ*, X, Y
        PRINT*, 'X = ', X
        PRINT*, 'Y = ', Y
C
C TEST AND SWITCH IF NECESSARY
        IF (Y .GT. X) THEN
           TEMP = X
           X = Y
           Y = TEMP
        ENDIF
C
C PRINT RESULTS
        PRINT*, 'LARGER = ', X
        PRINT*, 'SMALLER = ', Y
C
        STOP
        END


*** PROGRAM EXECUTION OUTPUT ***

X =    3.500000000000
Y =    7.200000000000
LARGER =   7.200000000000
SMALLER =   3.500000000000
```

**Fig. 3.6**  FORTRAN program and sample output for Problem 3B.

### 3.2.2  Single-Alternative Decision with One Dependent Statement

Very often, the True Task of a single-alternative decision structure will consist of only one FORTRAN statement. In such cases, the shorter logical IF statement can be used as shown next.

**Example 3.1:**   The flow diagram below

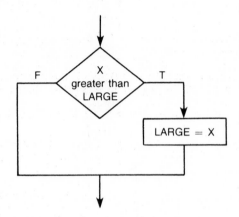

can be implemented in FORTRAN using the single-alternative decision structure

```
IF (X .GT. LARGE) THEN
    LARGE = X
ENDIF
```

or the equivalent logical IF statement

```
IF (X .GT. LARGE) LARGE = X
```

The logical IF statement is described in the next display.

---

**Logical IF Statement**

IF (*condition*) *dependent statement*

**Interpretation:** The *condition* is evaluated. If it is true, the *dependent statement* is executed; otherwise, the *dependent statement* is not executed.
*Notes:* Only one FORTRAN statement may follow the *condition*. The *dependent statement* may be any simple executable statement such as assignment, STOP, READ or PRINT. It may not be another logical IF statement or a control structure.

---

**Exercise 3.1:**   Trace the program in Fig. 3.6 for X = 7.2, Y = 3.5 and X = Y = 6.2.

**Exercise 3.2:**   Locate and correct the syntax errors in the following statements:

a) ```
IF (X = 0.0) XZER =
   XZER + 1
```

b) ```
IF (X .GT. 0.0)
   Y = Y + 1
   PRINT*, 'X IS POS'
ENDIF
```

c) ```
IF (X .GT. 0.0) THEN Y = Y + L
```

d) ```
IF (X .GT. 0.0) IF (X .LT. Y)
   Y = Y - 1
```

e) ```
IF (X .GT. Y) THEN
   XBIG = XBIG + 1
ELSE YBIG = YBIG + 1
ENDIF
```

**Exercise 3.3:**   Flow diagram the decisions stated below, using either single- or double-alternative decision structures.

a)   Read a number into the variable NMBR. If this number is positive, add one to the contents of NPOS. If the number is not positive, add one to the contents of NNEG.

b)   Read a number into NMBR. If NMBR is zero, add one to the contents of NZERO.

c)   (A combination of (a) and (b)). Read a number into NMBR. IF NMBR is positive, add one to NPOS; if NMBR is negative, add one to NNEG; and if NMBR is zero, add one to NZERO.

**Exercise 3.4:**   The True Task in the decision structure of the program for Problem 3B contains three statements and uses an additional variable TEMP. Could we have accomplished the same task performed by this group with either of the statement groups

a)  ```
X = Y
Y = X
```
  or   b)  ```
TEMP = Y
X = TEMP
Y = X
```

What values would be stored in X and Y after these statement groups execute? Modify statement group (b) so that it works properly.

**Exercise 3.5:**   Convert the following English descriptions of algorithms to flow diagrams and FORTRAN statement groups, using the single- and double-alternative decision structures or the conditional statement.

i)   If the remainder (REM) is equal to zero, then print N.

ii)   If the product (PROD) is equal to N, then print the contents of the variable DIV and read a new value into N.

iii)   If the number of traffic lights (NBRLTE) exceeds 25.0, then compute the gallons required (GALREQ) as total miles (MILES) divided by 14.0. Otherwise compute GALREQ as MILES divided by 22.5.

## 3.3   THE WHILE LOOP STRUCTURE

### 3.3.1   Introduction

We now turn our attention to two control structures that are useful for describing loops. The WHILE loop structure is more general and will be studied first. A second loop structure, the DO loop, will be introduced later in this chapter.
```

If you review the previous discussion about loops, you will note that all loops consist of two parts: the *loop body,* or sequence of algorithm steps that must be repeated, and *loop control.* The loop control is used to specify the number of times a loop must be executed or, alternatively, to specify some condition under which the loop is to be repeated or terminated.

For example, in the algorithm for Problem 2E (shown again in Fig. 3.7), there are two steps (Steps 1.3 and 1.4) comprising the loop body. The first step reads a data item into the variable ITEM, and the second step adds ITEM to SUM. The loop control in Fig. 3.7 consists of the steps to initialize COUNTR to zero, increment COUNTR by one, and compare the value of the counter to the number of data items. The loop is repeated as long as (WHILE) the loop repetition test "COUNTR is less than NRITMS" is true.

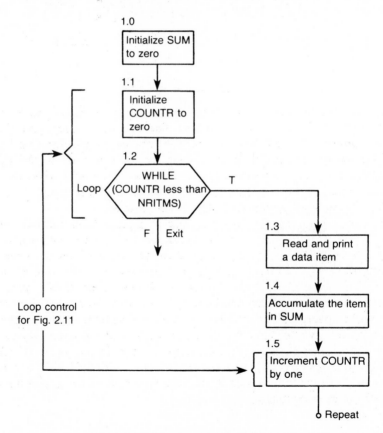

**Fig. 3.7**  The loop pattern for Problem 2E.

### 3.3.2  WHILE Loop Flow Diagram Pattern

The loop control pattern shown in Fig. 3.7 is often called a WHILE loop. The flow diagram for a WHILE loop was given in Chapter 2; it is repeated here for convenience, with the word WHILE added to the loop repetition test box (see Fig. 3.8).

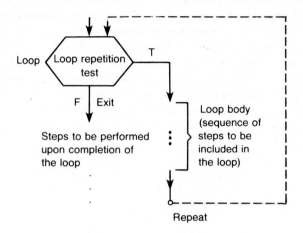

**Fig. 3.8**   The WHILE loop pattern.

The repeat condition is evaluated prior to the execution of the loop body. As long as (WHILE) the condition is true, the loop body is repeated. Once the condition becomes false, the loop is exited, and the next algorithm step is executed. It is important to remember that the loop repetition test for the WHILE loop is at the beginning of the loop, and loop exit can occur only when this test is evaluated. The loop body will not be executed at all if the repeat condition is false the first time that the repetition test is encountered.

Normally, when a WHILE loop is used in a program, a particular variable may be identified as the loop control variable. The loop control variable is the variable whose value is continually tested in the repeat condition. In order for the repeat condition to determine correctly the number of executions of the loop body, the loop control variable must be initialized prior to the start of the WHILE structure. Also, within the loop body, the value of the loop control variable should be updated (modified) with each loop execution; otherwise, the repeat condition will always produce the same test result, and loop exit may never occur. The updating of the loop control variable is usually the last step within the loop body. In Fig. 3.7, the loop control variable is COUNTR. It is initialized prior to entry into the loop, tested in the loop repetition test box, and updated in the last step of the loop body.

### 3.3.3   Implementing the WHILE Loop in FORTRAN

Unfortunately, there is some disagreement among FORTRAN experts regarding the implementation of the WHILE loop. For this reason, the WHILE loop control structure was not included in the standard version of FORTRAN 77. However, many computer science educators support the use of the WHILE loop control structure, and it has been incorporated in at least two FORTRAN compilers (WATFIV, developed at the University of Waterloo (Ontario, Canada) and M77, developed at the University of Minnesota).

We shall formally describe the WHILE loop control structure that is implemented in both of these compilers in the next section. If you are able to use this structure, by all means do so; it will simplify the translation of your flow diagrams into programs. However, you should be aware that this feature is nonstandard and, hence, is not available on all FORTRAN compilers. We shall also show how to implement the WHILE loop pattern in standard FORTRAN in Section 3.3.5. Your instructor will tell you which of these sections is most relevant for your compiler.

### 3.3.4  The WHILE Loop Control Structure (Nonstandard)

The WHILE loop control structure is shown in the display below.

---

**WHILE Loop Structure (Nonstandard)**

WHILE (*condition*) DO

```
-------
-------
-------  } Loop body
-------
```

ENDWHILE

**Interpretation:** The *condition* is evaluated first. If the *condition* is true, then the *loop body* will be executed. This sequence is repeated as long as the *condition* evaluates to true. If the *condition* is false, the *loop body* will be skipped, and execution will continue with the statement following the ENDWHILE.

---

This loop control structure can be used to implement the flow diagram pattern of Fig. 3.7, as shown next.

```
SUM = 0.0
COUNTR = 0
WHILE (COUNTR .LT. NRITMS) DO
    READ*, ITEM
    PRINT*, ITEM
    SUM = SUM + ITEM
    COUNTR = COUNTR + 1
ENDWHILE
```

**Fig. 3.9**  Using the nonstandard WHILE loop structure.

The statement beginning with WHILE is the *loop header statement*. It specifies the condition that must be true in order for the loop body to be executed (e.g., COUNTR less than NRITMS). The end of the loop body is marked by the ENDWHILE statement, the *loop terminator*. We will always indent the loop body so that it can be recognized easily in a program.

**Exercise 3.6:** Use a WHILE loop structure and write the flow diagram and program for a loop that will find the largest cumulative product of the numbers 1, 2, 3, 4, . . . that is smaller than 10,000. *Hint:* The idea is to compute 1 * 2 * 3 * 4 * . . . and continue while the resulting product is less than 10,000. Then the last product computed that was less than 10,000 is the one you want to print. Use an additional program variable to "remember" this value.

### 3.3.5   Implementing the WHILE Loop in Standard FORTRAN: Transfer Statements

If your compiler does not support the WHILE loop structure, then you will have to implement the structure using the *GO TO*, or *transfer statement*. A transfer statement changes the order in which instructions are executed. The WHILE loop shown earlier (Fig. 3.9) is implemented in Fig. 3.10 using the GO TO statement.

```
                    SUM = 0.0
        repeat      COUNTR = 0
        loop   ┌─►13 IF (COUNTR .GE. NRITMS) GO TO 19──┐
        test   │        READ*, ITEM                    │
               │        PRINT*, ITEM                   │
               │        SUM = SUM + ITEM               │
               │        COUNTR = COUNTR + 1            │
               └─── GO TO 13                           │
                 19 CONTINUE◄───────────────────────────┘
                    .                    exit
                    .                    loop
                    .
```

       **Fig. 3.10**   Implementing the WHILE loop in Standard FORTRAN.

There are two transfer statements in Fig. 3.10. The statement

```
        GO TO 13
```

follows the last statement in the loop body. As shown by the dashed line, it causes a transfer back to the statement associated with label 13 (the loop header).

A *label* in FORTRAN is a string of one to five digits. A label is associated with an executable statement if it appears in the *label field* (columns 1 through 5) of that statement. Each label must be associated with exactly one executable statement.

The loop header

```
        13 IF (COUNTR .GE. NRITMS) GO TO 19
```

is a *conditional transfer statement.* It causes a transfer to label 19 only when the condition is true; otherwise, the next statement in sequence (**READ*, ITEM** in Fig. 3.10) is executed.

The statement

```
        19 CONTINUE
```

is used simply as a loop terminator to mark the end of the loop. Although a

CONTINUE statement is considered an executable FORTRAN statement, it neither performs an operation nor affects the program behavior in any way.

The standard FORTRAN implementation of the WHILE loop in Fig. 3.10 works as follows. The *loop exit condition* (COUNTR .GE. NRITMS) is tested; if this condition is true, a transfer is made out of the loop to label 19 and the first instruction following the loop will be executed. If the loop exit condition is false, there is no transfer and the loop body is executed instead. When the transfer statement at the end of the loop (GO TO 13) is reached, control is transferred back to the loop header and the loop exit condition is retested. Consequently, the loop body will be executed as long as (WHILE) the loop exit condition remains false. For now, you should use transfer statements only to implement the WHILE loop if your compiler does not support one.

Notice that the loop exit condition (COUNTR .GE. NRITMS) is the *complement* of the loop repetition condition (COUNTR .LT. NRITMS) for the original WHILE loop structure. The complement of a condition is defined to be true when the condition is false, and vice versa. Thus, the complement of the condition "X is equal to Y" is "X is not equal to Y". Similarly, the complement of "COUNTR is less than NRITMS" is "COUNTR is greater than or equal to NRITMS". Table 3.1 shows the complements of the FORTRAN relational operators.

| Relational operator | Complement |
|:---:|:---:|
| .EQ. | .NE. |
| .GT. | .LE. |
| .LT. | .GE. |
| .NE. | .EQ. |
| .LE. | .GT. |
| .GE. | .LT. |

**Table 3.1 Relational operator complements.**

The GO TO statement and its use in implementing the WHILE loop are described in the displays that follow.

---

**The GO TO (Transfer) Statement**

                                GO TO *lab*

**Interpretation:** This statement causes an immediate transfer of control to the executable statement associated with label *lab*. The label *lab* must be a string consisting of 1 to 5 digits and must be placed in positions 1 through 5 of a single executable statement.

*Notes:* If a label *lab* appearing in a transfer statement is not attached to an executable program statement, an

                    UNDEFINED STATEMENT LABEL *lab*

error will occur.

If a label *lab* is attached to more than one statement, a

                    DUPLICATE STATEMENT LABEL *lab*

error will occur.

---

---

**WHILE Loop (Standard FORTRAN Implementation)**

*head*   IF (*loop exit condition*) GO TO *term*

$$\left.\begin{array}{l}\text{------}\\\text{------}\\\text{------}\end{array}\right\}\;\textit{loop body}$$

        GO TO *head*
*term*  CONTINUE

**Interpretation:** The *loop body* is repeated as long as the *loop exit condition* is false. The GO TO statement following the *loop body* causes a transfer back to the loop header (identified by the label *head*). The *loop exit condition* is tested before each execution of the *loop body*. A transfer to the loop terminator (the CONTINUE statement associated with label *term*) occurs when the *loop exit condition* becomes true; the loop is exited and execution continues with the first statement following the terminator.

---

**Program Style**

*Guidelines for choosing GO TO labels*

The statement labels (13 and 19) in Fig. 3.10 were chosen arbitrarily; any other labels could have been used instead. However, we shall base the label of each loop header (conditional transfer statement) on the step number of the loop repetition test in the flow diagram (Step 1.3 in Fig. 3.7). The label of the loop terminator (CONTINUE statement) will have the same first digit (or digits) as the loop header and will always end with the digit 9.

---

Throughout this text, we will illustrate both forms of the WHILE loop control structure whenever it appears in a program. Since the loop body is the same for both forms, it will only be necessary to specify the loop control statements. The nonstandard WHILE loop form will always be shown in the program; the corresponding control statements for the standard FORTRAN implementation will be shaded at the right of the program, as shown next. Ignore these statements if your compiler supports the WHILE loop. Remember, each statement label must be placed in columns 1 through 5.

```
      SUM = 0.0
      COUNTR = 0
      WHILE (COUNTR .LT. NRITMS) DO      13 IF (COUNTR .GE.
         READ*, ITEM                        Z NRITMS) GO TO 19
         PRINT*, ITEM
         SUM = SUM + ITEM
         COUNTR = COUNTR + 1
      ENDWHILE                               GO TO 13
                                         19 CONTINUE
```

**Exercise 3.7:**  Write the complements of the following conditions.

a) X .LT. Y                         c) ITEM .NE. SENVAL
b) COUNTR .EQ. NRITMS               d) 2.0 .GT. 3.0

## 3.4 APPLICATIONS OF THE WHILE LOOP STRUCTURE

### 3.4.1 Controlling Loop Repetition with Computational Results

The WHILE loop structure is well suited for use in controlling loop repetition in which the repetition condition involves a test of values that are computed in the loop body. For example, in processing checking account transactions, we might want to continue processing transactions as long as the account balance is positive or zero, and stop and print a message when the balance becomes negative.

In problems of this sort, the loop control variable serves a dual function. It is used for storage of a computational result as well as for controlling loop repetition. Occasionally, more than one computed value will be involved in the repetition test, as illustrated in the following problem.

**Problem 3C:**  Two cyclists are involved in a race. The first has a headstart because the second cyclist is capable of a faster pace. We will write a program that will print out the distance from the starting line each cyclist has traveled as the race progresses. These distances will be printed for each half-hour of the race, beginning when the first cyclist departs, and continuing as long as the first cyclist is still ahead of the second cyclist.

**Data Table for Problem 3C**

*Program Parameters*

INTERV = 0.5, the time interval

*Input variables*

SPEED1: Average speed of first cyclist in mph

SPEED2: Average speed of second cyclist in mph

HEADST: Headstart expressed in hours

*Output variables*

TIME: Elapsed time from start of first cyclist in hours

DIST1: Distance travelled by first cyclist

DIST2: Distance travelled by second cyclist

**Discussion:**  This problem illustrates the use of the computer to *simulate* what would happen in a real-world situation. We can get an estimate of the progress of the cyclists before the race even begins and perhaps use this information to set up monitoring or aid stations.

The loop repetition test will involve a comparison of the distances travelled by each cyclist. The initial flow diagram and first refinement are shown in Fig. 3.11a. The values of DIST1 and DIST2 are initially zero. The loop repetition test involves a comparison of the two output variables DIST1 and DIST2, both of which are updated at the end of the loop (Steps 2.6 and 2.7).

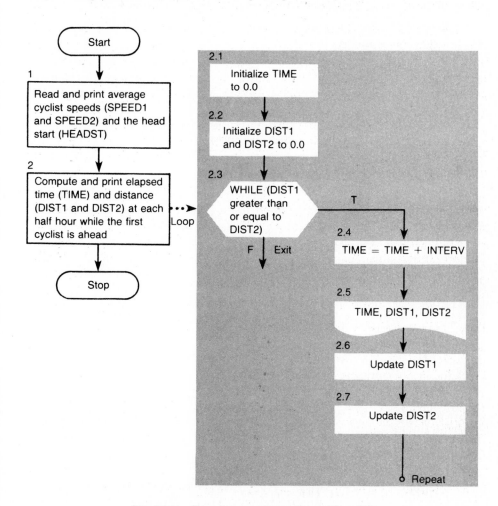

**Fig. 3.11a**  Flow diagrams for cyclist problem (3C).

To refine Step 2.6, we can use the familiar formula

$$\text{distance} = \text{speed} \times \text{time}$$

For the first cyclist, this will be implemented as the assignment statement

```
DIST1 = SPEED1 * TIME
```

The second cyclist does not depart until the elapsed time (TIME) exceeds the time of the headstart (HEADST). The time of travel for the second cyclist will always be reduced by the amount of the headstart. These considerations are reflected in the refinements of Step 2.7, shown in Fig. 3.11b. The expression involves two arithmetic operators and is evaluated as specified by the parenthesization (subtraction first).

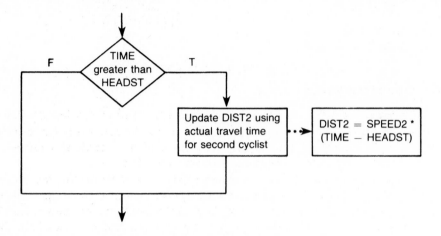

**Fig. 3.11b**    Refinement of Step 2.7 of Fig. 3.11a.

The program for Problem 3C is shown in Fig. 3.12 and the output is shown on page 100.

```
C CYCLE RACE PROBLEM
C
C PROGRAM PARAMETERS
      REAL INTERV
      PARAMETER (INTERV = 0.5)
C
C VARIABLES
      REAL SPEED1, SPEED2, HEADST
      REAL DIST1, DIST2, TIME
C
C READ IN SPEEDS AND HEADSTART
      READ*, SPEED1, SPEED2, HEADST
      PRINT*, 'FIRST CYCLIST SPEED = ', SPEED1
      PRINT*, 'SECOND CYCLIST SPEED = ', SPEED2
      PRINT*, 'FIRST CYCLIST HEADSTART (IN HOURS) = ', HEADST
C
C COMPUTE AND PRINT DISTANCES PER HALF HOUR
      PRINT*, ' '
      PRINT*, '        TIME            DISTANCE 1            DISTANCE 2'
      TIME = 0.0
      DIST1 = 0.0
      DIST2 = 0.0
```

*(Continued)*

```
        WHILE (DIST1 .GE. DIST2) DO        23 IF (DIST1 .LT. DIST2)
           PRINT*, TIME, DIST1, DIST2         Z GO TO 29
           TIME = TIME + INTERV
           DIST1 = SPEED1 * TIME
           IF (TIME .GT. HEADST) THEN
              DIST2 = SPEED2 * (TIME - HEADST)
           ENDIF
        ENDWHILE                               GO TO 23
  C                                         29 CONTINUE

        STOP
        END
```

**Fig. 3.12**   Program for Problem 3C.

### 3.4.2   Use of the Sentinel Card

Often, when we need to perform a task such as the one in Problem 2E (form the sum of a set of data items), we do not know exactly how many items there are to be processed. We might be handed a stack of data cards and asked to count them in order to determine the number of input data items.

One way to avoid this trying task is to append a *sentinel card* to the end of the stack of data cards. A sentinel card can be used to signal the program that all of the data cards have been read into the computer memory and processed. A sentinel card contains a value that would not normally occur as a data item for the program. When that value is read as a data item, it can be recognized by the program as an indication that all of the actual data items have been processed. The value on the sentinel card is called the *sentinel value*.

The concept of a sentinel card can be incorporated in the WHILE loop pattern as shown in Fig. 3.13.

The variable into which each data item is read acts as a loop control variable (LCV). The first read step (Step 2.1) is executed only once; it is used to initialize the LCV prior to loop entry so that the loop repetition test may be performed. After each data item is processed, the second read step (Step 2.5) enters the next data item. Every data item read is tested; if it is not the sentinel value, the loop body is repeated. We illustrate these and other points concerning the use of the sentinel card in the following problem.

**Problem 3D:**   Write a program that will read all of the scores for a course examination and compute and print the largest of these scores, as well as the number of scores processed.

**Discussion:**   In order to gain some insight into a solution of this problem, we should consider how we would go about finding the largest of a long list of numbers without the computer. Most likely we would read down the list of numbers, one at a time, and remember or "keep track of" the largest number that we had found at each point. If, at some point in the list, we should encounter a number, n, that is larger than the largest number found prior to that point, then we would make n the new largest number, and remember it rather than the previously found number.

**Program Style**

*Printing results in tabular form*

The output from the program shown in Fig. 3.12 would appear as follows:

```
*** PROGRAM EXECUTION OUTPUT ***

FIRST CYCLIST SPEED =    19.00000000000
SECOND CYCLIST SPEED =    27.00000000000
FIRST CYCLIST HEADSTART (IN HOURS) =    2.000000000000

            TIME              DISTANCE 1            DISTANCE 2
             0                    0                     0
    .5000000000000       9.500000000000               0
    1.000000000000       19.00000000000               0
    1.500000000000       28.50000000000               0
    2.000000000000       38.00000000000               0
    2.500000000000       47.50000000000        13.50000000000
    3.000000000000       57.00000000000        27.00000000000
    3.500000000000       66.50000000000        40.50000000000
    4.000000000000       76.00000000000        54.00000000000
    4.500000000000       85.50000000000        67.50000000000
    5.000000000000       95.00000000000        81.00000000000
    5.500000000000       104.5000000000        94.50000000000
    6.000000000000       114.0000000000        108.0000000000
    6.500000000000       123.5000000000        121.5000000000
```

The statements

```
PRINT*, ' '
PRINT*, '      TIME          DISTANCE 1          DISTANCE 2'
```

that precede the **WHILE** loop cause two character strings to be printed. Since the first string is a blank, a single line of spaces will be inserted in the program output. The second string contains three words that serve as column headings for the table of values to be printed. The spacing between these words may have to be adjusted slightly so that each word appears above its corresponding column of numbers.
The statement

```
                PRINT*, TIME, DIST1, DIST2
```

inside the **WHILE** loop prints each set of values for **TIME**, **DIST1** and **DIST2** on a single line allowing us to display the program results in an easy-to-read tabular form consisting of three columns of numbers. Unfortunately, not all compilers align numbers in columns as shown above.

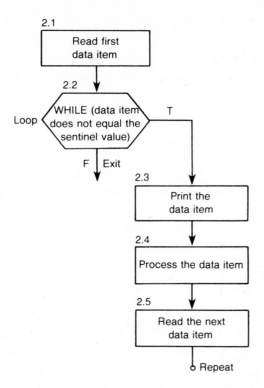

**Fig. 3.13** Use of the sentinel card in a WHILE loop pattern.

An example of how this might proceed is given in the monologue shown in Fig. 3.14.

| Test Scores | Effect of Each Score |
|---|---|
| 35 | "Since this is the first number, we will consider it to be the largest number initially. |
| 12 | "This is smaller than 35, so 35 is still largest. |
| 68 | "This is larger than 35. Therefore, 35 cannot be the largest item. Forget it and remember 68. |
| 8 | "This is smaller than 68, so 68 is still the largest. |
| −1 | "There are no more numbers, so 68 is the value we seek." |

**Fig. 3.14** Finding the largest of four numbers (a monologue).

We can use this procedure as a model for constructing an algorithm that solves Problem 3D. We will use a loop to instruct the computer to process a single score at a time, and to keep track of the largest score it has processed at any given point during the execution of the loop. The variable LARGE will be used to store this largest value.

In order to terminate the loop repetition, we will use a sentinel value of $-1$, which is not within the possible range of scores for the exam. The use of the sentinel value is desirable since we do not know beforehand how many test scores are to be processed. The sentinel value will be associated with the program parameter, SENVAL.

In addition, if we wish to keep track of the number of scores processed (al-

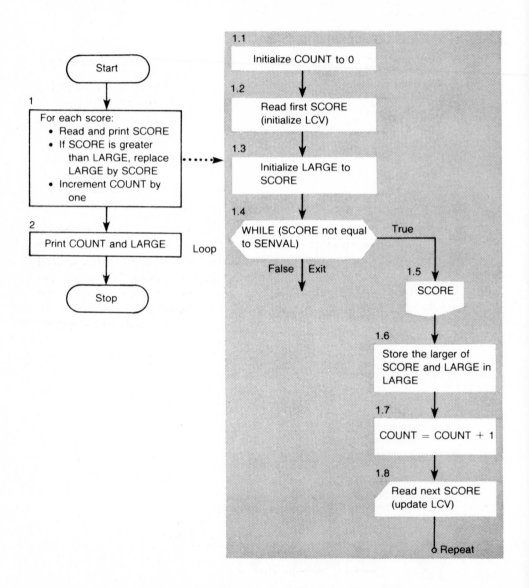

**Fig. 3.15**  Flow diagram for largest value problem (3D).

ways a good idea if this number is not known beforehand), we must introduce an output variable, COUNT, for this purpose. COUNT must be increased by one after each score is processed.

The data table follows; the algorithm is shown in Fig. 3.15.

**Data Table for Problem 3D**

*Program Parameters*

SENVAL = −1, sentinel value

| *Input variables* | *Output variables* |
|---|---|
| SCORE: The exam score currently being processed | LARGE: The largest of all scores processed at any point |
| | COUNT: Represents a count of the number of scores processed at any point |

From the refinement of Step 1, we see that SCORE is the variable used for loop control (this information should be included in the data table entry for SCORE). Each time a score is read in, it must first be compared to the sentinel value in order to determine whether loop execution is complete. The repeat condition is "SCORE not equal to SENVAL". Prior to performing this test for the first time, we must initialize the loop control variable SCORE via a READ statement (Step 1.2). Finally, at the end of the loop, we must update SCORE (also via a READ statement, Step 1.8).

Step 1.6 is refined as the single-alternative decision shown below.

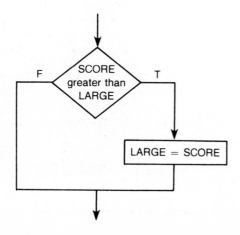

In order for the conditional test above to be meaningful, LARGE must be defined before the loop is entered. Since exam scores are normally positive, we could initialize LARGE to 0; however, the algorithm will be more general if we initialize LARGE to the first exam score, as done in Step 1.3 of Fig. 3.15.

The program is shown in Fig. 3.16, along with sample output for the data

$$\left.\begin{array}{r} 35 \\ 12 \\ 68 \\ 8 \end{array}\right\} \text{ exam scores}$$

$-1$    (sentinel value)

### 3.4.3  Reminders for Algorithm Development Using the WHILE Loop

The flow diagram pattern for the loop in Problem 3D is identical to the patterns developed earlier. This pattern may be characterized as shown in Fig. 3.17.

In addition, the steps leading to the construction of all of the loops seen so far are the same. These steps are summarized in the following list.

1. Complete a description of what must be done in the loop (the loop body).
2. Identify the loop control variable. This variable may already be a part of the loop body (such as SCORE), or it may need to be added (such as COUNTR in Problem 2E).
3. Set up the loop control variable test to be performed at the head of the loop body.
4. Initialize the loop control variable just prior to the test.
5. Update the loop control variable as the last step of the loop.

Not all loops will fit the category just described by the above pattern and loop-construction steps. However, a significant percentage of the loops you will write do fit into this category, so you should familiarize yourself with both the pattern and the construction steps.

**Exercise 3.8:**   What would happen in the execution of the program in Fig. 3.16 if we accidentally omitted all data except the sentinel value?

**Exercise 3.9:**   Modify the data table, flow diagram, and program for the largest value problem, to print the average of scores processed.

**Exercise 3.10:**   Modify the largest-value problem flow diagrams and data table so that the smallest score (SMALL) and the largest score are found and printed. Also, compute the range of the grades (RANGE = LARGE − SMALL).

```
C FIND THE LARGEST OF A COLLECTION OF EXAM SCORES
C
C PROGRAM PARAMETERS
      INTEGER SENVAL
      PARAMETER (SENVAL = -1)
C
C VARIABLES
      INTEGER SCORE, LARGE, COUNT
C
C PROCESS EACH EXAM SCORE TO FIND THE LARGEST VALUE AND THE COUNT
      PRINT*, 'LIST OF SCORES PROCESSED'
      COUNT = 0
      READ*, SCORE
      LARGE = SCORE
      WHILE (SCORE .NE. SENVAL) DO        13 IF (SCORE .EQ. SENVAL)
         PRINT*, SCORE                       Z GO TO 19
         IF (SCORE .GT. LARGE) LARGE = SCORE
         COUNT = COUNT + 1
         READ*, SCORE
      ENDWHILE                                  GO TO 13
C                                           19 CONTINUE
C PRINT THE COUNT AND THE LARGEST SCORE
      PRINT*, ' '
      PRINT*, COUNT, ' SCORES WERE PROCESSED'
      PRINT*, ' '
      PRINT*, LARGE, ' IS THE LARGEST SCORE'
C
      STOP
      END


*** PROGRAM EXECUTION OUTPUT ***

LIST OF SCORES PROCESSED
                35
                12
                68
                 8

            4 SCORES WERE PROCESSED

           68 IS THE LARGEST SCORE
```

**Fig. 3.16**   Program and sample output for Problem 3D.

**Exercise 3.11:**   On January 1, the water supply tank for the town of Death Valley contained 10,000 gallons of water. The town used 183 gallons of water a week, and it expected no rain in the foreseeable future. Write a loop to compute and print the amount of water remaining in the tank at the end of each week. Your loop should terminate when there is insufficient water to last out a week.

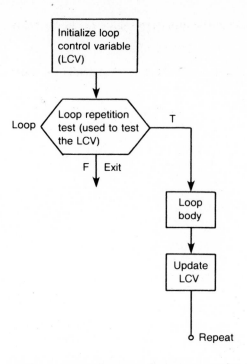

**Fig. 3.17**   Loop pattern for Problems 2E and 3D.

## 3.5   COUNTER-CONTROLLED LOOPS AND THE DO LOOP

In the last section, we showed how to use a sentinel value to control loop repetition when the number of data items to be processed was not known beforehand. In Chapter 2, we used a counter to control the repetition of a loop which was to be repeated exactly two thousand times (see Problem 2E). Since counter-controlled loops (counting loops) are so common, FORTRAN provides a special control structure, the *DO loop,* to simplify their implementation.

**Example 3.2:**   The program in Fig. 3.18 uses a DO loop to compute the sum of the integers from one to N inclusive, where N is a program parameter (value of 100). The header statement

```
DO 10 NEXT = 1 , N
```

identifies NEXT as the loop control variable. The *loop parameters* 1, N specify that the loop body

```
SUM = SUM + NEXT
```

should be executed once for each integer value of NEXT between 1 and N (1,2, . . . ,100). The label 10 following the word DO refers to the loop terminator. Any label may be used, however, the same label must appear in the label field (columns 1 through 5) of the loop terminator as shown below.

```
10 CONTINUE
```

We will always use a labelled CONTINUE statement to terminate a DO loop.

```
C COMPUTE THE SUM OF THE FIRST N INTEGERS
C
C PROGRAM PARAMETERS
      INTEGER N
      PARAMETER (N = 100)
C
C VARIABLES
      INTEGER SUM, NEXT
C
C COMPUTE SUM
      SUM = 0
      DO 10 NEXT = 1, N
         SUM = SUM + NEXT
   10 CONTINUE
C
C PRINT SUM
      PRINT*, 'THE SUM OF THE FIRST ', N, ' INTEGERS IS ', SUM
C
      STOP
      END


*** PROGRAM EXECUTION OUTPUT ***

THE SUM OF THE FIRST                100 INTEGERS IS                5050
```

**Fig. 3.18**   Program to compute the sum of the first 100 integers.

    A simplified form of the DO loop is described in the next display. In Chapter 6, we shall study the general form of the DO loop.

---

**DO Loop (simplified form)**

$$\text{DO } term\ lcv = initial,\ final$$

```
----------
---------- } loop body
----------
```

        *term* CONTINUE

**Interpretation:** The loop body is executed once for each integer value of the loop control variable, *lcv*, between *initial* and *final*, inclusive. The loop parameters *initial* and *final* may be integer constants, integer variables, or integer expressions. The label *term* must appear in the statement field of the header statement (following DO) and be placed in the label field (columns 1 through 5) of the terminator statement.

*Note 1:* The value of *lcv* is automatically increased by 1 each time the *loop body* is executed. It may not be modified by any statement in the *loop body*.

*Note 2:* The value of *initial* and *final* are determined just prior to loop entry. Any subsequent change in these values during loop execution will have no effect on the number of loop repetitions.

*Note 3:* Upon exit from the DO loop, the value of *lcv* will be one more than its value during the last execution of the *loop body*.

*Note 4:* If *initial* is greater than *final* (e.g., DO 15 I = 5, 1), the *loop body* will not be executed.

*Note 5:* A comma may be used after the label *term* in the loop header.

We shall use the flow diagram form shown in Fig. 3.19 to represent a DO loop.

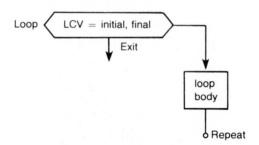

**Fig. 3.19**  Flow diagram of a DO loop.

**Example 3.3:**  The program in Fig. 3.20 prints a table showing the conversion from degrees Celsius to degrees Fahrenheit for temperatures from −5°C to +5°C.

The statement

        FAHREN = 1.8 * CELSIUS + 32.0

is the FORTRAN representation of the formula

        degrees fahrenheit = 1.8 × degrees celsius + 32

where the multiplication is performed first. We shall study additional examples of FORTRAN expressions with more than one operator in the next chapter.

The following problem also illustrates the use of the DO loop.

**Problem 3E:**  The banks in your area all advertise different interest rates for various kinds of long-term savings certificates. Usually the advertisements state the minimum investment term for the certificate (4 years, 6 years, etc.), and the annual interest rate. We will write a program which, given an investment term in years, a fixed annual interest rate, and an initial amount of deposit in dollars and cents, will compute and print the interest earned and the total value of the certificate at the end of each year of the investment term. We will assume that interest is compounded annually.

```
C CELSIUS TO FAHRENHEIT CONVERSION
C
C PROGRAM PARAMETERS
      INTEGER MINCEL, MAXCEL
      PARAMETER (MINCEL = -5, MAXCEL = 5)
C
C VARIABLES
      REAL FAHREN
      INTEGER CELSIUS
C
C COMPUTE AND PRINT TABLE
      PRINT*, '              CELSIUS      FAHRENHEIT'
      DO 20 CELSIUS = MINCEL, MAXCEL
         FAHREN = 1.8 * CELSIUS + 32.0
         PRINT*, CELSIUS, FAHREN
   20 CONTINUE
C
      STOP
      END


     *** PROGRAM EXECUTION OUTPUT ***

              CELSIUS    FAHRENHEIT
                 -5   23.00000000000
                 -4   24.80000000000
                 -3   26.60000000000
                 -2   28.40000000000
                 -1   30.20000000000
                  0   32.00000000000
                  1   33.80000000000
                  2   35.60000000000
                  3   37.40000000000
                  4   39.20000000000
                  5   41.00000000000
```

**Fig. 3.20**   Temperature conversion program and output.

**Discussion:**   An initial data table for this problem is shown below; the level one flow diagram is shown in Fig. 3.21a.

**Data Table for Problem 3E**

| *Input variables* | *Output variables* |
|---|---|
| CTTERM: Term of certificate | INTRST: Interest amount computed at the end of each year |
| RATE: Annual interest rate | |
| | VALUE: Certificate value at the end of each year |
| STVAL: Initial value of certificate | |

From the level one algorithm, it is clear that the repetition of a short sequence of steps is needed in the refinement of Step 2. The repetition can easily be

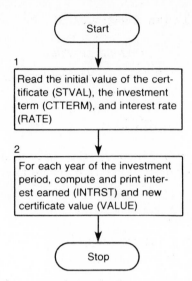

**Fig. 3.21a** Level one flow diagram for the bank certificate problem (3E).

controlled by using a counter, YEAR, that takes on successive integral values from 1 (first year) through CTTERM (last year). The new data table entry follows; the refinement of Step 2 is shown in Fig. 3.21b.

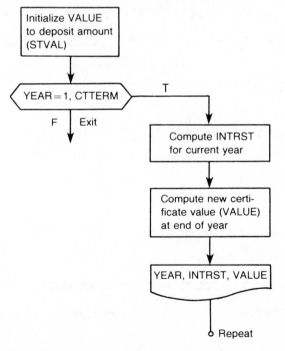

**Fig. 3.21b** Level two refinement of Step 2 of Fig. 3.21.

**Additional Data Table Entry for Problem 3E**

*Program variables*

YEAR: Loop control variable
of DO loop; initial
value 1, final value
**CTTERM**

The program is given in Fig. 3.22, along with sample output for an initial certificate value of $3000, a term of 10 years, and an interest rate of 12 percent.

```
C BANK CERTIFICATE PROBLEM
C
      REAL STVAL, RATE, INTRST, VALUE
      INTEGER CTTERM, YEAR
C
C READ AND PRINT THE INITIAL VALUE OF THE CERTIFICATE, THE
C INTEREST RATE AND THE TERM
      PRINT*, 'BANK CERTIFICATE ANALYSIS '
      PRINT*, ' '
      READ*, STVAL, RATE, CTTERM
      PRINT*, 'INITIAL VALUE OF CERTIFICATE = $', STVAL
      PRINT*, 'INTEREST RATE = ', RATE
      PRINT*, 'TERM OF CERTIFICATE = ', CTTERM, ' YEARS'
C
C COMPUTE AND PRINT INTEREST AND VALUE AFTER EACH YEAR OF THE TERM
      PRINT*, ' '
      PRINT*, '           YEAR      INTEREST           TOTAL VALUE'
      VALUE = STVAL
      DO 10 YEAR = 1, CTTERM
         INTRST = VALUE * RATE
         VALUE = VALUE + INTRST
         PRINT*, YEAR, INTRST, VALUE
   10 CONTINUE
C
      STOP
      END


*** PROGRAM EXECUTION OUTPUT ***

BANK CERTIFICATE ANALYSIS

INITIAL VALUE OF CERTIFICATE = $  3000.000000000
INTEREST RATE =    .1200000000000
TERM OF CERTIFICATE =              10 YEARS
```

*(Continued)*

| YEAR | INTEREST | TOTAL VALUE |
|------|----------|-------------|
| 1 | 360.0000000000 | 3360.000000000 |
| 2 | 403.2000000000 | 3763.200000000 |
| 3 | 451.5840000000 | 4214.784000000 |
| 4 | 505.7740800000 | 4720.558080000 |
| 5 | 566.4669696000 | 5287.025049600 |
| 6 | 634.4430059520 | 5921.468055552 |
| 7 | 710.5761666662 | 6632.044222218 |
| 8 | 795.8453066662 | 7427.889528884 |
| 9 | 891.3467434661 | 8319.236272351 |
| 10 | 998.3083526821 | 9317.544625033 |

**Fig. 3.22**    Program and sample output for Problem 3E.

**Exercise 3.12:**    Implement the counting loop shown in Fig. 3.22 using a WHILE loop.

**Exercise 3.13:**    Implement the counting loop shown in Fig. 3.18 using a WHILE loop.

**Exercise 3.14:**    Modify the program for Problem 3E to read in the starting year of the certificate (e.g., 1981) as a fourth input value and to print the actual years of the life of the certificate (e.g., 1981, 1982, 1990) rather than the integers 1,2, . . .10.

## 3.6   THE WIDGET INVENTORY CONTROL PROBLEM

We will now turn our attention to the solution of a problem that illustrates the use of most of the structures introduced in the chapter.

**Problem 3F:**    The Widget Manufacturing Company needs a simple program to help control the manufacturing and shipping of widgets. Specifically, the program is to process orders for the shipment of new widgets and check each order to see whether an inventory sufficient to fill the order is in stock. The message FILLED or NOT FILLED should be printed next to each shipment request. After all orders have been processed, the program should print out the final value of the inventory, the number of widgets shipped, and the number of additional widgets that must be manufactured to fill all outstanding orders. The program data must start with the initial inventory value followed by all orders for widgets, ending with a sentinel value.

**Discussion:**    A loop will be needed to process all orders. The loop exit will occur when the sentinel value is read. After each order is read in, it must be compared to the widget inventory and processed. Since inventory records and orders involving a piece or fractional part of a widget are impossible, we may treat all data to be processed by this program as type INTEGER. The data table follows.

**Data Table for Problem 3F**

*Program Parameters*

SENVAL = 0, sentinel value

| *Input variables* | *Output variables* |
|---|---|
| OLDINV: Initial inventory at start of processing | CURINV: Used to keep track of inventory as orders are processed |

ORDER: Contains each
  order as it is being
  processed

ADWID: Additional
  widgets required
  to fill outstanding
  orders

SHIP: Number of widgets
  shipped

    The input information for this problem is the initial inventory (OLDINV) and each widget order (ORDER). The output information will be the remaining widget inventory (CURINV), the total number of additional widgets required to fill the outstanding orders (ADWID) and the number of widgets shipped (SHIP). The value of SHIP can easily be computed as the difference of OLDINV and the final value of CURINV. The flow diagrams are drawn in Fig. 3.23a.

    As each order is processed (Step 2.3), it should be filled if it is less than or equal to the current inventory (CURINV) and the value of CURINV must be decreased by the amount of the order (ORDER). If the order cannot be com-

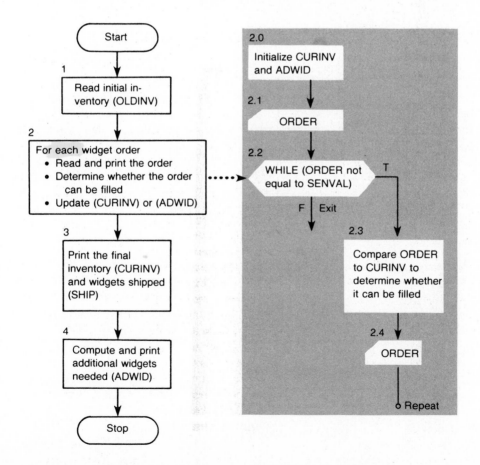

**Fig. 3.23a**  Flow diagrams for widget inventory problem (3F).

pletely filled, the order amount should be added to ADWID. Before entering the loop, CURINV should be initialized to the starting inventory value (OLDINV) and ADWID must be initialized to zero. If ADWID does not remain zero, then more widgets are needed. This amount is determined by subtracting the final value of CURINV (after all orders are processed) from ADWID. The refinements of Steps 2.3 and 4 are shown in Fig. 3.23b. The program is given in Fig. 3.24, along with the output for the sample data below.

$$
\begin{array}{ll}
75 & \text{(initial inventory)} \\
\left.\begin{array}{l}
20 \\
50 \\
100 \\
3 \\
15 \\
2
\end{array}\right\} & \text{widget orders} \\
0 & \text{(sentinel value)}
\end{array}
$$

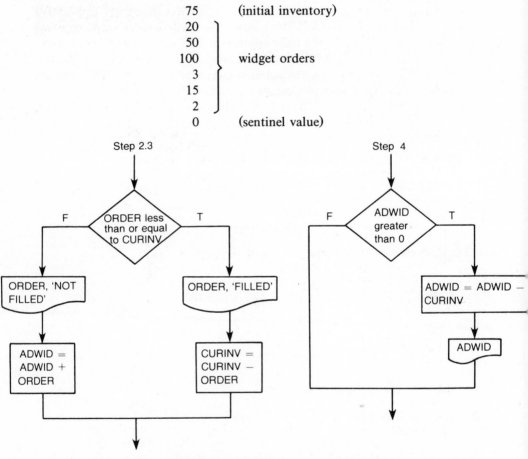

**Fig. 3.23b**   Additional flow diagram refinements for the widget inventory problem (3F).

```
C WIDGET INVENTORY CONTROL PROBLEM
C
C PROGRAM PARAMETERS
      INTEGER SENVAL
      PARAMETER (SENVAL = 0)
C
C VARIABLES
      INTEGER OLDINV, CURINV, ORDER, ADWID, SHIP
C
```

(*Continued*)

```
C READ STARTING INVENTORY
      READ*, OLDINV
      PRINT*, 'INITIAL INVENTORY = ', OLDINV
C
C READ AND PROCESS EACH ORDER
      CURINV = OLDINV
      ADWID = 0
      PRINT*, 'LIST OF ORDERS'
      READ*, ORDER
      WHILE (ORDER .NE. SENVAL) DO        22 IF (ORDER .EQ. SENVAL)
         IF (ORDER .LE. CURINV) THEN        Z GO TO 29
            PRINT*, ORDER, ' FILLED'
            CURINV = CURINV - ORDER
         ELSE
            PRINT*, ORDER, ' NOT FILLED'
            ADWID = ADWID + ORDER
         ENDIF
         READ*, ORDER
      ENDWHILE                            GO TO 22
C                                         29 CONTINUE
C PRINT FINAL INVENTORY AND COUNT OF WIDGETS SHIPPED
      PRINT*, ' '
      PRINT*, 'FINAL INVENTORY = ', CURINV
      PRINT*, ' '
      SHIP = OLDINV - CURINV
      PRINT*, SHIP, ' WIDGETS SHIPPED'
C
C PRINT COUNT OF ADDITIONAL WIDGETS NEEDED (IF ANY)
      IF (ADWID .GT. 0) THEN
         ADWID = ADWID - CURINV
         PRINT*, ADWID, ' NEW WIDGETS NEEDED'
      ENDIF
C
      STOP
      END


*** PROGRAM EXECUTION OUTPUT ***

INITIAL INVENTORY =                  75
LIST OF ORDERS
                20 FILLED
                50 FILLED
               100 NOT FILLED
                 3 FILLED
                15 NOT FILLED
                 2 FILLED

FINAL INVENTORY =                    0

                75 WIDGETS SHIPPED
               115 NEW WIDGETS NEEDED
```

Fig. 3.24  Program and sample output for Problem 3F.

---

**Program Style**

*Sequences and lists of control structures*

If you follow the suggestions for algorithm design presented in this chapter and Chapter 2, your level one flow diagrams will consist of a sequence of major steps. Some of these steps may be refined as decision steps and still others may be refined as loops. When your algorithm is translated into FORTRAN, there will be a group of program statements or a control structure corresponding to each algorithm step or its refinement in the flow diagram. These statements and structures will be in the same order as their corresponding flow diagram steps.

Occasionally, one control structure will be *nested*, or wholly contained, in another. This will happen if the refinement of a step in a control structure is itself a control structure. An example of this is the refinement of Step 2.3 as a decision structure (See Fig. 3.23b). Since Step 2.3 is part of the loop body, the IF-THEN-ELSE control structure corresponding to Step 2.3 is nested in the WHILE loop as shown in Fig. 3.24. This means that the IF-THEN-ELSE structure is executed during each repetition of the loop and the message 'FILLED' or 'NOT FILLED' is printed along with each order amount.

---

**Exercise 3.15:**   Is it possible for an order for widgets to be filled even if the one before it was not? Hand-trace the execution of this program for the data 75, 20, 50, 100, 3, 15, 2, 0.

## 3.7  DEBUGGING PROGRAMS

### 3.7.1  Introduction

It is very rare that a program runs correctly the first time it is typed or keypunched. Often, one spends a considerable amount of time in removing errors or *bugs* from programs.

The process of locating and removing errors from a program is called *debugging.* You will find that a substantial portion of your time is spent debugging. The debugging time can be reduced if you follow the algorithm and program development steps illustrated in the text without taking any shortcuts.

This approach requires a careful analysis of the problem description, the identification in a data table of the input and output data for the problem, and the careful development of the algorithm for the problem solution. The algorithm development should proceed on a step-by-step basis, beginning with an outline of the algorithm in the form of a level one list of subtasks. Additional algorithm detail (refinements) should be provided as needed, until enough detail has been added so that writing the program is virtually a mechanical process. The data table should be updated during the refinement process, so that all variables introduced in the algorithm are listed and clearly defined in the table.

Once the algorithm and data table are complete, a systematic hand simulation (or trace) of the algorithm, using one or two representative sets of data, can help eliminate many bugs before they show up during the execution of your program. When the hand trace is complete, the program may be written, using the data table and the algorithm refinements.

There are two general categories of errors that you may encounter when running programs:

- syntax errors
- run-time and logic errors.

### 3.7.2   Syntax Errors

We discussed syntax errors in Chapter 1. A syntax error results when an illegal statement is present in the program. If a statement cannot be translated because it does not follow the FORTRAN syntax rules, the compiler will print a diagnostic message indicating the illegal statement and the type of error.

Unfortunately, the diagnostic messages printed by the compiler often seem vague or confusing. Sometimes careful interpretation and clever detective work are needed to correlate the message with the actual error. A further complication is that an error in one program statement is likely to "confuse" the compiler and cause it to detect errors in subsequent program statements that are actually correct. For this reason, it is usually wisest to concentrate on the first few diagnostics generated by the compiler, particularly if they relate to your declarations. Correcting these early mistakes will often drastically reduce the number of diagnostic messages printed in the next program run. After all syntax errors have been eliminated, the program will then execute, although it may still contain bugs.

### 3.7.3   Run-time and Logic Errors

Errors that occur during program execution (*run-time errors*) are normally the result of programmer carelessness, or errors in logic. They are not severe enough to prevent the compiler from translating the program; however, they will prevent the program from executing correctly. Depending on the severity of the error, program execution may be terminated.

A common run-time error is caused by referencing a variable before its value has been defined. For example, the assignment statement

$$X = Y * Z$$

cannot be executed in a meaningful manner unless the values of Y and Z have been previously defined.

Another example of a run-time error is the failure to provide enough data values for a program. After the last data value is read, a diagnostic message of the form

```
ATTEMPT TO READ BEYOND END OF INPUT FILE
```

would be printed if another READ statement is reached. This type of message would also be printed if a loop containing a READ statement did not terminate when expected. In this case, the loop would continue to execute and attempt to read data after all data values had been entered. This diagnostic message is often the only symptom of an improper loop repetition test or missing sentinel value.

If your program executes but doesn't produce the desired results, there may be an error in logic. If there was not enough output information printed in your first run, it is often worthwhile to make an extra debugging run in which all pertinent variable values are printed at different steps in the execution of your program.

The PRINT statements used for debugging should be added with some care and thought. Since your program is organized as a linear sequence of steps, it is desirable to print the values of variables that are changed in each step before and after execution of that step. You should carefully simulate the execution of each step to verify that the values printed are correct.

Often the first debugging run will do no more than tell you what step is in error. At this point you should concentrate on that step and insert additional PRINT statements to trace its execution. Be careful when inserting extra PRINT statements in a loop as a new set of values will be printed with each repetition of the loop. This may waste considerable paper and time.

For decision and loop steps, it is a good idea to print the values of variables involved in the conditional test. These values should be printed just before execution of the test. For example, if the conditional test has the form

```
COUNTR .LT. NRITMS
```

then you might insert the statement

```
PRINT*, 'COUNTR = ', COUNTR
```

just before the condition and as the last step in a loop. If a loop is used to accumulate a result, it is useful to print the accumulated value just after loop exit.

For example, the PRINT statements in the following loop would show that the loop control update step was accidently omitted from the loop. Since the value of the loop control variable I never changes, this loop will execute "forever," or until the program uses up all of the time allotted for this execution.

```
READ*, N
PRINT*, 'N = ', N
I = 1
SUM = 0
PRINT*, 'INITIAL I = ', I
WHILE (I .LE. N) DO
        SUM = SUM + I
        PRINT*, 'NEXT I = ', I
ENDWHILE
PRINT*, 'SUM = ', SUM
```

If N = 4, the output below would be printed, clearly illustrating the error.

```
N = 4
INITIAL I = 1
NEXT I = 1
NEXT I = 1
NEXT I = 1
NEXT I = 1
    .
    .
    .
```

To correct this error, a loop control variable update step such as

$$I = I + 1$$

must be inserted in the loop body.

Once you have located an error, go back to your algorithm, modify the steps that you believe are in error and then completely retrace the modified algorithm. This last step is extremely important and one that is often overlooked. Each algorithm change may have important side effects that are difficult to anticipate. Making what seems to be an obvious correction in one step of the algorithm may introduce new errors into other algorithm steps. The only way to establish that there are no side effects is to retrace the revised algorithm.

Once the revised algorithm has been carefully checked out, write the new program statements that are needed, and correct and rerun your program.

## 3.8 TESTING PROGRAMS

### 3.8.1 Preparing Test Data Sets

Once all errors have been removed and the program has executed to normal completion, the program should be tested as thoroughly as possible. In Section 2.5.1, we discussed tracing an algorithm and suggested that enough sets of test data be provided to ensure that all paths through the algorithm are traced.

In a similar way, the final program should be tested with a variety of data. For example, to test the widget inventory program (Fig. 3.24), one data set should be provided that contains orders that are not filled as well as orders that are filled. As suggested in Exercise 3.15, this data set should be prepared so that some orders are filled, even though earlier orders were not. A second data set should also be provided that only contains orders that are filled. (Why?)

### 3.8.2 Handling the Exceptions

As you get more experienced in programming, you will begin to write programs that not only handle all "normal" situations, but also the unexpected as well. For example, what would the widget inventory program do if the initial inventory value or any widget order happened to be negative? As currently implemented, all negative orders would automatically be filled. A better solution would be to ignore a negative order and print an "error message" indicating that an invalid order was received.

You are right if you are thinking that no reasonable person would request a negative order; however, users of programs often do not know what is reasonable and what is not. This is particularly likely to be true if the program user is different from the program designer. Consequently, experienced programmers practice "defensive programming" so that their programs will behave in a reasonable way even when the data is unreasonable.

Wherever possible, your programs should test for the occurrence of unusual data values and print an error message when they occur. Very often, these unusual values will eventually result in a run-time error. Whether this happens or not, your program should print an error message to warn the user of a potential problem and its source.

## 3.9   COMMON PROGRAMMING ERRORS

In using control structures, there are many pitfalls to avoid. The header and terminator statements of each control structure must satisfy the syntax rules. Care should be taken to ensure that header and terminator statements are included for each structure, and that the header and terminator match the corresponding flow diagram step.

Missing terminators can be detected easily by the compiler, which will print a diagnostic indicating that the terminator statement is missing. If the terminator statements for two nested control structures such as a decision structure and a loop structure are interchanged (see Fig. 3.25), the compiler will not be able to translate these structures properly. Consequently, it will print a diagnostic message indicating that the structures overlap or are not terminated in the proper sequence. Use of the wrong structure (for example, using a decision structure when a loop is required) will often go undetected until execution of the program.

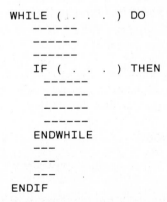

```
WHILE ( . . . ) DO
    ------
    ------
    ------
    IF ( . . . ) THEN
      ------
      ------
      ------
      ------
    ENDWHILE
    ---
    ---
    ---
ENDIF
```

**Fig. 3.25**   Overlapping structures.

A common error in using loop structures involves the specification of too many or too few loop repetitions. You should check carefully that all loop repetition counters are initialized properly and that the loop repetition condition specifies the exact number of loop repetitions. A common error is performing one more loop repetition than desired. For example, this would happen if the loop repetition condition

$$(COUNTR\ .LE.\ NRITMS)$$

was substituted for

$$(COUNTR\ .LT.\ NRITMS)$$

It is always essential to verify that either your input data or some computation will eventually cause a loop repetition condition to become false. This is especially important when the condition involves a test for equality or inequality. In Problem 3D, the loop repetition condition (SCORE .NE. SENVAL) will become false only if a data value of $-1$ is read into SCORE. Otherwise, the loop will not terminate when it should.

*Nonterminating loops* can cause a great waste of computer paper and time. Eventually, a program with a nonterminating loop will either exceed the amount of computer time allocated to it, or will run out of input data. The latter problem will be indicated by a diagnostic message indicating "an insufficient amount of data," or an "attempt to read past the end of the input file."

## 3.10  SUMMARY

In this chapter we have discussed some of the control structures that are available in FORTRAN. Two decision structures and two loop structures were described.

FORTRAN implementations of each of these structures were provided. They all begin with a header statement and end with a special terminator statement. Two forms of the WHILE loop were given as it is not part of standard FORTRAN.

The header statement is used to distinguish each structure from the others and to indicate the type of the structure. The IF-THEN header indicates a decision structure; the WHILE or DO header indicates a loop. Each header has its own meaning, and this meaning is defined by the way in which the structure is translated by the compiler into machine-language.

Terminator statements serve as end markers, indicating to the compiler where the physical end of a structure is in the program. The ENDIF indicates the physical end of a decision structure; the ENDWHILE or CONTINUE marks the end of a loop.

In formulating your solution to a programming problem, it is essential that you think in terms of the structures and their functions. The emphasis should be on how the structures affect what is to be done, and not on the various transfers of control that are inserted by the compiler during translation. To emphasize this point further, we urge you to review the interpretation of the IF-THEN, IF-

THEN-ELSE and WHILE and DO loop structures, as given in Sections 3.2 and 3.3 and summarized in Figs. 3.26 and 3.27.

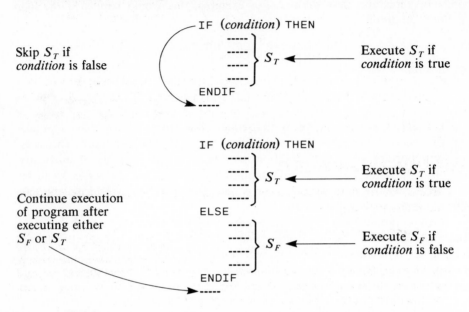

Skip $S_T$ if *condition* is false

IF (*condition*) THEN

$S_T$    Execute $S_T$ if *condition* is true

ENDIF

Continue execution of program after executing either $S_F$ or $S_T$

IF (*condition*) THEN

$S_T$    Execute $S_T$ if *condition* is true

ELSE

$S_F$    Execute $S_F$ if *condition* is false

ENDIF

**Fig. 3.26**  Graphical summary of the effect of the decision-control structures.

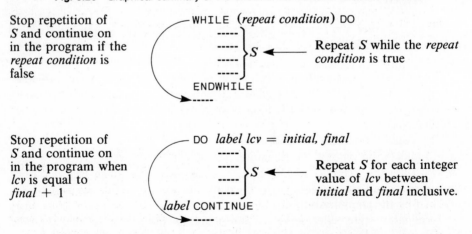

Stop repetition of $S$ and continue on in the program if the *repeat condition* is false

WHILE (*repeat condition*) DO

$S$    Repeat $S$ while the *repeat condition* is true

ENDWHILE

Stop repetition of $S$ and continue on in the program when *lcv* is equal to *final* + 1

DO *label lcv = initial, final*

$S$    Repeat $S$ for each integer value of *lcv* between *initial* and *final* inclusive.

*label* CONTINUE

**Fig. 3.27**  Graphical summary of the effect of the WHILE and DO loop structure.

Unfortunately, the WHILE loop form shown in Fig. 3.27 is not part of standard FORTRAN and, hence, is not universally available. The standard FORTRAN form shown below tests the loop exit condition that is the complement of the repeat condition.

Stop repetition of *S* and continue on in the program if the *exit condition* is true

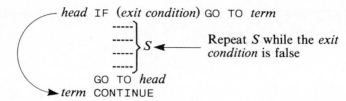

```
      head IF (exit condition) GO TO term
           -----
           -----
           ----- } S ←——— Repeat S while the exit
           -----                condition is false
      GO TO head
 term CONTINUE
```

Repeat *S* while the *exit condition* is false

We recommend the use of the structures introduced in this chapter whenever possible in the formulation and implementation of algorithms. Careful understanding and use of the structures will pay off in programs that are easier to write and have fewer and more easily detected errors.

## PROGRAMMING PROBLEMS

(Unless indicated otherwise, a data table and flow diagram should be provided for each problem.)

**3G** Write a program for the following problem: Read in a list of integer data items punched one to a card, and find and print the index of the first card containing the number 12. Use a sentinel value of 0. Your program should print the message 'NOT FOUND' if the number is not found. (The index is the location of the card containing 12 in the data-card section. For example, if the 11th card read in contains 12, then the index value 11 should be printed.)

**3H** Write a program to read in a collection of exam scores ranging in value from 1 to 100, and count and print the total number of scores, and the number of outstanding scores (90–100), the number of satisfactory scores (60–89), and the number of unsatisfactory scores (1–59). Test your program on the following data:

| | | |
|---|---|---|
| 63 | 75 | 72 |
| 72 | 78 | 67 |
| 80 | 63 | 75 |
| 90 | 89 | 43 |
| 59 | 99 | 82 |
| 12 | 100 | |

**3I** Write a program to process weekly employee time cards for all employees of an organization. Each employee will have one card indicating an identification number (IDN, an integer), the hourly wage rate (RATE), and the number of hours (HOURS) worked during a given week. Each employee is to be paid time-and-a-half for all hours worked over 40.0. A tax amount of 3.625 percent of gross salary (GROSS) will be deducted. The program output should show the employee's number and net pay (NET). An employee number of 0 will serve as the sentinel value.

**3J** Complete the flow diagram, and write the program for Exercise 3.11. Have your program indicate whether or not the town has enough water to last for one year.

**3K** Suppose you own a beer distributorship that sells Piels (ID number 1), Coors (ID2), Bud (ID3), and Iron City (ID number 4) by the case. Write a program to (a) read in the case inventory for each brand for the start of the week; (b) process all weekly sales and purchase records for each brand; and (c) print out the final inventory. Each case transaction will consist of the brand identification number (an integer) and the amount purchased (a positive integer value) or the amount sold (a negative integer value). The weekly inventory for each brand (for the start of the week) will consist of the identification and inventory amount for that brand. For now, you may assume that you always have sufficient foresight to prevent the complete depletion of your inventory for any brand. *Hint:* Your data-card deck should begin with four cards representing the case inventory. These should be followed by all the transaction cards, followed by a sentinel card.

**3L**  Write a program for the following problem: Read in a positive integer N and compute SUMSLO $= \sum_{i=1}^{N}$    $i = 1 + 2 + 3 + 4 + \cdots + N$ (the sum of all integers from 1 to N, inclusive). Then compute SUMFST $= (N \times (N + 1))/2$ and compare SUMFST and SUMSLO. Your program should print both SUMSLO and SUMFST and indicate whether or not they are equal. (You will need a loop to compute SUMSLO and three arithmetic statements to compute SUMFST. Which computation method is preferable?

To verify your hypothesis of the relationship between SUMSLO and SUMFST, modify your program so that it will process a collection of numbers (use a loop that will terminate when a value of 0 is read).

**3M**  Write a program to find the largest value in a collection of N numbers, where the value of N will be the first data card read into the program. If the value of N were unknown, could a sentinel value be used to terminate execution of the main program loop for this problem?

**3N**  Write a program to process a collection of checking-account transactions (deposits or withdrawals) for Mr. Shelley's account. Your program should begin by reading in the previous account balance (OLDBAL), and then process each transaction (TRANS), computing the new balance (NEWBAL). Use a 0.00 transaction as your sentinel value. Your output should appear in three columns, with withdrawals on the left, deposits in the middle, and the new balance (after each transaction) on the right. To accomplish this, use the print statement

```
PRINT*, ZERO, TRANS, NEWBAL
```

if the transaction is a deposit (a positive number), and

```
PRINT*, TRANS, ZERO, NEWBAL
```

if the transaction is a withdrawal (a negative number). The value of the variable ZERO should be 0.0. Test your program with the following data.

Old balance = 325.50

Transactions:    25.00, −79.25, −60.00, 16.75,
                          −259.47, 42.00, −5.50

**3O**  Modify the data table, flow diagram, and program of Problem 3N to compute and print the following additional information:

The number of withdrawals; The number of deposits;
The number of transactions;
The total sum of all withdrawals;
The total sum of all deposits.

**3P**   Following the processing of the transaction −259.47 in Problem 3N (or 3O), the value of NEWBAL was negative, indicating that Mr. Shelley's account was over-drawn. Modify your data table, flow diagram, and program for Problem 3N (or 3O) so that the resulting new program will test for withdrawal amounts that are not covered. Have your program completely skip processing each such withdrawal, and instead use the following print statement to indicate an overdrawn account:

PRINT*, TRANS, '*WITHDRAWAL NOT COVERED AND NOT PROCESSED** '

The value of NEWBAL should not be altered by withdrawals that are not covered. Your program should count the number of such withdrawals and print a total at the end of execution. (Note that in Problem 3N or 3O Mr. Shelley's final balance was positive. This indicates that he made a deposit during the current time period to cover the $259.47 withdrawal. What could be done to prevent such a transaction from being considered as overdrawn as long as the final account balance for the current period is positive?)

**3Q**   Write a program to compute and print the fractional powers of two (1/2, 1/4, 1/8, 1/16, . . .) in decimal form. Your program should print two columns of information, as shown below:

| Power | Fraction |
|-------|----------|
| 1 | 0.5 |
| 2 | 0.25 |
| 3 | 0.125 |
| 4 | 0.0625 |
| . | . |
| . | . |
| . | . |

The program should terminate when the decimal fraction becomes less than or equal to 0.000001. Draw a flow diagram first.

**3R**   Modify the program for Problem 3Q to accumulate and print the sum of the fractions computed *at each step*. Add a third column of output containing the accumulated sum.

| Sum |
|-----|
| 0.5 |
| 0.75 |
| 0.875 |
| 0.9375 |
| . |
| . |
| . |

Explain the results in this column. Could this value ever reach 1.0?

**3S**   The trustees of a small college are considering voting a pay raise for the 12 full-time faculty members. They want to grant a 9 1/2 percent pay raise. However, be-fore doing so, they want to know how much additional cash this will cost the college. Write a program that will provide this information. Test your program for the following salaries:

| | |
|---|---|
| $12,500.00 | $14,029.50 |
| $16,000.00 | $13,250.00 |
| $15,500.00 | $12,800.00 |
| $20,000.50 | $18,900.00 |
| $13,780.00 | $17,300.00 |
| $14,120.25 | $14,100.00 |

**3T**     The assessor in the local township has punched, one value per card, the estimated value of all 14 properties in the township. Properties are assessed a flat tax rate of 125 mils per dollar of assessed value, and each property is assessed at only 28% of its estimated value. Write a program to compute the total amount of taxes that will be collected on the 14 properties in the township. (A mil is equal to 0.1 of a penny). The estimated values of the properties are:

| | |
|---|---|
| $50,000.00 | $48,000.00 |
| $45,500.00 | $67,000.00 |
| $37,600.00 | $47,100.00 |
| $65,000.00 | $53,350.00 |
| $28,000.00 | $58,000.00 |
| $52,250.00 | $48,000.00 |
| $56,500.00 | $43,700.00 |

**3U**     Write a program that will read in a positive real number and determine and print the number of digits to the left of the decimal point. *Hint:* Repeatedly divide the number by 10.0 until it becomes less than 1.0. Test the program with the following data:

| | |
|---|---|
| 4703.62 | 0.01 |
| 0.47 | 57642.00 |
| 10.12 | 4000.00 |

# DATA TYPES

## 4.1   INTRODUCTION

While writing earlier programs you may have thought about, and perhaps even written, FORTRAN assignment statements containing parentheses and more than one arithmetic operator. You may have wondered whether or not FORTRAN could be used to instruct the computer to manipulate something other than numbers and, if so, how?

In this chapter, we will see that FORTRAN can be used to manipulate character strings, logical values (true and false), as well as numbers. We will learn how to form FORTRAN assignment statements of greater complexity than those used so far to specify numeric data manipulations, and we will introduce some simple manipulations of character and logical data. Additional features of character and logical data manipulation will be described in Chapter 9. All of these features will make it still more convenient to program in FORTRAN.

## 4.2   DATA TYPES, DECLARATIONS AND CONSTANTS

### 4.2.1   External Representations of Different Data Types

In all our programming so far, we have manipulated only numeric information. This is, in fact, an unnecessary restriction since it is possible to write FORTRAN programs that manipulate a number of different types of data. In this text, we will discuss four *data types*: real, integer, character and logical.

In order to best understand the differences among the four data types, it is helpful to examine the differences in their external representations. The *external representation* of a data item is the representation that is used in writing constants in FORTRAN programs, and in keypunching information that is to be read by a program. It is also the form in which results of program computations are printed for us to read.

The program shown in Fig. 4.1 is identical to the program for Problem 3E (the bank certificate problem), except that the input variable NAME (used to contain the name of the certificate holder) has been added. The program illustrates the use of three types of data: real, integer and character. The first statement in the program

```
CHARACTER * 24 NAME
```

is a *character type declaration* that identifies NAME as a variable used for storage of character data.

A brief study of this program will aid your understanding of the differences between the data types. The sample output shown with the program was generated using the input

```
'EMMYLOU HARRIS'
3000.0   0.12    10
```

These data illustrate the differences between the external representation of integer, real and character data. Real and integer data consist of a string of decimal digits possibly preceded by a sign (+ or −). Real data (such as 3000.0 and 0.12)

must contain a decimal point, whereas integer data must not. Character data items (such as 'EMMYLOU HARRIS') consist of a string of characters enclosed in apostrophes. Any character in the FORTRAN character set (see Fig. 1.5) may be included in the string.

```
C BANK CERTIFICATE PROBLEM
C
      CHARACTER * 24 NAME
      REAL STVAL, RATE, INTRST, VALUE
      INTEGER CTTERM, YEAR
C
C READ AND PRINT THE NAME OF THE CERTIFICATE OWNER, THE INITIAL VALUE
C OF THE CERTIFICATE, THE INTEREST RATE AND THE TERM
      READ*, NAME
      PRINT*, 'BANK CERTIFICATE ANALYSIS FOR ', NAME
      PRINT*, ' '
      READ*, STVAL, RATE, CTTERM
      PRINT*, 'START VALUE OF CERTIFICATE = $', STVAL
      PRINT*, 'INTEREST RATE = ', RATE
      PRINT*, 'TERM OF CERTIFICATE = ', CTTERM, ' YEARS'
      PRINT*, ' '
      PRINT*, '             YEAR     INTEREST          TOTAL VALUE'
C
C COMPUTE AND PRINT INTEREST AND VALUE AFTER EACH YEAR OF THE TERM
      VALUE = STVAL
      DO 10 YEAR = 1, CTTERM
         INTRST = VALUE * RATE
         VALUE = VALUE + INTRST
         PRINT*, YEAR, INTRST, VALUE
   10 CONTINUE
C
      STOP
      END

*** PROGRAM EXECUTION OUTPUT ***

BANK CERTIFICATE ANALYSIS FOR EMMYLOU HARRIS

START VALUE OF CERTIFICATE = $  3000.000000000
INTEREST RATE =    .1200000000000
TERM OF CERTIFICATE =                 10 YEARS

            YEAR      INTEREST          TOTAL VALUE
              1   360.0000000000      3360.000000000
              2   403.2000000000      3763.200000000
              3   451.5840000000      4214.784000000
              4   505.7740800000      4720.558080000
              5   566.4669696000      5287.025049600
              6   634.4430059520      5921.468055552
              7   710.5761666662      6632.044222218
              8   795.8453066662      7427.889528884
              9   891.3467434661      8319.236272351
             10   998.3083526821      9317.544625033
```

**Fig. 4.1** An illustration of the use of real, integer and character data.

The program output shows the external form of these constants, except that character data are printed without the enclosing apostrophes. The strings

```
'INTEREST RATE = '
'TERM OF CERTIFICATE = '
```

are also examples of character data; they are used to annotate the program output. More complete descriptions of the external representation of all four types of data (including logical) are given in the next display.

---

### External Data Representations in FORTRAN

*Integer data* Type integer data must be represented as a string of decimal digits 0–9, prefixed by an optional $+$ or $-$ sign. No other character is permitted in the representation of an integer. (*Examples*: 0, $-9$, $+12$, 068, 123456)

*Real data* Type real data may be represented using the digits 0–9 and a decimal point, prefixed by an optional $+$ or $-$ sign. The decimal point must always be used for type real data. It is the sole means available for the compiler to distinguish between real and integer constants. (*Examples*: 0., 12.0, $-6.325$, .00625, $+7.2$)

A FORTRAN *scientific notation* for real data may also be used. A real data item written in FORTRAN scientific notation consists of a sign followed by a standard real or integer constant, followed by the letter E, another sign, and an integer constant ($+$ signs may be omitted). (*Examples*: $-0.325E6$, $325E+5$, $13764.25E-10$, $-110.E02$) The value of any number written in the above form is determined as follows:

> Multiply the first constant times $10^n$, where n is the integer constant following E. (*Example*: $-0.325E6 = -0.325 \times 10^6 = 325000$)

*Character data* Character data items consist of any sequence of legal FORTRAN characters enclosed in apostrophes. *Examples*:

```
'A'        'JIMMY'        'HARRIET  BEECHER  STOWE'
'SOON, I WILL LEAVE'      '3*X*2-6+X+4'        '*/,+-W)X$'
```

Two adjacent apostrophes are used inside a string to indicate contraction or possession:

```
'I''M DREAMING'   'JOE''S HAT'
```

When character data are printed, the enclosing apostrophes are omitted.

*Logical data* There are only two data values of type logical. They are true and false, and are represented in FORTRAN as .TRUE. and .FALSE. (or .T. and .F.). When logical data are printed, the letters T and F, representing .TRUE. and .FALSE., respectively, are printed without the enclosing periods.

## Example 4.1:

a) The following are legal integer data:

```
 -7                -456789
2463                 13
   0                +32
```

The following are illegal integer data:

```
-27E3       Use of exponential form is not legal with integer.
 32.0       Decimal point indicates real.
```

b) The following are legal real data:

```
 3.14159         0.1E-6
 0.0            -22.3E12
-22.0            0.3889E-10
 0.000031        57E18
```

The following are illegal real data:

```
6382         No decimal point indicates integer.
.TRUE.       Logical data.
57E18.0      No decimal point allowed after the E.
```

c) The following are legal character data items:

```
'THE'            'ITSY, BITSY'
'4*AC'           '679'
'HIT IS'         '72.H'
'.TRUE.'         '.0025E-02'
```

The following are illegal character data:

```
'IT'S'       Extra (or missing) apostrophe.
'HERMAN      Missing apostrophe.
 3.0'        Missing apostrophe.
.TRUE.       Logical constant.
```

d) The following are legal logical data:

```
.TRUE.
.FALSE.
```

The following are illegal logical data:

```
'.TRUE.'     Character string data.
1            Integer data.
TRUE         Missing periods; treated as a variable name.
```

**Exercise 4.1:**   Indicate which of these data items are legal FORTRAN data and identify the type (real, integer, character, or logical).

a) 37.86            f) -18.E-3
b) 219-40-0677      g) 18+4
c) .FLAG.           h) 'WHAT, FOR'
d) 'OOPS'           i) $16.27
e) -64              j) WHYNOT

### 4.2.2  Internal Representation of Data

Just as each data type has a unique external form, each type has its own *internal form* as well. This is the form used by the computer when storing data in memory. Although the precise details of the internal form of data will vary from computer to computer, there are some common attributes which will be described in this section.

All computers use strings of 0's and 1's (*binary strings*) to represent data in memory. Consequently, each data item must be converted from its external form to its internal representation (a binary string) before it can be stored.

The internal forms of real and integer data items are called *floating point* and *fixed point*, respectively. Floating point form is analogous to *scientific notation*. Recall that the number 6,850,000,000 can be written very compactly in scientific notation as $0.685 \times 10^{10}$, where multiplying by $10^{10}$ is equivalent to shifting the decimal point ten positions to the right. Similarly, a very small number such as $-.0000000012$ can be written as $-0.12 \times 10^{-8}$, where multiplication by $10^{-8}$ implies that the decimal point should be shifted eight positions to the left. In the previous example, the fraction $-0.12$ is called the *mantissa* and the integer $-8$ is called the *exponent*.

The internal, floating point form of a real number consists of three separate binary strings, representing the exponent, mantissa and the sign of the number. The internal, fixed point form of an integer, however, consists only of a sign, followed by the binary representation of the magnitude of the number. These differences are illustrated below.

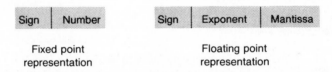

Fixed point                         Floating point
representation                      representation

The most important thing to remember from this picture is that a memory cell has one numerical value when interpreted as being in fixed point form (integer variable) and a different numerical value when interpreted as being in floating point form (real variable). For this reason, the internal format of a data item stored in a variable must always be consistent with the declared type of the variable.

In addition to the capability of storing fractions, real variables can be used to represent a considerably larger range of numbers than can integers. For example, on Control Data Corporation computers, positive numbers from $10^{-294}$ (a very small fraction) to $10^{+322}$ may be stored in real variables; whereas the range of positive integers extends from 1 to approximately $10^{15}$. On the other hand, operations involving integers are always exact, whereas there is often a small numerical error when real variables are manipulated. Also, on most computers, operations involving integers are considerably faster than operations involving real variables.

Some compilers use binary strings that differ only in one position to differentiate between the logical values .TRUE. and .FALSE.. For example, a string of 0's ending in a 1 might represent .TRUE. and a string of all 0's might represent .FALSE.. On other compilers, the value .TRUE. might be represented as a string of all 1's and .FALSE. as all 0's.

The representation of character data is somewhat more complicated. In most computers, the internal representation of a single character is given in terms of a unique binary string called a *character code*. The internal representation of character data consists of a *concatenation* (joining) of these individual codes to form a longer binary string.

**Example 4.2:**   The character codes shown below are used on IBM computers:

|  *letter*  |  *code*  |
|:---:|:---:|
| A | 11000001 |
| B | 11000010 |
| C | 11000011 |

Consequently, the character string 'CAB' would be represented as the binary string

$$11000011\,11000001\,11000010$$
$$\quad\ \ \text{C}\qquad\ \ \text{A}\qquad\ \ \text{B}$$

This string consists of 24 binary bits (0 or 1).

### 4.2.3   Type Declarations

The main rationale for allowing different types of data in FORTRAN is for our convenience. It is much easier to prepare, read and modify our programs and data using familiar external forms than it would be using binary strings. Internally, however, the computer cannot distinguish among the different data types since all information is stored in its memory as binary strings. Therefore, the programmer must instruct the computer to *interpret* the data stored in memory in a manner that is consistent with its type.

We can tell the compiler the type of data to be stored in each memory cell through the use of explicit *type declarations*. We have already seen how to declare the data type of variables used for storage of integer and real numbers. The type declaration for logical variables is similar except that the word LOGICAL is used instead of INTEGER or REAL. The program in Fig. 4.1 contains an example of the declaration of a fourth variable type, CHARACTER, which is shown in general form in the following display.

Type declarations must precede all executable statements in a program, although the declarations themselves can appear in any order. The proper typing of all variables used in a program is essential if the program is to produce the de-

---

**CHARACTER Type Declaration**

CHARACTER * len var-list

**Interpretation:** *Var-list* is a list of variable names. The integer constant *len* denotes the length of each variable specified in the list. Each character variable will be allocated sufficient memory storage space to accommodate a character string of the specified length (consisting of *len* characters).
*Note:* If the * *len* is omitted, then the length is assumed to be 1 and only a single character may be stored in each variable listed.

---

sired results. The type of data stored in each variable and the operations performed on that data must be consistent with the declared data type. The rule for storing data is summarized in the following display.

---

**Data Storage Rule**

A data item must always be stored in a variable of the same type.

---

Violations of this rule will normally be recognized by the compiler. The operations used with each data type are the subject of the rest of this chapter.

**Example 4.3:**

a) Some legal type declarations:

```
REAL X, ALPHA, LAX, NIMBLE
LOGICAL FLAG, SWITCH
INTEGER BETA, LOON, DOPE
CHARACTER * 4 NAME, FLOWER
CHARACTER * 2 N25, N26
CHARACTER * 1 B, C, D
```

b) Some legal storage specifications, given the declarations in a.):

```
ALPHA = LAX
LAX = 35.7
NIMBLE = LAX
BETA = 25
LOON = 4
FLAG = .TRUE.
READ*, LAX          data:  47.2
READ*, LOON         data:  3
NAME = 'MIKE'
NAME = 'FLAG'
READ*, N25, B, FLOWER    data: 'XY' 'B' 'ROSE'
```

c) Illegal storage specifications, given the declarations in a.):

```
ALPHA = .FALSE.
BETA = 'B'

FLAG = 35

READ*, LOON   data: 68.4
```
(logical data item assigned to a real variable)
(character data item assigned to an integer
   variable)
(integer data item assigned to a logical
   variable)
(real data read into an integer variable)

**Exercise 4.2:**   What type declarations would you use to declare the following variables? Defend your choice.

|            |                                                            |
|------------|------------------------------------------------------------|
| a) GROSS   | Employee gross salary                                      |
| b) HOURS   | Employee hours worked                                     |
| c) NOFAIL  | Number of failures in a class examination                 |
| d) SOLD    | Indicates whether or not a house in a multiple real estate listing has been sold |
| e) COLOR   | The color of a car                                        |
| f) CURSET  | The current assets of a corporation                       |
| g) ID      | A student identification number                           |
| h) SSNO    | Your Social Security number                               |
| i) GPA     | Your grade point average                                  |
| j) TOOBIG  | Indicates whether or not your new car will fit into your old garage |

**Exercise 4.3:**   Which of the following storage assignments are legal? Which are illegal and why? Assume the following declarations:

```
CHARACTER * 24 NAME
INTEGER AGE
REAL PAY
LOGICAL EMPLYD
```

a) NAME = 'HENRY'
b) PAY = 'ABC'
c) READ*, NAME        *data*:
   'ABCDEFGHIJKLMNOPQRSTUVWXYZ'
d) AGE = 47.6
e) PAY = 127.27

f) EMPLYD = .TRUE.
g) EMPLYD = 'NO'
h) AGE = 36
i) PAY = $400
j) NAME = .FALSE.

## 4.2.4  Implicit Typing of Variable Names

In FORTRAN, there is an implicit "first letter" typing convention for all variable names that are not explicitly declared through a type declaration. We recommend that you always explicitly declare the names used in your programs, but you should be aware of the existence of the implicit typing convention and how it works.

---

**Implicit Typing Convention**

Variable names that are not declared in explicit type declarations are typed implicitly by the FORTRAN compiler as follows:
- If the variable name begins with any one of the letters I through N, it is typed as INTEGER.
- If the variable name begins with any one of the letters A through H, or O through Z, it is typed as REAL.
- No variable names are implicitly typed as LOGICAL or CHARACTER.

---

Any variable name that does not appear in a type declaration is automatically associated with the address of a computer memory cell when it is first encountered in a FORTRAN program. The type of such a name (REAL or INTEGER) is determined according to the first letter typing convention described above.

## 4.3   ARITHMETIC EXPRESSIONS

### 4.3.1   Introduction

In Chapter 1 we mentioned that FORTRAN was particularly well suited to the programming of mathematical problems, because it permits a convenient representation of mathematical formulas. Even though FORTRAN is used for solving a wide variety of problems, the formula specification facility is one of the most important aspects of the language. In this section, we will describe in some detail the FORTRAN features for the manipulation of integer and real data.

In FORTRAN, formulas are expressed primarily in terms of assignment statements. The general form of the FORTRAN assignment statement is repeated below.

---

**Assignment Statement**

$$result = expression$$

**Interpretation:** This statement is used to assign a particular value (indicated by the *expression*) to the variable indicated by *result*.

---

The assignment statement itself is quite simple and rather unexciting. In fact, we have used many assignment statements in earlier programs. For example:

```
SUM = 0.0
GROSS = RATE * HOURS
VALUE = VALUE + INTRST
```

It is really the expression part of the assignment statement that is of primary interest. Until now, we have been limited when writing expressions in FORTRAN. However, there is a good deal of variation actually permitted.

An *arithmetic expression* is a grouping of real or integer variables and constants, the arithmetic operators and parentheses. In FORTRAN, it is possible to write arithmetic expressions containing more than one operator and nested parentheses. For example, expressions such as

```
HOURS * RATE - TAX
B * B - 4.0 * A * C
(40.0 + 1.5 * (HOURS - 40.0)) * RATE
(N / DIV) * DIV
```

are all legal in FORTRAN.

Since the computer is capable of performing only a single basic operation at a time, the compiler must translate more complicated expressions such as these into an equivalent sequence of basic operations. We must specify each expression in a precise form or the compiler translation may not conform to our expectations. For example, there must be no confusion in our minds concerning the meaning of expressions such as A + B * C or X / Y * Z. In the former, do we mean (A + B) * C or A + (B * C)? Do we intend to compute (X / Y) * Z or X / (Y * Z) in the latter?

In a programming language such as FORTRAN, the meaning of all expressions is completely determined by the rules of translation that are followed by the compiler. That is, the meaning is determined by the order in which the compiler specifies that the list of basic operations will be carried out. In general, the rules followed by the compiler correspond to the rules of algebra.

**Example 4.4:**   The following diagram illustrates how the statement

$$X = A / (B + C)$$

is translated into a sequence of operations to be carried out, one at a time, by the computer. For this example, we assume that A = 14.0, B = 4.0, C = 3.0, and X is initially undefined.

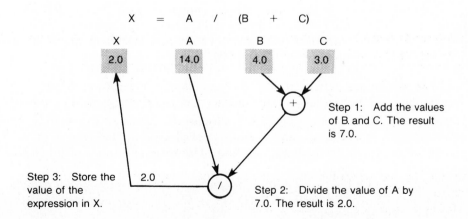

Step 1:   Add the values of B. and C. The result is 7.0.

Step 3:   Store the value of the expression in X.

Step 2:   Divide the value of A by 7.0. The result is 2.0.

**Example 4.5:**   In addition to the basic operators $(+, -, *, \text{and } /)$, there is another arithmetic operator, **. This is the exponentation operator and it raises its first operand to the power indicated by the second. Thus, $X^3$ may be written in FORTRAN as X ** 3.

**Exercise 4.4:**   Show by example that (X / Y) * Z and X / (Y * Z) do indeed compute different values. Which interpretation do you think the compiler gives to the expression X / Y * Z?

### 4.3.2   Evaluating FORTRAN Arithmetic Expressions

We must understand the way expressions are evaluated in FORTRAN in order to write expressions that produce the desired results. Unfortunately, the algorithms used by most compilers to analyze an expression and to specify the list of basic operations indicated by the expression are beyond the scope of this text. However, it is possible to formulate a set of *rules of expression evaluation* which, if followed, will produce the same computational results as the rules followed by the compiler. These rules, which are based upon the algebraic rules of *operator precedence*, are summarized in the following display.

---

**Rules of Expression Evaluation**

a) All parenthesized subexpressions must be evaluated first. Nested parenthesized subexpressions must be evaluated inside-out, with the innermost expression evaluated first.

b) Operators in the same subexpression are evaluated according to the following hierarchy:

- exponentiation, **                                    first
- multiplication and division, *, /                     next
- addition and subtraction, +, −                        last

c) Operators in the same subexpression and at the same hierarchical level (such as + and −) are evaluated left to right. The only exception to this rule is that consecutive exponentiation operators are evaluated right to left (thus X ** Y ** Z is evaluated as X ** (Y ** Z)).

---

To illustrate the application of the rules for expression evaluation, we provide diagrams like the one in Fig. 4.2. Each operator is enclosed in a circle. The order of operator evaluation and the rule applied at each step (rule a, b, or c above) are indicated alongside the operator. The lines are used to connect each operator with its operands.

Fig. 4.2 shows the FORTRAN representation of the expression for computing the area of a circle ($\pi r^2$). If RADIUS is 4.0, then the expression value is 3.14159 * 16.0, or 50.26544.

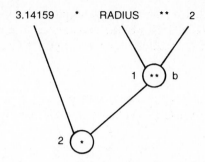

**Fig. 4.2**  An example of expression evaluation.

**Example 4.6:**  The expression below illustrates all three evaluation rules.

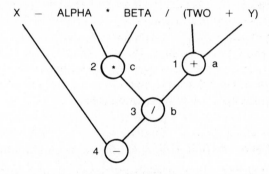

The operator $+$ is evaluated first as it is inside parentheses (rule a); the operator $*$ is evaluated before the operator $/$ (rule c); the higher precedence operator $/$ is evaluated before the operator $-$ (rule b).

**Example 4.7:**  The formula for computing the amount on deposit in a savings account after a fixed number of days is given by

$$\text{amount} = \text{deposit} (1 + \text{rate}/365)^{\text{days}}$$

where deposit is the initial deposit and rate is the yearly interest rate. This formula is written and evaluated in **FORTRAN** as

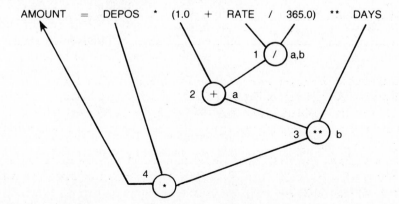

**Example 4.8:**

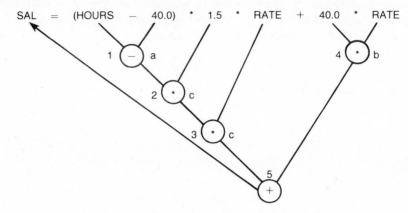

This expression could be used to calculate the salary for an employee who has worked more than forty hours, and is paid time and a half for all hours worked over forty. The subexpression in parentheses is evaluated first, according to rule (a). All multiplications are performed before the addition because of rule (b). The multiplications are performed in left-to-right order according to rule (c). Finally, the addition operation is performed and the assignment is carried out. HOURS, RATE, and SAL should be real variables.

It should be clear that inserting parentheses in an expression will affect the order of operator evaluation. If you are in doubt as to the order of evaluation that will be followed by the FORTRAN compiler, you should use parentheses to clearly specify the intended order of evaluation.

**Exercise 4.5:**   Write the following formula in FORTRAN and draw a diagram indicating its order of evaluation:

$$A = C + \frac{S}{Q} + \frac{H \times Q}{2.0 \times R}$$

### 4.3.3   Rules for Writing Arithmetic Expressions

There are three inherent difficulties in representing a mathematical formula in FORTRAN: one concerns multiplication, another concerns division, and the other concerns expressions with adjacent operators. Multiplication can often be implied in a mathematical formula by writing the two items to be multiplied next to each other; e.g., a = bc. In FORTRAN, however, the * operator must always be used to indicate multiplication:

$$A = B * C$$

The second difficulty arises in formulas involving division. We normally write these with the numerator and denominator on separate lines:

$$m = \frac{y - b}{x - a}$$

In FORTRAN, all assignment statements must be written on a single line; consequently, parentheses are often needed to separate the numerator from the denominator, and to indicate clearly the order of evaluation of the operators in the expression. The formula above would be written as

$$M \ = \ (Y \ - \ B) \ / \ (X \ - \ A)$$

The mathematical formula

$$f = g^{-h}$$

illustrates the third problem encountered in writing expressions in FORTRAN: it is illegal for two operators to appear side by side in an expression. Thus, this formula must be written as

$$F \ = \ G \ ** \ (-H)$$

where the left parenthesis separates the operators ** and —. The consecutive asterisks (**) represent a single operator (exponentiation).

**Example 4.9:** This example illustrates how several mathematical formulas can be implemented in FORTRAN using expressions involving multiple operators and parentheses. In this example, A, B, C, D, X, and Y are assumed to be real; K is assumed to be an integer.

| Mathematical formula | FORTRAN expression |
|---|---|
| a. $b^2 - 4ac$ | B ** 2 - 4.0 * A * C |
| b. $a + b - c$ | A + B - C |
| c. $\dfrac{a + b}{c + d}$ | (A + B) / (C + D) |
| d. $a^b$ | A ** B |
| e. $\dfrac{1}{1 + x^2}$ | 1.0 / (1.0 + X ** 2) |
| f. $xy - \dfrac{a}{d^5}$ | X * Y - A / D ** 5 |
| g. $1 + x^{-k}$ | 1.0 + X ** (-K) |

The points just illustrated are summarized in the following display.

---

### Rules of Formation for FORTRAN Expressions

- Always specify multiplication explicitly by using the operator * where needed (see Example 4.9a and f).
- Use parentheses when required to control the order of operator evaluation (see Example 4.9c and e).
- Never write two arithmetic operators in succession; they must be separated by an operand or parentheses (see Example 4.9g).

---

**Exercise 4.6:**   Write the mathematical equivalents of the following FORTRAN expressions:

  a) (W + X) / (Y + Z)
  b) G * H - F * W
  c) A ** (B ** 2)
  d) (B ** 2 - 4.0 * A * C) ** .5
  e) (X * X - Y * Y) ** .5
  f) X * 2 + R / 365.0 ** N
  g) P2 - P1 / T2 - T1

**Exercise 4.7:**   Let X = 2.0, Y = 3.0, and Z = 5.0. What are the values of the following FORTRAN expressions?

  a) (X + Y) / (X + Z)
  b) X + Z / X / Z
  c) X + Y * Z
  d) X / Y * Z
  e) X ** (Y - Z)
  f) X ** Y - Z

**Exercise 4.8:**   Write FORTRAN assignment statements for the following mathematical expressions. Do not mix types, except for exponentiation. State your assumption concerning the type of each variable used in your expressions.

  a)   $c = (a^2 + b^2)^{1/2}$
  b)   $y = 3x^4 + 2x^2 - 4$
  c)   $z = 3k^4(7k + 4) - k^3$
  d)   $x = \dfrac{a^2(b^2 - c^2)}{bc}$
  e)   $d = (a^2 + b^2 + c^2)^{1/2}$
  f)   $z = \pi r^2$ (use $\pi = 3.14159$)
  g)   $r = 6.27 \times 10^{15}s$
  h)   $p = c_0 + c_1 x - c_2 x^2 - c_3 x^3 - c_4 x^4$
  i)   $b = a^{-5}$

**Exercise 4.9:**   Let A, B, C, and X be the names of four variables of type real, and I, J, K the names of three type integer variables. Each of the statements below contains a violation of the rules of formation of arithmetic expressions. Rewrite each statement so that it is consistent with these rules.

  a) X = 4.0 A * C                    d) K = 3(I + J)
  b) A = AC                           e) X = A/BC
  c) I = 2 * -J                       f) I = J3

**Exercise 4.10:**   Write the FORTRAN equivalents for the following arithmetic expressions. Each symbolic name consists of a single letter. Write the constants involving powers of 10 in scientific notation.

  a) $x(y + w)z$                      b) $x^2 + y^2$

  c) $(xy)^2$                         d) $\dfrac{ax + bg}{aw + by}$

  e) $3.0(xy + wz) - 12.0$            f) $(3.2 - 7.5y)^2$

  g) $3.14 \times 10^6 - .013 \times 10^{-4}(x + y)$     h) $\dfrac{18.0x - 12.5 \times 10^3 x}{x^{(n-3)} + y^{(n-2)}}$

### 4.3.4    Mixed-types and Integer Division

A *mixed-type expression* is one involving the use of both integer and real data. Although the FORTRAN rules for writing arithmetic expressions do not specifically prohibit them, we suggest you avoid the use of mixed-type expressions whenever possible. The only exception to this is in expressions of the form

$$X \ ** \ P$$

where P may be of type integer, even though X is of type real. In fact, you should use an integer type exponent wherever possible, since this is more efficient and likely to produce more accurate results (e.g., X ** 3 is preferred to X ** 3.0 even when X is real). In all other operations involving a real and integer operand, the real equivalent of the integer operand must be obtained before the operation can be performed.

**Example 4.10:** If X is a type real variable (value 3.9) and NRITMS is a type integer variable, the assignment statements

```
NRITMS = 2 * X
X = NRITMS / 2
```

are considered *mixed assignment statements* as they assign a value of one numeric type to a variable of the other numeric type. In evaluating the mixed expression 2 * X, the value of X (3.9) is multiplied by the real equivalent of 2 (2.0). The result (7.8) must be converted to an integer before it can be stored in NRITMS. In doing this conversion, the compiler *truncates*, or removes, the fractional part and assigns the integer 7 to NRITMS.

In performing the second assignment above, the expression NRITMS / 2 must first be evaluated. Since both operands of the division operator are type integer, this is an example of *integer division*. The result of any arithmetic operation involving two integer operands must be an integer; consequently, the remainder from the integer division 7 / 2 is lost and the expression value is 3. Finally, the real equivalent of the integer 3 (3.0) is assigned to X.

While the truncation resulting from integer division or the assignment of a real value to an integer variable can be useful in some programming problems, it can have disastrous effects when it occurs by accident (because of programmer error). We therefore suggest that you always explicitly specify the conversions involved in a mixed expression or assignment (integer to real or real to integer), and we will show how this can be done in Section 4.6.

## 4.4    MANIPULATING CHARACTER DATA

### 4.4.1    Introduction

Prior to this chapter, we used character data (strings of characters enclosed in apostrophes) only for annotating program output. However, in the program in Fig. 4.1 (Sec. 4.2) we used a character variable (NAME) for storage of a character string ('EMMYLOU HARRIS'). The following examples illustrate the use of the assignment statement and the READ statement for storing character data in character variables. The character data are enclosed in apostrophes.

**Example 4.11:** The variable values shown on the right result from the execution of the statements shown at the left. The symbol □ represents a blank.

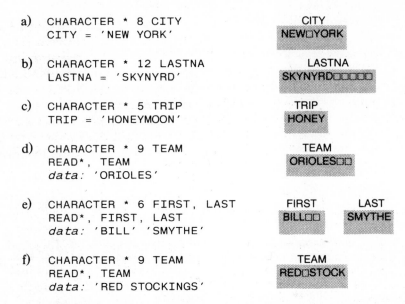

a)  CHARACTER * 8 CITY                    CITY
    CITY = 'NEW YORK'                      NEW□YORK

b)  CHARACTER * 12 LASTNA                  LASTNA
    LASTNA = 'SKYNYRD'                     SKYNYRD□□□□□

c)  CHARACTER * 5 TRIP                     TRIP
    TRIP = 'HONEYMOON'                     HONEY

d)  CHARACTER * 9 TEAM                     TEAM
    READ*, TEAM                            ORIOLES□□
    *data:* 'ORIOLES'

e)  CHARACTER * 6 FIRST, LAST       FIRST          LAST
    READ*, FIRST, LAST              BILL□□         SMYTHE
    *data:* 'BILL' 'SMYTHE'

f)  CHARACTER * 9 TEAM                     TEAM
    READ*, TEAM                            RED□STOCK
    *data:* 'RED STOCKINGS'

For the moment, we will only use simple assignment statements, as illustrated in Example 4.11, parts a through c. The use of more complicated character assignment statements, involving character operators and expressions, is deferred until Chapter 9.

In a character assignment, the number of characters that will be stored is the same as the declared length of the variable receiving the character data. If the character string is too short, then blanks are added to the right end of the string (see Example 4.11 b, d and e). This technique is often referred to as "padding the string on the right." If the character string is too long, any extra characters at the right end of the string are lost, as illustrated in Examples 4.11c and f.

When a character variable is printed, the number of characters displayed is the same as the declared length of the variable. Hence, any blanks added for padding will be printed. Naturally, any characters that were lost during storage will not be printed.

It makes very little sense and, in fact, is illegal, to specify arithmetic operations with character data. Nevertheless, there are some simple and useful things that can be done with character strings, as illustrated in the next section.

**Example 4.12:**   Given the declarations

```
CHARACTER * 11 SSNO
REAL GROSS, PAY
CHARACTER * 6 LASTNA, FRSTNA
```

and the data

```
'219-40-0677' 'MONK' 'THEO' 500.00 417.26
```

the READ statement

```
READ*, SSNO, LASTNA, FRSTNA, GROSS, PAY
```

will place the data items in memory as shown below.

| SSNO | LASTNA | FRSTNA | GROSS | PAY |
|------|--------|--------|-------|-----|
| 219-40-0677 | MONK□□ | THEO□□ | 500.00 | 417.26 |

The PRINT statement

```
PRINT*, FRSTNA, LSTNA, PAY
```

would produce an output line of the form:

```
THEO  MONK  417.2600
```

**Exercise 4.11:**   Write an appropriate sequence of type declarations and READ statements for entering the data shown here. Use meaningful names. The first line of data contains the ID number, name, and class of a student. The last three lines contain the course ID, credit hours, and a grade for each of three courses taken by the student. These three lines should be processed by a READ statement within a program loop.

```
700007    'J.A. OSHEA'     'SENIOR'
 'S101'    3     'A'
 'M130'    3     'B'
 'C224'    4     'A'
```

### 4.4.2   Character String Comparison

Although somewhat restricted, it is possible to compare character data using the *relational operators* (.GT., .LT., .LE., .GE., .EQ., .NE.). Recall from the discussion in Section 4.2.2 that a character string is internally represented as a binary string formed from individual character codes. This string is treated by the computer as a single binary number. Thus, the comparison of two character strings actually involves a comparison of two numbers. If the character strings are not the same size, then the shorter string is padded with blanks on the right before the comparison is performed.

The binary code for each character is not the same on all computers. Consequently, comparing the same two character strings on different computers will often yield different results. However, FORTRAN requires the following *collating sequence* (ordering of characters) for all compilers.

---

**Collating Sequence**

The character blank precedes (is less than) all of the digits (0, 1, 2, . . . , 9) and all of the letters (A, B, C, . . . , Z).

The letters are ordered lexigraphically (in dictionary sequence). That is, A precedes B, B precedes C, . . . , Y precedes Z.

The digits follow their normal numeric sequence. That is, the character 0 precedes 1, 1 precedes 2, . . . , 8 precedes 9.

---

The FORTRAN collating sequence specifies that the conditions

```
'ACE'   .LT.  'BAT'
'ACES'  .GT.  'ACE'
```

will be true on all computers. In other words, the result of comparing two words (strings of letters only) is based upon the dictionary order of these words. In evaluating the second condition, the equal length strings 'ACES' and 'ACE ' are actually compared. The binary codes for the letter S (in 'ACES') and the blank (in 'ACE ') are, of course, different. The condition is true since the blank character precedes the letter S in the collating sequence.

The conditions

```
'1234'  .LT.  '3333'
'9999'  .GT.  '9998'
```

will be true for all compilers. The result of comparing two equal length *numeric strings* (strings of digits) corresponds to their numeric relationship. However, the result of comparing two numeric strings that are not the same length does not always correspond to the numeric relationship between these strings. For example, the condition '3' .GT. '12' is true since the character '3' follows '1' in the collating sequence. (The equal length strings '3 ' and '12' are actually compared.)

Similarly, two character variables containing letters only, or digits only, may be compared. The result of these comparisons can also be determined from the collating sequence, as previously illustrated.

It is also possible to compare a character variable to a character string as shown in the conditions

```
(IDNUM .NE.  '9999')
(CARD  .EQ.  'ACE')
```

These comparisons are legal only if IDNUM and CARD are declared to be character variables.

The collating sequence for FORTRAN 77 does not specify any relationship between the letters and digits. It also makes no statement concerning the relationship among special characters, or between the special characters and the letters, digits, and the blank. Consequently, in comparing strings containing both letters and digits, or strings containing special characters, the results may vary from computer to computer. In such cases, we suggest that you restrict yourself to the use of the relational operators .EQ. (equal) and .NE. (not equal) until you learn more about the complete collating sequence used by your compiler.

**Example 4.13:**   The conditions specified below will be true on all computers.

| | | |
|---|---|---|
| 'C' .LT. 'G' | | '*/.' .EQ. '*/.' |
| '2' .LT. '9' | | 'A14B' .NE. '2A4B' |
| 'ABC' .LT. 'CBA' | | '1234' .LT. '34' |
| 'BETA' .GT. 'ANT' | | '  ' .LT. 'A' |
| 'ANSI' .LT. 'ANTS' | | '  ' .LT. 'O' |
| 'BELL' .LT. 'BELLTOWER' | | '134' .GE. '1226' |

**Exercise 4.12:**   Describe the effect of the following groups of statements.

a)
```
CHARACTER * 11 SSNO
INTEGER HOURS
REAL RATE
READ*, SSNO, HOURS, RATE          data: '219-40-0677' 30 7.25
```
b)
```
CHARACTER * 6 FIRST, LAST
FIRST = 'FUDGE'
LAST = 'RIPPLE'
PRINT*, FIRST, LAST
```
c)
```
CHARACTER * 4 WORD              data: 'ALL'
READ*, WORD                           'FOR'
WHILE (WORD .NE. 'DONE') DO           'NOUGHT'
   PRINT*, WORD                       'IS'
   READ*, WORD                        'DONE'
ENDWHILE
PRINT*, 'ALL ', WORD
```

### 4.4.3   A Sample Problem—The Registered Voters List

The next problem illustrates the reading, printing and comparison of character data.

**Problem 4A:**   For the local election next Tuesday, town officials have decided to use three clerks, Abraham, Martin and John, to verify that each resident wishing to vote is legally registered and votes only once. The officials decided that in order to distribute the registered voters fairly evenly among the three clerks, they would assign Abraham to check voters with last names beginning with A through I, Martin to check the voters with last names beginning with J through R, and John to check voters with last names beginning with S through Z.

We will write a program to print a complete voter list with the correct clerk assignment for each voter. The program should read the voter name, house number and street name for each registered voter in the township. The program will then print a master list with the house number and street name printed first, followed by the voter name, and the clerk assigned to the voter printed last. We will assume that all names are entered correctly, with the last name entered first. The number of voters assigned to each clerk should be printed at the end.

**Discussion:**   The data table for this problem is shown below. As illustrated, from now on, we will always indicate the type of each entry listed in a data table. The level one flow diagram is shown in Fig. 4.3a.

**Data Table for Problem 4A**

*Input variables*

N: Number of voters
   (integer)

HSENBR: House number
   (integer)

STREET: Street name
   (character)

VOTER: Voter name
   (character)

*Output variables*

CLERK: Clerk
   assigned to a voter
   (character)

N1, N2, N3: Count
   of voters assigned
   to Abraham, Mar-
   tin and John (in-
   teger)

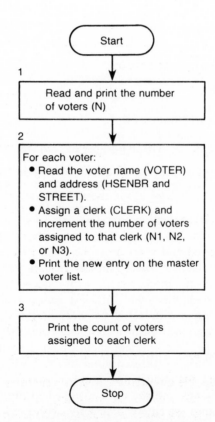

**Fig. 4.3a**   Level one flow diagram for voter registration problem (4A).

The refinement of Step 2 requires a loop in which the data for a voter are read (Step 2.3), the clerk is designated (Step 2.4), and an additional entry in the master voter list is printed (Step 2.5). Since the number of loop repetitions is known beforehand (the loop is repeated once for each of the N voters), a counting loop (DO loop) may be used. The additional data table entry for the loop control variable is shown next; the flow diagram refinement for Step 2 is given in Fig. 4.3b. As is always the case for variables used for counting or accumulating results, N1, N2, and N3 must be initialized to zero prior to entering the loop.

**Additional Data Table Entry for Problem 4A**

*Program variable*

I: Counter and loop control variable (integer)

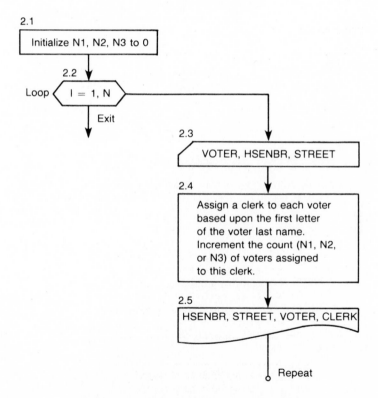

**Fig. 4.3b**   Refinement of Step 2 in Fig. 4.3a.

A nesting of two double alternative decision structures (Fig. 4.3c) is used to refine Step 2.4. The conditions (VOTER less than 'J', VOTER less than 'S') are used to determine which of the clerks should be assigned to each voter. The program is shown in Fig. 4.4, the output for a sample execution is shown in Fig. 4.5.

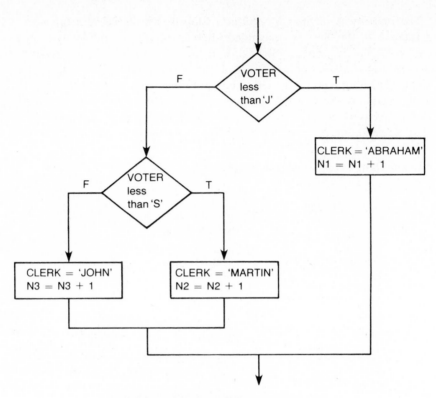

**Fig. 4.3c**  Refinement of Step 2.4 of Fig. 4.3b.

```
C TOWNSHIP VOTER CLERK ASSIGNMENT LIST
C PRINTS MASTER LIST OF EACH VOTER AND ASSIGNED CLERK
C ALSO COUNTS NUMBER OF VOTERS ASSIGNED TO EACH CLERK
C     VOTERS A-I ASSIGNED TO CLERK ABRAHAM (COUNT IN N1)
C     VOTERS J-R ASSIGNED TO CLERK MARTIN (COUNT IN N2)
C     VOTERS S-Z ASSIGNED TO CLERK JOHN (COUNT IN N3)
C
      CHARACTER * 10 CLERK
      CHARACTER * 20 VOTER, STREET
      INTEGER N, N1, N2, N3, HSENBR, I
C
C READ IN AND PRINT NUMBER OF REGISTERED VOTERS.
C INITIALIZE COUNTS
      READ*, N
      PRINT*, 'THE NUMBER OF REGISTERED VOTERS IS ', N
      PRINT*, ' '
      N1 = 0
      N2 = 0
      N3 = 0
C
C LOOP TO READ VOTER NAME, HOUSE ADDRESS AND STREET
C ASSIGN CLERK TO EACH VOTER
C UPDATE COUNTER FOR THE CLERK
C PRINT HOUSE ADDRESS, STREET, VOTER, NAME AND CLERK
```

*(Continued)*

```
        PRINT*, '               REGISTERED VOTER ADDRESS',
      Z '     NAME                CLERK'
        DO 10 I = 1, N
            READ*, VOTER, HSENBR, STREET
C           SEE IF CLERK IS ABRAHAM
            IF (VOTER .LT. 'J') THEN
               CLERK = 'ABRAHAM'
               N1 = N1 + 1
            ELSE
C              SEE IF CLERK IS MARTIN OR JOHN
               IF (VOTER .LT. 'S') THEN
                  CLERK = 'MARTIN'
                  N2 = N2 + 1
               ELSE
                  CLERK = 'JOHN'
                  N3 = N3 + 1
               ENDIF
            ENDIF
            PRINT*, HSENBR, ' ', STREET, VOTER, CLERK
   10   CONTINUE
C
C PRINT COUNTS
        PRINT*, ' '
        PRINT*, 'THE NUMBER OF VOTERS ASSIGNED TO ABRAHAM IS ', N1
        PRINT*, 'THE NUMBER OF VOTERS ASSIGNED TO MARTIN IS ', N2
        PRINT*, 'THE NUMBER OF VOTERS ASSIGNED TO JOHN IS ', N3
C
        STOP
        END
```

**Fig. 4.4**   Program for Problem 4A.

*Input:*

```
6
'ADAMS, JOHN'    125    'ABBOT STREET'
'ADAMS, MARY'    125    'ABBOT STREET'
'WASHINGTON, GEORGE'   137    'MOUNT VERNON AVE.'
'KING, MARTIN L.'   270    'PEACHTREE LANE'
'JEFFERSON, THOMAS'   112    'XAVIER RD.   '
'KOFFMAN, ELLIOT'   322    'SPEAKMAN HALL'
```

*Output:*

```
THE NUMBER OF REGISTERED VOTERS IS              6

          REGISTERED VOTER ADDRESS     NAME           CLERK
             125 ABBOT STREET        ADAMS, JOHN       ABRAHAM
             125 ABBOT STREET        ADAMS, MARY       ABRAHAM
             137 MOUNT VERNON AVE.   WASHINGTON, GEORGE JOHN
             270 PEACHTREE LANE      KING, MARTIN L.   MARTIN
             112 XAVIER RD.          JEFFERSON, THOMAS MARTIN
             322 SPEAKMAN HALL       KOFFMAN, ELLIOT   MARTIN

THE NUMBER OF VOTERS ASSIGNED TO ABRAHAM IS          2
THE NUMBER OF VOTERS ASSIGNED TO MARTIN IS           3
THE NUMBER OF VOTERS ASSIGNED TO JOHN IS          1
```

**Fig. 4.5**   Sample input and output for Problem 4A.

## 4.5   USING LOGICAL DATA

### 4.5.1   Syntax Rules for Logical Expressions

The data type LOGICAL is the fourth type discussed in Section 4.2. Type LOGICAL variables may be assigned either of the logical values, .TRUE. or .FALSE..

Since the beginning of the text, we have been using logical expressions to specify the conditions in decision structures and WHILE loops. Examples of such expressions are listed below.

```
ORDER .LE. CURINV
SCORE .NE. SNTVAL
COUNTR .LT. NRITMS
VOTER .LT. 'S'
```

As shown in these examples, logical expressions may consist of two arithmetic or two character operands connected by the relational operators (.LT., .GT., .LE., .GE., .EQ., .NE.).

Logical expressions may also consist of logical constants (.TRUE. or .FALSE.) or logical variables used by themselves. For example, if ERRMSG is a logical variable, then the statement

```
IF (ERRMSG) THEN
    PRINT*, 'ERROR FOUND. EXECUTION TERMINATED.'
    STOP
ENDIF
```

will print the indicated message and terminate program execution if ERRMSG has the value .TRUE..

We may also form *compound logical expressions* from the simpler expressions just described using the *logical operators* .AND., .OR. and .NOT.. In Chapter 9 we will learn more about the use and evaluation of logical expressions and operators. However, we will provide a few examples now so that you may begin using these operators.

**Example 4.14:**   Either of the following two decision structures may be used to determine if an integer value (SCORE) falls in the range zero to one hundred $(0 \leq \text{SCORE} \leq 100)$.

```
      IF (SCORE .GE. O .AND. SCORE .LE. 100) THEN
C         PROCESS A VALID SCORE
          .  .  .  .
      ELSE
          PRINT*, SCORE, ' OUT OF RANGE AND IS IGNORED.'
      ENDIF

      IF (SCORE .LT. O .OR. SCORE .GT. 100) THEN
          PRINT*, SCORE, ' OUT OF RANGE AND IS IGNORED.'
      ELSE
C         PROCESS A VALID SCORE
          .  .  .  .
      ENDIF
```

Note that the form of the expressions on both sides of the .AND. (.OR.) operator is the same. Parentheses are not needed around these expressions since the relational operators are evaluated before the logical operators. The item (SCORE) being compared must be written twice as shown in order to form a legal logical expression. A logical expression formed using the .AND. operator is true only when the expressions on the left *and* right are both true. A logical expression formed using the .OR. operator is true if either the left expression *or* the right expression (or both) is true. The flow diagram pattern for the decision structure that uses the .AND. operator is shown in Fig. 4.6.

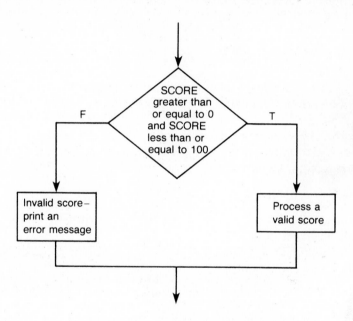

**Fig. 4.6**  Use of the AND operator.

**Example 4.15:**  The following are illustrations of very common errors made in writing conditions such as the ones shown in Example 4.14.

```
(SCORE .GE. 0 .AND. .LE. 100)
(SCORE .LT. 0 .OR. .GT. 100)
```

Both of these expressions are illegal since there is no operand between the logical operator and the second relational operator.

**Example 4.16:**  The logical expression

```
.NOT. FOUND
```

is true only if the logical variable FOUND is false. .NOT. is a logical operator with only one logical operand; it forms the *complement,* or negation, of its operand. The program segment shown next will print a message only when FOUND is false.

```
IF (.NOT. FOUND) THEN
    PRINT*, 'KEY NOT FOUND'
ENDIF
```

The logical operators are described in the next display.

---

### Logical Operator .AND.

$$log\ expr_1\ .\text{AND}.\ log\ expr_2$$

**Interpretation:** The compound logical expression above is true only if both *log expr₁* and *log expr₂* are true.

### Logical Operator .OR.

$$log\ expr_1\ .\text{OR}.\ log\ expr_2$$

**Interpretation:** The compound logical expression above is true if either *log expr₁* or *log expr₂* is true. The compound logical expression is false only when both *log expr₁* and *log expr₂* are false.

### Logical Operator .NOT.

$$.\text{NOT}.\ log\ exp$$

**Interpretation:** The logical expression above is true when *log exp* is false; it is false when *log exp* is true.

---

**Exercise 4.13:** Implement both conditions shown in Example 4.14 using nested decision structures with simple conditions (without using the .AND. and .OR. operators).

### 4.5.2  Program Flags

The prime numbers have been studied by mathematicians for many years. A *prime number* is an integer greater than 1 that has no divisors other than 1 and itself (e.g., 2, 3, 19, 37). In order to demonstrate that a number is not prime, we have to identify one or more divisors. In writing a program to solve this problem, we shall introduce the use of a logical variable as a program flag. We shall also make use of the remainder truncation resulting from integer division.

**Problem 4B:** Find and print all exact divisors of an integer N other than 1 and N itself. If there are no divisors, print out the message "N is a prime number." The value of N will be provided as a data item to be read in by the program.

**Discussion:** The general approach we will take is to see whether we can find an integer, DIVISR, which divides N evenly (with no remainder). We shall print any exact divisors.

The initial data table is shown below. The level one flow diagram is shown on the left in Fig. 4.7.

**Data Table for Problem 4B**

*Input variables*

N: Number to be tested
for prime property
(integer)

*Output variables*

DIVISR: The value of
each divisor of N
(integer)

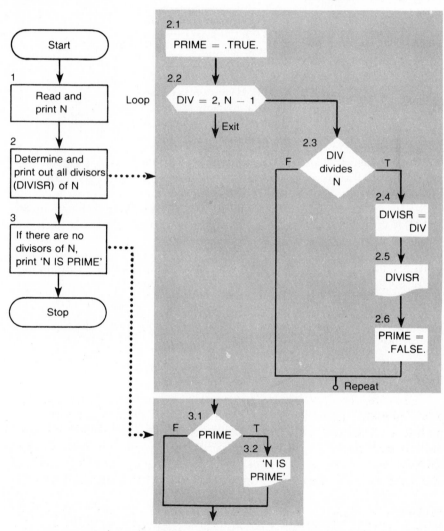

**Fig. 4.7** Flow diagrams for prime number problem (4B).

Steps 2 and 3 of the level one diagram require additional detail as shown on the right in Fig. 4.7. In order to carry out Step 3, we need to know whether any divisors were found in Step 2. Consequently, in Step 2 we must test each integer between 2 and $N - 1$ to see whether it divides N. We will use a logical variable **PRIME** to summarize the results of this test. The value of **PRIME** will indicate whether N is still considered a prime number (PRIME equal to .TRUE.) or is known to be not prime (PRIME equal to .FALSE.).

**Additional Data Table Entries for Problem 4B**

*Program variables*

PRIME: Indicates whether
N is considered prime
(.TRUE.) or a divisor
was found (.FALSE.)
(logical)

DIV: Each number tested
as a divisor; used as
loop control variable
(integer)

The logical variable **PRIME** is called a *program flag*. A program flag is a variable that is used to communicate to one program step the result of computations performed in another step. Initially, we will consider N to be prime; hence, **PRIME** is set to an initial value of .TRUE.. If a divisor is found in Step 2, **PRIME** will be reset to .FALSE.. The program will test **PRIME** at Step 3 to determine whether or not N is still considered prime.

As shown in the flow diagram, a loop is used to check all integers between 2 and N − 1 inclusive as possible divisors of N. Step 2.3, the divisibility test, is performed by evaluating the condition

```
(N / DIV) * DIV .EQ. N
```

The first step is to calculate the quotient of N divided by DIV (N / DIV). If there is no remainder (DIV is a divisor of N), then when the quotient is multiplied by DIV, the result should be N as shown below for N = 4, DIV = 2.

```
(4 / 2) * 2 = 2 * 2 = 4
```

However, if DIV does not divide N exactly, the remainder is lost (remember, all of the variables involved in this computation are integers). Consequently, when the quotient is multiplied by DIV, the result will be less than N as shown below for N = 5, DIV = 2.

$$(5 \ / \ 2) \ * \ 2 = 2 \ * \ 2 = 4 \neq 5.$$

The program for Problem 4B is shown in Fig. 4.8. Note that the logical variable **PRIME** is used as the condition in the logical IF statement

```
IF (PRIME) PRINT*, N, ' IS PRIME'
```

This is legal since the value of **PRIME** can only be .TRUE. or .FALSE.. The condition

$$(PRIME \ .EQ. \ .TRUE.)$$

would be invalid since the relational operators cannot be used with logical operands.

```
C PRIME NUMBER PROBLEM
C
      INTEGER N, DIV, DIVISR
      LOGICAL PRIME
C
C READ AND PRINT N AND VALIDATE N
      READ*, N
      IF (N .LE. 1) THEN
         PRINT*, 'N IS LESS THAN OR EQUAL TO 1 --'
         PRINT*, 'PROGRAM EXECUTION TERMINATED.'
         STOP
      ENDIF
C
C TEST ALL POSSIBLE DIVISORS OF N
      PRINT*, 'LIST OF DIVISORS OF ', N
      PRIME = .TRUE.
      DO 20 DIV = 2, N - 1
C        PRINT OUT EACH DIVISOR
         IF ((N / DIV) * DIV .EQ. N) THEN
            DIVISR = DIV
            PRINT*, DIVISR
            PRIME = .FALSE.
         ENDIF
   20 CONTINUE
C
C TEST FOR PRIME
      IF (PRIME) THEN
         PRINT*, 'NO DIVISORS -- ', N, ' IS PRIME'
      ENDIF
C
      STOP
      END


*** PROGRAM EXECUTION OUTPUT ***

LIST OF DIVISORS OF                     195
               3
               5
              13
              15
              39
              65
```

Fig. 4.8   Program for Problem 4B.

---

**Program Style**

*Validating input data*

In Fig. 4.8, a decision structure has been inserted that prints an error message if N is less than or equal to one. Values of N satisfying this condition are considered to be invalid as the prime number property is not defined for these values. The loop in Fig. 4.8 would not be executed for invalid data values as the program terminates before reaching it. (If the program did not terminate, these values would be classified incorrectly as prime numbers.) It is always a good idea to test for invalid input data whether or not this is requested in the problem statement.

Note that the loop body is also not executed when the value read into N is 2 (N $-$ 1 is 1). However, in this case, the program would correctly determine that 2 is a prime number.

---

**Exercise 4.14:** *For the more mathematically inclined.* The program shown in Fig. 4.8 tests all integer values between 2 and N $-$ 1 inclusive to see if any of them divide N. This is, in fact, quite inefficient, for we need not test all of these values. Revise the algorithm shown in Fig. 4.7 to minimize the number of possible divisors of N that must be tested to determine whether or not N is prime. Make certain that your improved algorithm still works. *Hints:* If 2 does not divide N, no other even number will divide N. If no integer value between 2 and N / 2 divides N, then no integer value between N / 2 + 1 and N $-$ 1 will divide N. (In fact, we can even compute a smaller maximum test value than N / 2. What is it?)

## 4.6   LIBRARY FUNCTIONS

### 4.6.1   Using Functions

Library functions are a feature of FORTRAN that are of considerable help in specifying computations that produce a single result. These functions are referenced directly in an expression: the value computed by the function is then substituted for the function reference.

**Example 4.17:** SQRT is the name of a library function that computes the square root of a non-negative value. Consider the statement

$$Y = 5.5 + \text{SQRT}(20.25)$$

This statement uses the SQRT function to compute the square root of 20.25, the *function argument.* The value computed by the function reference SQRT(20.25) is 4.5; the result of the evaluation of the addition operation (5.5 + 4.5) is 10.0, which is stored in the variable Y.

There are a number of numerically-oriented operations, such as finding the square root or logarithm of a number, or computing trigonometric functions, that many users of FORTRAN need to perform. These operations are not among the basic operations of most computers. However, they are provided as part of a library of FORTRAN-related programs at each computer installation. Table 4.1 provides a summary of the names and descriptions of some of the most commonly used *FORTRAN library functions.*

| Name | Form | Operation | Type of Arguments | Result |
|------|------|-----------|------------------|--------|
| INT | INT (*a*) | Convert *a* to type integer | Real | Integer |
| REAL | REAL (*a*) | Convert *a* to type real | Integer | Real |
| NINT | NINT (*a*) | Round *a* to nearest integer | Real | Integer |
| IABS<br>ABS | IABS (*a*)<br>ABS (*a*) | Return the absolute value of *a* | Integer<br>Real | Integer<br>Real |
| MOD | MOD (*a1, a2*) | Return remainder of *a1/a2* | Integer | Integer |
| MAXO*<br>AMAX1 | MAXO (*a1, ..., an*)<br>AMAX1 (*a1, ..., an*) | Select largest value from list | Integer<br>Real | Integer<br>Real |
| MINO*<br>AMIN1 | MINO (*a1, ..., an*)<br>AMIN1 (*a1, ..., an*) | Select smallest value from list | Integer<br>Real | Integer<br>Real |
| SQRT | SQRT (*a*) | Return $\sqrt{a}$ for $a > 0$ | Real | Real |
| EXP | EXP (*a*) | Return e**a | Real | Real |
| ALOG | ALOG (*a*) | Return $\log_e(a)$ or $\ln(a)$ for $a > 0$ | Real | Real |
| ALOG10* | ALOG10 (*a*) | Return $\log_{10}(a)$ for $a > 0$ | Real | Real |
| SIN | SIN (*a*) | Return sine of *a* | Real (radians) | Real |
| COS | COS (*a*) | Return cosine of *a* | Real (radians) | Real |
| TAN | TAN (*a*) | Return tangent of *a* | Real (radians) | Real |

*The last character of this function name is a zero.

**Table 4.1 Table of FORTRAN library functions.**

A function may be thought of as an operator that performs a transformation on its input data and produces a single output value. The input data and the resulting output value have a particular relationship that is determined by the *function definition.* Input is provided to a function through its argument(s); most functions have only a single argument, but the use of multiple arguments (separated by commas in an *argument list*) is allowed. In computing an output value, a function will not alter the value of its input argument(s).

To reference or call a function, we simply write its name followed by its argument(s), enclosed in parentheses, in an expression. The type of the *result returned* (function output) for each library function in Table 4.1 is based on the implicit typing convention (Sec. 4.2.4): function names that begin with I - N return an integer result; otherwise, the result is real.

A function call can be part of a complicated expression and may have expressions as well for its arguments. The statement

```
R1 = (-B + SQRT(B ** 2 - 4.0 * A * C)) / (2.0 * A)
```

is the FORTRAN form of the formula for a root of a quadratic equation

$$R1 = \frac{-B + \sqrt{B^2 - 4AC}}{2A}$$

Here the function SQRT would be called after the evaluation of its argument B ** 2 − 4.0 * A * C was performed. Then the quantity −B would be added to the result (the square root), and the entire quantity would finally be divided by the product 2.0 * A.

**Example 4.18:**   The ABS, SQRT, and INT functions are illustrated in the program of Fig. 4.9. Each line of the sample output (bottom of Fig. 4.9) shows the value of X and the three functions applied to X.

The PRINT statement in the loop shows that it is permissible to include a function reference in an output list

```
PRINT*, X, ABS(X), SQRT(ABS(X)), INT(X)
```

The third item in the output list above, SQRT(ABS(X)), is an example of a nested function reference. Since the square root of a negative number is mathematically undefined, we should first compute the absolute value of X, ABS(X), before computing the square root.

The definition of the absolute value function ABS may be stated as follows:

If the argument (X in this case) is negative, the result is the negation of the argument (−X);

If the argument is not negative, the result is identical to the argument (no computation is necessary in this case).

The INT function is used to convert its real argument to an integer value. The fractional part of the argument is truncated and the result is the integral part of the argument.

```
C AN ILLUSTRATION OF SQRT, ABS, AND INT
C
      REAL X
      INTEGER I
C
C READ FIVE VALUES OF X AND PRINT EACH FUNCTION VALUE
      PRINT*, '        X              ABS(X)',
     Z '        SQRT(ABS(X))        INT(X)'
      DO 40 I = 1, 5
         READ*, X
         PRINT*, X, ABS(X), SQRT(ABS(X)), INT(X)
   40 CONTINUE
C
      STOP
      END
```

\*\*\* PROGRAM EXECUTION OUTPUT \*\*\*

| X | ABS(X) | SQRT(ABS(X)) | INT(X) |
|---|---|---|---|
| -6.300000000000 | 6.300000000000 | 2.509980079602 | -6 |
| 0 | 0 | 0 | 0 |
| 19.80000000000 | 19.80000000000 | 4.449719092257 | 19 |
| -7.750000000000 | 7.750000000000 | 2.783882181415 | -7 |
| 20.25000000000 | 20.25000000000 | 4.500000000000 | 20 |

**Fig. 4.9** The ABS, SQRT and INT functions.

The next example illustrates the conversion of formulas from physics into FORTRAN statements. You need not be concerned if the formulas are unfamiliar; the main point of the example is to illustrate the implementation of these formulas in a FORTRAN program.

**Example 4.19:** Prince Valiant is trying to rescue Rapunzel by shooting an arrow with a rope attached through her tower window which is 100 feet off the ground. We will assume that the arrow travels at a constant velocity. Hence, the time it takes to reach the tower is given by the formula

$$T = \frac{D}{V \cos(\theta)}$$

where D is the distance Prince Valiant is standing from the tower, V is the velocity of the arrow and $\theta$ (THETA) is its angle of elevation.

Our task is to determine whether or not the Prince's arrow goes through the window by computing its distance off the ground when it reaches the tower as given by the formula

$$H = VT \sin(\theta) - \frac{GT^2}{2}$$

For the arrow to go through the window, H should be between 100 and 110 feet. We will print out an appropriate message to help Prince Valiant correct his aim.

The program is shown in Fig. 4.10, along with a sample output. The assignment statement

$$RADIAN = THETA * (PI / 180.0)$$

is used to convert the angle THETA from degrees to radians as required for input to the SIN and COS functions.

```
C PRINCE VALIANT TAKES AIM AT RAPUNZEL
C
C PROGRAM PARAMETERS
      REAL PI, G
      PARAMETER (PI = 3.14159, G = 32.17)
C
C VARIABLES
      REAL H, T, V, THETA, D, RADIAN
C
C READ AND PRINT INPUT DATA
      READ*, D, V, THETA
      PRINT*, 'DISTANCE FROM TOWER = ', D
      PRINT*, 'VELOCITY OF ARROW = ', V
      PRINT*, 'ANGLE OF ELEVATION = ', THETA, 'DEGREES'
C
C THE FORTRAN TRIG FUNCTIONS REQUIRE INPUT ANGLES
C IN RADIANS.  CONVERT THETA FROM DEGREES TO RADIANS.
      RADIAN = THETA * (PI / 180.0)
C COMPUTE TRAVEL TIME OF ARROW
      T = D / (V * COS(RADIAN))
C COMPUTE ARROW HEIGHT
      H = V * T * SIN(RADIAN) - G / 2.0 * T ** 2
C PRINT MESSAGE TO CORRECT PRINCE'S AIM
      IF (H .LT. 100.0) THEN
         IF (H .LT. 0.0) THEN
            PRINT*, 'ARROW DID NOT REACH THE TOWER'
         ELSE
            PRINT*, 'ARROW WAS TOO LOW, HEIGHT WAS ', H
         ENDIF
      ELSE
         IF (H .GT. 110.0) THEN
            PRINT*, 'ARROW WAS TOO HIGH, HEIGHT WAS ', H
         ELSE
            PRINT*, 'GOOD SHOT PRINCE'
         ENDIF
      ENDIF
C
      STOP
      END

*** PROGRAM EXECUTION OUTPUT ***

DISTANCE FROM TOWER =    100.0000000000
VELOCITY OF ARROW =    500.0000000000
ANGLE OF ELEVATION =     47.00000000000    DEGREES
GOOD SHOT PRINCE
```

**Fig. 4.10**  Prince Valiant rescues Rapunzel.

**Example 4.20:** The library function MOD can be used to test whether its first argument is divisible by its second argument. The value returned by the function call

$$MOD(N,DIV)$$

is the remainder of the division of N by DIV. For example, if N = 17 and DIV = 3, then the value 2 should be returned as illustrated below.

$$3 \overline{)\ 17 \atop \phantom{)}\ \underline{15}} \atop \phantom{)}\ \ 2 \leftarrow \text{remainder}$$

If the value returned is 0 (no remainder), then N is evenly divisible by DIV. Thus, the loop that tests all possible divisors of N in the prime number problem (Fig. 4.8) can be rewritten using the MOD function as shown next.

```
DO 20 DIV = 2, N - 1
   IF (MOD(N,DIV) .EQ. 0) THEN
      DIVISR = DIV
      PRINT*, DIVISR
      PRIME = .FALSE.
   ENDIF
20 CONTINUE
```

**Exercise 4.15:** Write a program that displays the sine and cosine of angles between 0° and 180° in increments of 10°.

**Exercise 4.16:** The roots of an equation of the form

$$y = Ax^2 + Bx + C$$

where a, b, and c are real numbers may be computed as follows.

$$r1 = \frac{-B + \sqrt{B^2 - 4AC}}{2A}$$

$$r2 = \frac{-B - \sqrt{B^2 - 4AC}}{2A}$$

Write a program to read in three values A, B, and C, and determine and print r1 and r2. Test your program with the following values for A, B, and C.

| A | B | C |
|---|---|---|
| 1 | 1 | −6 |
| 1 | −8 | 16 |
| 1 | 0 | −1 |
| 1 | 0 | 1 |
| 15 | −2 | −1 |
| 0 | 0 | 0 |

If $B^2 - 4AC$ is less than zero, print an appropriate message.

**Exercise 4.17:**   Modify the Prince and Rapunzel program so that the velocity of the arrow will automatically be increased by 10 feet/sec if the arrow is too low and decreased by 8 feet/sec if the arrow is too high. This repetition should terminate when the arrow enters the window. *Hint*: Use a program flag as the loop control variable.

### 4.6.2  Type Conversion Functions

In Sec. 4.3.4, we urged you to avoid using mixed-type expressions involving integers and reals in your programs. This can be done most easily by using the three library functions REAL, INT, and NINT for performing *type conversion*. The function REAL takes an integer argument (in fixed point form) and provides as output an equivalent real value (in floating point form). The functions INT and NINT take a real argument and provide as output an integer value. The value provided by INT corresponds to the integral part of the argument only (the fractional part is truncated), whereas NINT provides the nearest rounded integer as its value.

**Example 4.21:**   The functions NINT and INT are illustrated below.

| function call | result | function call | result |
|---|---|---|---|
| NINT(16.8) | 17 | INT(16.8) | 16 |
| NINT(−7.8) | −8 | INT(−7.8) | −7 |
| NINT(16.2) | 16 | INT(16.2) | 16 |
| NINT(−7.5) | −8 | INT(−7.5) | −7 |

The functions INT, NINT and REAL may also be used to avoid mixed-type assignment statements, such as

```
XN = N
 I = PSI
```

where XN and PSI are real, and N and I are integers. The assignment statements below are preferred since they explicitly specify the type conversion that is to take place.

```
   XN = REAL(N)
    I = INT(PSI) (truncates PSI)
or  I = NINT(PSI) (rounds PSI)
```

**Example 4.22:**   The program in Fig. 4.11a computes and prints a table showing the conversion from degrees Celsius to degrees Fahrenheit for temperatures ranging from 0°C to 100°C. Several lines of the table are shown following the program. The computation for CDEGR = 26 is illustrated in Fig. 4.11b.

```
C COMPUTE AND PRINT TABLE OF CELSIUS TO FAHRENHEIT
C CONVERSION FOR 0 TO 100 DEGREES CELSIUS
C
      INTEGER CDEGR, FDEGR
C
```

*(Continued)*

```
      PRINT*, '              CELSIUS              FAHRENHEIT'
      PRINT*, '              DEGREES                DEGREES  '
      DO 10 CDEGR = 0, 100
         FDEGR = NINT(1.8 * REAL(CDEGR) + 32.0)
         PRINT*, CDEGR, FDEGR
   10 CONTINUE
C
      STOP
      END
```

*** PROGRAM EXECUTION OUTPUT ***

```
          CELSIUS            FAHRENHEIT
          DEGREES              DEGREES
             0                   32
             1                   34
             2                   36
             3                   37
             4                   39
             5                   41
             6                   43
             7                   45
             8                   46
```

**Fig. 4.11a**  Converting Celsius to Fahrenheit degrees.

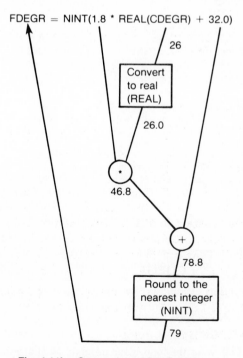

**Fig. 4.11b**  Conversion of 26° Celsius.

**Exercise 4.18:**   Let X = 10.0, START = 3.0, and YNEW = 0.0, as shown in the program below. What will be the value of YNEW, as printed by this program? (Show all intermediate output from your hand simulation. Keep your computations accurate to 4 decimal places. Two repetitions of the loop should be sufficient.) *Hint:* This program computes an approximation to the square root of X.

```
REAL X, Y, YNEW, START, ERRLIM
PARAMETER (ERRLIM = 0.001)
X = 10.0
START = 3.0
Y = START
YNEW = (Y + X / Y) / 2.0
WHILE (ABS(YNEW - Y) .GT. ERRLIM) DO
    Y = YNEW
    YNEW = (Y + X / Y) / 2.0
ENDWHILE
PRINT*, 'THE SQUARE ROOT OF ', X, ' IS ', YNEW
STOP
END
```

**Exercise 4.19:**   Write an assignment statement that rounds a positive real number to two significant decimal places. *Hint:* You will have to multiply by 100.0. Generalize your expression to round a positive real number accurate to N decimal places, where N is any positive integer. *Hint:* 100.0 = 10.0**2

**Exercise 4.20:**   Assume that you have a program in which the following declarations and data statements appear.

```
INTEGER COLOR, LIME, STRAW, YELLOW, RED, ORANGE
REAL BLACK, WHITE, GREEN, BLUE, PURPLE, CRAYON
DATA COLOR, BLACK, CRAYON/2, 2.5, -1.3/
DATA STRAW, RED, PURPLE/1, -3, +0.3E1/
```

For each of the following assignment statements, eliminate all mixed-mode expressions and assignments and then determine the value that is assigned to the variable appearing on the left-hand side of the equal sign.

```
a) WHITE  = COLOR * 2.5/PURPLE
b) GREEN  = COLOR/PURPLE
c) ORANGE = COLOR/RED
d) BLUE   = (COLOR+STRAW)/(CRAYON+0.3)
e) LIME   = PURPLE**COLOR + RED/CRAYON
f) YELLOW = 1.0 * STRAW/COLOR+0.5
```

**Exercise 4.21:**   The numeric constant e is known as Euler's Number. It has had an established place in mathematics alongside the Archimedean number $\pi$ for over 200 years. The approximate value of e (accurate to 9 decimal places) is 2.71828183. Write a FORTRAN program that prints $e^x$ and ln(x) for integer values of x from 1 to 10.

**Exercise 4.22:**   Let N = 1000. Write a program to compute P = 0.1 * N and S = $\sum_{i=1}^{N}$ 0.1 (where $\sum_{i=1}^{N}$ 0.1 is equal to 0.1 added to itself 1000 times). You will need a loop to compute S. Following the loop, your program should print the values of S and P and also print an appropriate message indicating whether or not S and P are equal.

**Exercise 4.23:** The formula for the velocity of a body dropped from rest is v = gt, where g is the acceleration due to gravity, and t is time (air resistance is ignored here). Write a loop to compute v at 10-second intervals (starting with t = 0) for a pickle dropped from a building that is 600 meters tall, with g = 9.81 meters/second. *Hint:* Use the formula t = $\sqrt{2s/g}$ to determine the time, T1, it takes for the pickle to hit the ground (s equals 600). Use T1 to limit the number of repetitions of the loop that produces the table.

## 4.7  INTRODUCTION TO FORMATTED OUTPUT*

### 4.7.1  Format Specification

You have probably noticed that you have very little control over the exact appearance of an output line. This is because the compiler uses a predetermined number of columns to print each real number.

Rather than have the compiler dictate the appearance of program output, it is possible for the programmer to indicate the exact form of any output line through the use of *formatted output.* Formatted input and output will be discussed in detail in Chapter 8; however, we will provide a brief introduction to formatted output in this section.

When using formatted output, the programmer describes the exact form of an output line by using a *format specification.* A format specification is a list of *edit descriptors* enclosed in parentheses as shown below.

```
(A, A, 3X, F7.2, 2X, I2)
```

Each edit descriptor either provides information about spacing, or specifies how an item in the output list should be displayed. There are six edit descriptors in the format specification above. Three of these (A, F7.2, I2) describe how values should be printed. The two descriptors ending in X (3X and 2X) specify horizontal spacing, or spacing between output values. We shall illustrate the use of this format specification in the next example.

**Example 4.23:** Given the variables shown below

| NAME | GROSS | DEPEND |
|------|-------|--------|
| HARRISON | 347.5783 | 3 |

the PRINT statement

```
PRINT '(A, A, 3X, F7.2, 2X, I2)',
Z         ' ', NAME, GROSS,      DEPEND
```

generates the output line.

```
HARRISON      347.58      3
     A      3X   F7.2   2X I2
```

---

*This section may be omitted for now if your instructor prefers to wait until Chapter 8 to cover this material.

The labeled brackets indicate the *output field* specified by each edit descriptor as explained next.

The format specification is written as a character string (enclosed in apostrophes) and is separated by a comma from the output list. The Z indicates that the PRINT statement is continued on the next program line which contains the output list.

The compiler associates each edit descriptor that begins with A, F or I with a data item (variable or constant) in the output list. The correspondence is established by working from left to right in each list. The first edit descriptor A is associated with the string ' ' (a blank); the second A is associated with NAME, F7.2 is associated with GROSS, and I2 is associated with DEPEND. In our sample PRINT statement, we have positioned each output list item under its corresponding edit descriptor, although this is clearly not required. In fact, the output list can be written on the same line as the format specification.

Each edit descriptor that begins with A, F or I specifies how its associated output list item will be printed. The effect of each edit descriptor in the example above is described next.

| Edit Descriptor | Output List Item | Meaning |
| --- | --- | --- |
| A | ' ' | This descriptor specifies that the current line should be printed directly under the previous line using single line spacing. |
| A | NAME | The character variable NAME is printed. The number of print positions used is determined by the declared length of NAME. |
| 3X | | Three blanks are "printed." |
| F7.2 | GROSS | The value of the real variable GROSS is rounded to two fractional digits and printed right-justified in seven columns. |
| 2X | | Two blanks are "printed." |
| I2 | DEPEND | The value of the integer variable DEPEND is printed right-justified in two columns. |

Table 4.2 describes the general form of the FORTRAN edit descriptors used so far. In this table, n, w and d represent integer constants.

Table 4.3 describes the characters used to specify vertical spacing between lines, or for line control. Each output list should begin with a character string. The first character in this string is used only for line control and is not printed. This vertical spacing method normally works only on line printers.

| Edit descriptor | Effect |
| --- | --- |
| A | Used for printing a character data item. The field width used is determined by the length of the data item. |
| Aw | Used for printing a character data item in a field of width w. The string is printed left-justified. |

(Continued)

| Edit descriptor | Effect |
|---|---|
| Fw.d | Used for printing a real data item in a field of width w rounded to d fractional digits (w includes d, a decimal point, and a sign). The number is printed right-justified. |
| Iw | Used for printing an integer data item in a field of width w (w includes the sign). The number is printed right-justified. |
| nX | Used for "printing" n blanks, or skipping n spaces. |

**Table 4.2 Some FORTRAN edit descriptors.**

| Line control character | Effect |
|---|---|
| blank | Single line space. |
| 0 | Double line space. |
| 1 | Skip to the top of a new page. |
| + | Suppress line spacing (no space). |

**Table 4.3 Line control characters.**

It is essential that each edit descriptor and its corresponding output list item agree in type as indicated in Table 4.2. Any discrepancy will result in an error diagnostic during program execution.

**Example 4.24:**   The statements

```
      PRINT '(A, A)', ' EMPLOYEE IS ', NAME
      PRINT '(A, F6.0, 3X, A, I2)',
    Z        '0SALARY = $', GROSS, 'DEPENDENTS = ', DEPEND
```

generates the output lines below.

```
 EMPLOYEE IS HARRISON

SALARY = $  348.   DEPENDENTS =  3
```

The first character string in each output list above begins with the carriage control character for that line (blank in ' EMPLOYEE IS ' and 0 in '0SALARY = $'). The first PRINT statement displays a character string followed by a character variable. The second PRINT statement causes a line to be skipped before printing two character strings and a real and integer value. The format descriptor F6.0 specifies that the value of GROSS is to be rounded to the nearest dollar.

**Example 4.25:**   The statement below could be used to print a table heading at the top of a page:

```
      PRINT '(A, 10X, A, 10X, A)'
    Z        '1NAME', 'SALARY', 'DEPENDENTS'
```

Each of the column headings (a string) is printed at the top of a page with ten spaces between column headings. The PRINT statement

```
     PRINT '(A, A10, 4X, F7.2, 14X, I2)',
   Z         '0', NAME, GROSS, DEPEND
```

could be used inside a loop to print a table of double-spaced values under the column headings. An example of how the table might look follows.

| NAME | SALARY | DEPENDENTS |
|------|--------|------------|
| HARRISON | 347.82 | 3 |
| JACKSON | 299.63 | 10 |

The character variable NAME is printed left justified in a field of width ten (edit descriptor A10). The numeric values are printed right-justified in their respective fields.

The descriptors used for horizontal spacing (4X and 14X) were selected so that the second and third columns of values would be centered under their respective column headings. These descriptors should be fixed after the table heading has been laid out and the number of positions needed to print each data value is determined. These points are discussed in the next section.

### 4.7.2  Width of a Field

Each edit descriptor specifies the size of the field in which its corresponding data item will be printed. Obviously, the field should be wide enough to accommodate the data items. For integer data, this means that the field width should be the same as the number of digits in the largest absolute value to be printed. Any integer with a smaller absolute value will be printed right-justified in its field. If the value to be printed is larger than expected, then one or more asterisks may be printed instead. No other error indication will be given. If the value to be printed may be negative, an additional print position must be added to accommodate the sign.

In determining the field width required for a real value, one space should be allocated for the sign and one for the decimal point. If d fractional digits are desired, then at least $d + 2$ print positions are needed. The required field width may be determined by adding the largest number of digits expected to precede the decimal point (the integral part) to $d + 2$. As with integer values, a smaller number will be printed right-justified. A larger number may be printed in scientific notation, or one or more asterisks may be printed instead, depending on the compiler.

In general, it is better to specify a field width that is too large rather than too small. If the field width is too large, any extra blanks can normally be ignored. However, if the field width is too small, a data value may not be printed as expected.

The field width for a character data item does not need to be specified. If the descriptor A is used, then the field width will automatically correspond to the length of the character data item. However, the programmer may specify the field width by using a descriptor of the form Aw. In this case, the character data item is printed left-justified if w is larger than the length of the character data item. If w is smaller than the length of the character data item, then only the first w characters will be printed.

**Example 4.26:**  The statement below prints the string 'CASE' in three adjacent fields of different widths.

```
          PRINT '(A, A10, A2)', ' CASE', 'CASE', 'CASE'
```

**Exercise 4.24:**  Which of the following format specifications are correct, and which contain syntax errors? Correct the syntax errors that you find.

a) (A, 3X, I4, F12.2)
b) (A, 4I, 2X, F12.1)
c) (A, F16.35X)
d) (A, 2A, F3.3, X6)
e) (A, I4, 4X, F4.2)

**Exercise 4.25:**  Let K contain the value 1234, and ALPHA contain the value 555.4567. What would be printed by each of the statements below.

a) PRINT '(A, I4, F12.4)', ' ', K, ALPHA
b) PRINT '(A, I4, 4X, F8.4)', ' ', K, ALPHA
c) PRINT '(A, I4)', ' K = ', K
   PRINT '(A, F8.2)', 'OALPHA = ', ALPHA
d) PRINT '(A, I5, 10X, A, F10.3)',
   Z '1K = ', K, 'ALPHA = ', ALPHA

**Exercise 4.26:**  Consider the variable values shown below.

| SSNO1 | SSNO2 | SSNO3 | LAST | FIRST | HOURS | RATE | PAY |
|---|---|---|---|---|---|---|---|
| 219 | 40 | 9677 | DOG | HOT | 40.00 | 4.50 | 180.00 |

Write a segment of a FORTRAN program (including declarations) to produce the following output:

```
Line 1   SOCIAL SECURITY NUMBER 219-40-9677
Line 2
Line 3   DOG, HOT
Line 4
Line 5   HOURS RATE PAY
Line 6   40.00 4.50 180.00
```

## 4.8  NUMERICAL ERRORS*

All of the errors discussed in earlier chapters were programmer errors. However, even if a program contains no such errors, it is still possible to compute the wrong answer, especially in programs that involve extensive numerical computation. The cause of error is the inherent inaccuracy in the internal representation of floating point (real) data.

We stated earlier that all information is represented in the memory of the computer as a binary string. However, as shown in the next example, many decimal numbers do not have precise binary equivalents and, therefore, can only be approximated in the binary number system.

**Example 4.27:**   This example lists several binary approximations of the number 0.1. The precise decimal equivalent of the binary number being represented and the numerical error are also shown.

| Number of binary bits | Binary approximation | Decimal equivalent | Numerical error |
|---|---|---|---|
| 4 | .0001 | 0.0625 | 0.0375 |
| 5–7 | .0001100 | 0.09375 | 0.00625 |
| 8 | .00011001 | 0.09765625 | 0.00234375 |
| 9 | .000110011 | 0.099609375 | 0.000390625 |

We can see from this example that, as the number of binary bits used to represent 0.1 is increased, the precise decimal equivalent represented by the binary number gets closer and closer to 0.1. However, it is impossible to obtain an exact binary representation of 0.1, no matter how many bits are used. Unfortunately, the number of binary bits that can be used to represent a real number in the memory of the computer is limited by the size of a memory cell. The larger the cell, the larger the number of binary bits, and the greater the degree of accuracy that can be achieved.

The effect of a statistically small representational error can become magnified in a sequence of computations, or through repetition of an imprecise computation in a loop. Such magnification can sometimes be diminished by the use of functions and careful ordering of the computations. You should be aware that the problem exists, and that it may cause the same FORTRAN program to produce different results when run on computers having memory cells of different size.

**Example 4.28:**

a) The computation

$$B \;**\; 2 \text{ or } B \;*\; B$$

is likely to produce more accurate results than B**2.0, since many compilers will

---

*This section may be omitted.

use logarithms to perform the latter computation and multiplication for the former.

b) The computation

$$SQRT(X)$$

is likely to produce more accurate results than $X**.5$ since most square root functions produce more accurate results than the logarithm functions required in the latter computation.

c) If we have two real numbers A and B, whose difference is small, and a third number C that is relatively large (compared with $A - B$), then the calculation

$$(A - B) * C$$

may produce results that are less accurate than

$$A * C - B * C.$$

This is because the percentage of error is greater in a very small number such as $(A - B)$, and additional inaccuracy is introduced when a very small number is multiplied by one that is much larger.

## 4.9　COMMON PROGRAMMING ERRORS

Errors in type declarations are generally caused by spelling mistakes, missing commas or failure to declare a variable used in a program. Spelling errors and failure to declare variables will not be detected by most compilers, and missing commas between short variable names (such as IT and ARE) will also go unnoticed (in this case IT ARE, with a comma missing between the T and the A, will be interpreted as the single variable name ITARE). Such mistakes can often confuse the compiler and cause it to detect other errors in subsequent program statements.

The only way to avoid the grief that can be caused by type declaration errors is to check carefully all declarations that you write. There is also no substitute for complete data tables. Such tables can be used in writing program declarations, and can be checked for the correct spelling of variables during the writing of subsequent program statements.

A common input error is caused by attempts to read data of one type into a variable of a different type. Usually these errors are caused by one of the following:

- data out of order
- missing data
- incorrectly prepared data (such as character strings without apostrophes)

Most FORTRAN systems can detect and diagnose these errors, and print informative execution time diagnostics indicating which data items are in error.

Some of the more common programming errors involving expressions and assignment statements are listed below, along with their remedies. The compiler diagnostics for these errors will be similar in wording to the short descriptions that are given here.

*Mismatched or unbalanced parentheses.* The statement in error should be carefully scanned, and left and right parentheses matched in pairs, inside-out, until the mismatch becomes apparent. This error is often caused by a missing parenthesis at the end of an expression.

*Missing operator in an expression.* This error is usually caused by a missing multiplication operator, *. The expression in error must be scanned carefully, and the missing operator inserted in the appropriate position.

*Logical or character data used with arithmetic operator* or *logical data used with relational operator.* These errors are examples of mixed-type expressions; operators which can manipulate data of one type are being used with data of another type. A common error of this sort is a condition of the form (PRIME .EQ. .TRUE.) where PRIME is logical; the correct condition is simply (PRIME).

*Arithmetic underflow or overflow* or *division by zero attempted.* Another type of numerical error is caused by attempts to manipulate very large real numbers or numbers that are very close in value to zero. For example, dividing by a number that is almost zero may produce a number that is too large to be represented (*overflow*). You should check that the correct variable is being used as a divisor and that it has the proper value. On some compilers, a divisor that is undefined would be set to zero and would cause a *division by zero* diagnostic to be printed. Arithmetic *underflow* occurs when the magnitude of the result is too small to be represented in floating point form.

One type of programming error that cannot be detected by a compiler involves the writing of expressions that are syntactically correct, but do not accurately represent the computation called for in the problem statement. All expressions, especially long ones, must be carefully checked for accuracy. Often, this involves the decomposition of complicated expressions into simpler subexpressions producing intermediate results. Intermediate results should be printed and compared with hand calculations for a simple, but representative, data sample.

## 4.10  SUMMARY

The FORTRAN language provides a capability for manipulating a number of different *types* of data. We have introduced four of these types (integer, real, logical and character) in this chapter.

All information is stored in the computer as a binary string. The format of this string is determined by the type of data being represented. In order for the FORTRAN compiler to generate the correct machine language instructions for a program, it must know the types of all data being manipulated. Thus, the types of all constants and variable names used in a program must be clearly indicated to the compiler.

The types of variable names can be specified either by using type declaration statements, or by allowing the compiler to assign a type to each name according to the implicit typing convention (not recommended). Under this convention, all variable names beginning with the letters I through N are automatically typed *integer;* names beginning with A through H or O through Z are automatically typed *real.* Examples of the four type declarations introduced in this chapter are shown in Fig. 4.12.

| *Declaration* | *Use* |
|---|---|
| CHARACTER * 20 NAME | declares a variable for storing a character string of length 20 |
| CHARACTER LETTER | declares a variable for storing a character string of length 1 |
| REAL X, ALPHA | declares two variables for storing real data items |
| INTEGER I, COUNT | declares two variables for storing integer data items |
| LOGICAL FLAG | declares a variable for storing a logical data item (having a value of either .TRUE. or .FALSE.) |

**Fig. 4.12**  Examples of type declarations.

Constants are typed according to the way in which they are written. Real constants are numbers which contain a decimal point; integer constants do not contain a decimal point. Character string constants consist of a string of legal FORTRAN characters enclosed in apostrophes. The only two logical constants are .TRUE. and .FALSE..

In this chapter, we have also provided rules for forming and evaluating arithmetic expressions. This capability is most useful in numerically oriented problems. Knowledge of these rules will enable you to apply FORTRAN correctly to perform calculations. One useful guideline, which should always be kept in mind when transforming an equation or formula to FORTRAN, is "when in doubt of the meaning, insert parentheses."

Integer arithmetic was also discussed. Of the basic arithmetic operators +, *, /, and −, only the slash (division) produces different results when used with integers instead of reals. This is due to the fact that the internal fixed point format used for storage of integers does not permit the representation of a fractional remainder.

The use of functions in arithmetic expressions has also been introduced. Their use will enable you to more efficiently perform numerical calculations, since calling a function to compute a result is much faster and easier than trying to perform the calculation using the basic arithmetic operations.

Finally, we discussed formatted output and the use of the format specification to control the appearance of a printed line. Edit descriptors beginning with A, F and I are used for printing character, real and integer data, respectively.

The X descriptor is used for horizontal spacing in a line. Each output list should begin with one of the characters blank, 0, 1 or +; this character is used for vertical spacing, or line control.

## PROGRAMMING PROBLEMS

**4C**    Write a program to read in a collection of integers and determine whether each is a prime number. Test your program with the four integers 7, 17, 35, 96.

**4D**    Let $n$ be a positive integer consisting of up to 10 digits, $d_{10}\,d_9\ldots d_1$. Write a program to list in one column each of the digits in the number $n$. The rightmost digit $d_1$ should be listed at the top of the column. *Hint:* If $n = 3704$, what is the value of *digit* as computed according to the following formula?

$$\text{digit} = \text{MOD}(n, 10)$$

Test your program for values of $n$ equal to 6, 3704, and 1704985.

**4E**    An integer N is divisible by 9 if the sum of its digits is divisible by 9. Use the algorithm developed for Problem 4D to determine whether or not the following numbers are divisible by 9.

$$N = 154168$$
$$N = 62159382$$
$$N = 12345678$$

**4F**    Each month a bank customer deposits $50.00 in a savings account. The account earns 6.5 percent interest, calculated on a quarterly basis (one-fourth of 6.5 percent each quarter). Write a program to compute the total investment, total amount in the account, and the interest accrued, for each of 120 months of a 10-year period. You may assume that the rate is applied to all funds in the account at the end of a quarter regardless of when the deposits were made. The table printed by your program should begin as follows:

| MONTH | INVESTMENT | NEW AMOUNT | INTEREST | TOTAL SAVINGS |
|---|---|---|---|---|
| 1 | 50.00 | 50.00 | 0.00 | 50.00 |
| 2 | 100.00 | 100.00 | 0.00 | 100.00 |
| 3 | 150.00 | 150.00 | 2.44 | 152.44 |
| 4 | 200.00 | 202.44 | 0.00 | 202.44 |
| 5 | 250.00 | 252.44 | 0.00 | 252.44 |
| 6 | 300.00 | 302.44 | 4.91 | 307.35 |
| 7 | 350.00 | 357.35 | 0.00 | 357.35 |

Keep all computations accurate to two decimal places. How would you modify your program if interest were computed on a daily basis?

**4G**    Compute a table of values of $X/(1 + X^2)$ for integer values of $X = 1, 2, 3, \ldots,$ 50. Your table of values should be accurate to four decimal places and should begin as follows:

| X | $X / (1 + X^2)$ |
|---|---|
| 1 | .5000 |
| 2 | .4000 |
| 3 | .3000 |
| 4 | .2353 |
| 5 | .1923 |
| . | . |
| . | . |
| . | . |

**4H**    Write a program to compute the sum $1 + 2 + 3 + 4 + \ldots + N$ for any positive integer N; use a DO loop to accumulate this sum (SUM1). Then compute the value SUM2 by the formula

$$SUM2 = \frac{(N + 1) N}{2}$$

Have your program print both SUM1 and SUM2, compare them, and print a message indicating whether or not they are equal. Test your program for values of N = 1, 7, 25.

**4I**    The Hoidy Toidy baby furniture company has ten employees, many of whom work overtime (more than 40 hours) each week. They want a payroll program that reads the weekly time records (containing employee name, hourly rate (r), and hours worked (h) for each employee) and computes the gross salary and net pay as follows:

$$g = \text{gross salary} = \begin{cases} h \times r & (\text{if } h < = 40) \\ 1.5r(h - 40) + 40r & (\text{if } h > 40) \end{cases}$$

$$p = \text{net pay} = \begin{cases} g & (\text{if } g < = \$65) \\ g - (15 + .045g) & (\text{if } g > \$65) \end{cases}$$

The program should print a five column table listing each employee's name, hourly rate, hours worked, gross salary, and net pay. The total amount of the payroll should be printed at the end. It can be computed by summing the gross salaries for all employees. Test your program on the following data:

| name | rate | hours |
|------|------|-------|
| IVORY HUNTER | 3.50 | 35 |
| TRACK STAR | 4.50 | 40 |
| SMOKEY BEAR | 3.25 | 80 |
| OSCAR GROUCH | 6.80 | 10 |
| THREE BEARS | 1.50 | 16 |
| POKEY PUPPY | 2.65 | 25 |
| FAT EDDIE | 2.00 | 40 |
| PUMPKIN PIE | 2.65 | 35 |
| SARA LEE | 5.00 | 40 |
| HUMAN ERASER | 6.25 | 52 |

**4J**    The interest paid on a savings account is compounded daily. This means that if you start with X dollars in the bank, then at the end of the first day you will have a balance of

$$X \times (1 + \text{rate}/365)$$

dollars, where rate is the annual interest rate (0.06 if the annual rate is 6%). At the end of the second day, you will have

$$X \times (1 + \text{rate}/365) \times (1 + \text{rate}/365)$$

dollars, and at the end of N days you will have

$$X \times (1 + \text{rate}/365)^N$$

dollars. Write a program that will process a set of data cards, each of which contains values for X, rate, and N and computes the final account balance.

**4K**    The program written in Exercise 4.18 finds the square root of the number X (written $\sqrt[2]{X}$ or $\sqrt{X}$) accurate to 4 decimal places. This program can be adapted to find the Nth root $\sqrt[N]{X}$ of a number X, for any positive integer N. The major task required for this change is to generalize the expression for computing YNEW:

$$\text{YNEW} = \frac{1}{N}\left((N-1)\times Y + \frac{X}{Y^{N-1}}\right)$$

(*Note.* In Exercise 4.18, N was equal to 2.). Modify the program given in Exercise 4.18 to read in a number X, a value N, and a value for START, and compute $\sqrt[N]{X}$ accurate to 4 decimal places. Test your program on the following data.

| X | N | START |
|---|---|-------|
| 7.0 | 2 | 2.0 |
| 48.0 | 3 | 6.0 |
| 32.0 | 5 | 2.0 |

**4L**    Compute the monthly payment and the total payment for a bank loan, given:

1. the amount of the loan (LOAN),
2. the duration of the loan in months (MONTHS),
3. the interest-rate percent for the loan (RATE).

Your program should read in one card at a time (each containing a loan value months value, and rate value), perform the required computation, and print the values of the loan, months, rate, and the monthly payment (MPAYMT), and total payment (TOTPMT).

Test your program with at least the following data (and more if you want).

| Loan | Months | Rate |
|------|--------|------|
| 16000. | 300 | 6.50 |
| 24000. | 360 | 7.50 |
| 30000. | 300 | 9.50 |
| 42000. | 360 | 8.50 |
| 22000. | 300 | 9.50 |
| 300000. | 240 | 9.25 |

Don't forget to first read in a card indicating how many data cards you have.
*Notes.*
i) The formula for computing monthly payment is

$$\text{mpaymt} = \left[\frac{\text{rate}}{1200.}\times\left(1.+\frac{\text{rate}}{1200.}\right)^{\text{months}}\times\text{loan}\right]\Big/\left[\left(1.+\frac{\text{rate}}{1200.}\right)^{\text{months}}-1.\right].$$

ii) The formula for computing the total payment is

$$\text{totpmt} = \text{mpaymt}\times\text{REAL (months)}.$$

iii) Don't forget to declare the types of the variables LOAN and MPAYMT.
Also, you may find it helpful to introduce additional variables RATEM and EXPM defined below, and use these to simplify the computation of MPAYMT.

You can check the values of ratem and expm to see whether your program's computations are accurate.

$$\text{ratem} = \text{rate}/1200.$$
$$\text{expm} = (1. + \text{ratem})^{\text{months}}$$

**4M** The rate of radioactive decay of an isotope is usually given in terms of the half-life, HL (the time lapse required for the isotope to decay to one-half of its original mass). For the strontium 90 isotope (one of the products of nuclear fission), the rate of decay is approximately .60/HL. The half-life of the strontium 90 isotope is 28 years. Compute and print, in table form, the amount remaining after each year from an initial point at which 50 grams are present. *Hint:* For each year, the amount of isotope remaining can be computed using the formula

$$r = \text{amount} * C^{(\text{Year}/\text{HL})}$$

where amount is 50 grams (the initial amount), and C is the constant $e^{-0.693}$ ($e = 2.71828$).

**4N** Write a FORTRAN program to read in a 9-digit Social Security number, such as 219400677, and print this number in the form 219-40-0677. (The problem here is to break the Social Security number into three parts, SSNO1, SSNO2, SSNO3, and to print each part as a separate integer.

**4O** Write a program that will read a data record containing two words and store them in the character variables FIRST and LAST. The program will then process a collection of data items, each consisting of a single word, and print that word in column 1 if it precedes FIRST, column 2 if it lies between FIRST and LAST, and column 3 if it follows LAST. At the end, print the count of all words in each field. Assume a maximum length of 12 characters for your words.

**4P** An examination with nine questions is given to a group of 28 students. The exam is worth 10 points and everyone turning in an answer sheet receives at least 1 point. Each problem is graded on a no credit, half credit, full credit basis. An exam score (SCORE) and name (NAME) is entered for each student. Write a program to determine the rank for each score and print a three-column list containing the name, score and rank of each student. The ranks are determined as follows:

| Score | Rank |
|-------|------|
| 9.0–10.0 | GOOD |
| 6.0–8.5 | FAIR |
| 1.0–5.5 | POOR |

The program should also print the number of scores in each rank and the total number of scores.

**4Q** Write a program to simulate the tossing of a coin. Use the random number generator on your system, and consider any number less than 0.5 to represent tails, and any number greater than or equal to 0.5 to represent heads. Print the number produced by random number generator and its representation (heads or tails). *Hint:* Repeat the call to the random number generator 50 or 100 times. At the end, print a count of the number of heads versus the number of tails. Your instructor can tell you how to use the random number generator on your FORTRAN system.

# ARRAYS

5

## 5.1   INTRODUCTION

In many applications, we are faced with the problem of having to store and manipulate large quantities of data in memory. In our problems so far, it has been necessary to use only a few memory cells to process relatively large amounts of data. This is because we have been able to process each data item separately and then reuse the memory cell in which that data item was stored.

For example, in Problem 3D we computed the maximum value of a set of exam scores. Each score was read into the same memory cell, named SCORE, and then completely processed. This score was then destroyed when the next score was read for processing. This approach allowed us to process a large number of scores without having to allocate a separate memory cell for each one. However, once a score was processed, it was impossible to reexamine it later.

There are many applications in which we may need to save data items for subsequent reprocessing. For example, we might desire to write a program that computes and prints the average of a set of exam scores, and also the difference between each score and the average. In this case, all scores must be processed and the average computed before we can calculate the differences requested. We must, therefore, be able to examine the list of student exam scores twice, first to compute the average, and then to compute the differences. Since we would rather not have to read the exam scores twice, we will want to save all of the scores in memory during the first step, for reuse during the second step.

In processing each data item, it would be extremely tedious to have to reference each memory cell by a different name. If there were 100 exam scores to process, we would need to list each variable used to store a score in the input list of a READ statement. We would also need 100 assignment statements in order to determine the difference between each score and the average.

In this chapter, we will learn how to use a new feature of FORTRAN, called an *array*, for storing a collection of related data items. Use of the array will simplify the task of naming and referencing the individual items in the collection. Through the use of arrays, we will be able to enter an entire collection of data items with a single READ statement. Once the collection is stored in memory, we will be able to reference any of these items as often as we wish without ever having to reenter that item into memory.

## 5.2   DECLARING ARRAYS

In all prior programming discussed in this text, each symbolic name used in a program has always been used for storage of a single data item. An array is a collection of two or more adjacent memory cells, called *array elements,* that are associated with a single symbolic name. Whenever we wish to tell the compiler to associate two or more memory cells with a single name, we must use a declaration statement in which we state the name to be used, and designate the elements to be associated with this name. The *size* (number of elements) of the array is also specified by this declaration statement.

For example, the declaration statement

```
REAL X(8)
```

instructs the compiler to associate eight memory cells, designated as X(1), X(2), X(3), . . . , X(7), and X(8) with the name X (see Fig. 5.1). The array X is considered to be of size 8—i.e., to consist of 8 array elements. Each of the elements may contain a real number, as shown in Fig. 5.1.

Array X

| X(1) | X(2) | X(3) | X(4) | X(5) | X(6) | X(7) | X(8) |
|------|------|------|------|------|------|------|------|
| 16.0 | 12.0 | 6.0  | −2.0 | −12.0 | −24.0 | −38.0 | −54.0 |

First         Second       Third                                    Eighth
element     element     element              . . .              element

**Fig. 5.1**   The eight elements of the array X.

The *subscripted variable* X(1) can be used to reference the first element of the array X, X(2) the second element, and X(8) the eighth element. The integer enclosed in parentheses is the *array subscript.*

**Example 5.1:**   Let X be the real array shown in Fig. 5.1, and let SUM be a real variable containing the value 34.0 (the sum of the first three elements of X). Then the statement

$$SUM = SUM + X(4)$$

will cause the value −2.0 (the contents of the memory cell designated by X(4)) to be added to SUM. The statement

$$SUM = SUM + X$$

would be illegal since FORTRAN will not allow us to add the array X to SUM. An array name must always be used with a subscript in an assignment statement.

In the next section, we will study subscripts in more detail and we will see that integer constants are not the only form of a subscript that is allowed in FORTRAN. However, first we will describe a limited form of the FORTRAN array declaration.

---

**Array Declaration (Limited Form)**

*type name(size)*

**Interpretation:** *Type* may be any one of the four data types INTEGER, REAL, LOGICAL, or CHARACTER. The compiler will associate the number of memory elements indicated by *size* with the variable indicated by *name*. Each element is used for storage of a data item of the indicated type. The individual array elements will be referenced by the subscripted variables *name(1)*, *name(2), . . . , name(size)*, where the smallest legal subscript is 1, and the largest legal subscript is *size*. *Size* must be a positive integer constant or parameter.

*Note:* Arrays and simple variables may be declared in the same declaration statement; commas must be used between entries in the declaration.

**Example 5.2:** More than one array may appear in a declaration statement. The declarations

```
REAL CACTUS(5), NEEDLE, PINS(6)
INTEGER FACTOR(12), N, INDEX
```

will cause five array elements to be assigned to the real array CACTUS, six to the real array PINS, and twelve to the integer array FACTOR. In addition, separate memory cells will be allocated for storage of the real variable NEEDLE and the integer variables N and INDEX.

**Example 5.3:** The statements:

```
INTEGER MAX
PARAMETER (MAX = 20)
CHARACTER * 10 FIRNAM(MAX), LASNAM(MAX)
CHARACTER * 40 PARENT(15)
```

will allocate three arrays for storage of character data. Twenty array elements, each capable of storing a character string of length ten, will be assigned to the arrays FIRNAM and LASNAM; fifteen array elements, each capable of storing a character string of length forty, will be assigned to the array PARENT.

**Exercise 5.1:** Draw pictures similar to the one shown in Fig. 5.1 (complete with sample data) for the arrays declared below.

```
CHARACTER * 8 TEAM(6)
INTEGER RUNS(6), HITS(6), RBI(6)
REAL BA(6)
LOGICAL EASTDV(6)
```

## 5.3   ARRAY SUBSCRIPTS

In the preceding section, we introduced the array subscript as a means of differentiating among the individual elements of an array. We showed that an array element can be referenced by specifying the name of the array followed by a pair of parentheses enclosing a subscript.

FORTRAN allows a special class of arithmetic expressions to be used as the subscript of an array. The compiler can determine the particular array element referenced by evaluating the *subscript expression,* and using the result of this evaluation to indicate the element to be referenced. The rules for the specification and evaluation of array subscripts are summarized next.

---

**Array Subscripts**

*name(subscript)*

**Interpretation:** *Subscript* may be any integer expression whose value is not less than one and not greater than the size listed in the declaration for array *name*. During the execution of a program, the value of *subscript* determines which element of array *name* is referenced.

*Note:* If the value of *subscript* is not within the declared range, then an execution time error will result, and an *out-of-range subscript* message will be printed.

---

**Example 5.4:**   Let ISUB be a memory cell containing the value 3, and let X be
an integer array consisting of ten elements. Then:
X(ISUB) refers to the third element of the array X;
X(4) refers to the fourth element of the array X;
X(2*ISUB) refers to the sixth element of the array X;
X(5*ISUB−6) refers to the ninth element of the array X.

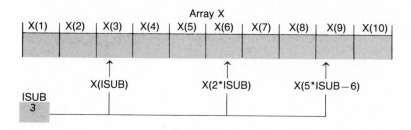

**Example 5.5:**   Let BALANC be an array of bank balances declared as

```
REAL BALANC(12)
```

a) The following program segment can be used to print the 12 values stored
in BALANC in a column (I is an INTEGER variable).

```
PRINT*, 'LIST OF BANK BALANCES'
DO 10 I = 1, 12
     PRINT*, BALANC(I)
10 CONTINUE
```

This loop executes once for each value of the loop variable I, starting with I = 1
and continuing until I = 12. Thus, one of the values BALANC(1), BALANC(2),
. . . , BALANC(11), BALANC(12) will be printed for each execution of the
loop.

b) The following program segment can be used to determine and print the
SUM (type REAL) of the balances in BALANC.

```
SUM = 0.0
DO 30 I = 1, 12
     SUM = SUM + BALANC(I)
30 CONTINUE
PRINT*, 'SUM = ', SUM
```

**Exercise 5.2:**   In Example 5.4, which elements of the array X are referenced if ISUB is
equal to 4 rather than 3?

**Exercise 5.3:**   Let I contain the integer 6 and let X be the array in Example 5.4. Which
of the following references to elements of X are within the range of X?

```
a) X(I)              e) X(4*I-12)
b) X(3*I-20)         f) X(I-2*1)
c) X(4+I)            g) X(30)
d) X(I*3-12)         h) X(I*I-1)
```

**Exercise 5.4:**   Let SCORES be an array declared as

                          INTEGER SCORES(6)

a) Illustrate the contents of SCORES following the execution of the following program segment, for the data on the right.

|                          |  *data:* 100 |
|--------------------------|--------------|
| DO 30 I = 1, 6           |           42 |
|     READ*, SCORES(I)     |           85 |
|     PRINT*, SCORES(I)    |           70 |
| 30 CONTINUE              |           58 |
|                          |           65 |

b) Write a program segment to compute and print the sum and average of the values in SCORES. Show a complete trace of the execution of your program. Your trace should include the values of I and the sum at each step, as well as an indication of which element of the array SCORES is being processed at each point.

c) Write a program segment to compute and print the absolute value of the differences between each element of SCORES and the average. Compute and print the sum of these "absolute differences" and the average absolute differences.

## 5.4   MANIPULATING ARRAY ELEMENTS

Array elements may be manipulated just as other variables are manipulated in FORTRAN statements. In most cases, we can only specify the manipulation of one array element at a time. Thus, for example, each use of an array name in a FORTRAN assignment statement or a condition must be followed by a subscript indicating the particular array element to be manipulated.

It is important to understand the distinction between the array subscript, the value of the subscript (sometimes called an *index* to the array), and the contents of the array element. The subscript is enclosed in parentheses following the array name. Its value is used to select one of the array elements for manipulation. The contents of that array element is either used as an operand or modified as a result of executing a FORTRAN statement.

**Example 5.6:**   Let G be a real array of 10 elements as shown below.

Array G

| G(1)  | G(2) | G(3) | G(4) | G(5) | G(6) | G(7) | G(8) | G(9) | G(10) |
|-------|------|------|------|------|------|------|------|------|-------|
| −11.2 | 12.0 | −6.1 | 4.5  | 8.2  | 1.3  | −.7  | 8.3  | 9.0  | −3.3  |

The following statements can be made about array G:
The contents of the second element (subscript value 2) in the array is 12.0.
The contents of the fourth element (subscript value 4) is 4.5.
The contents of the tenth element (subscript value 10) is −3.3.

Remember, the subscript value is used to select a particular array element, but it does not by itself tell us what is stored in that element.

**Example 5.7:**   The six assignment statements below involve the integer variables

M and N, the real variables X and Y, and the array G, shown in Example 5.6. Assume that $M = 2$, $N = 4$ and $X = 28.5$. Make sure you understand the differences between these statements as described below.

1. Y = REAL (M + N)               Y

                                   6.0

2. Y = G(M+N)                     Y

                                   1.3

3. Y = G(M) + G(N)                Y

                                   16.5

4. M + N = X                      illegal

5. G(M+N) = X                     G(6)

                                   28.5

6. G(M) + G(N) = X                illegal

Statement 1 assigns the value of $M + N$, or 6, to Y. This will be stored as the real number 6.0. Recall that REAL is the name of a function that converts an integer to a real value.

Statement 2 uses the expression $M+N$ as a subscript. The value of G(6), or 1.3, is assigned to Y.

Statement 3 computes the sum of array elements G(2) and G(4). This value, 16.5, is assigned to Y.

Statement 4 is illegal. (Why?)

Statement 5 assigns the value of X to the array element with subscript $M+N$. The value 28.5 is assigned to G(6); the previous value of G(6), or 1.3, is destroyed.

Statement 6 is illegal. (Why?)

**Example 5.8:** Let G be an array of size 10, as drawn in Example 5.6. Then the sequence of instructions

```
J = 1
I = 4
G(10) = 10.0
G(I) = 400.0
G(2*I) = G(I) + G(J)
```

will alter the contents of the 10th, 4th, and 8th elements of G, as shown in Table 5.1; the new array G is shown in Fig. 5.2.

| Statement | Subscript | Value of subscript | Effect |
|---|---|---|---|
| G(10) = 10.0 | 10 | 10 | Store 10.0 in G(10) |
| G(I) = 400.0 | I | 4 | Store 400.0 in G(4) |
| G(2*I) = | I | 4 | Add contents of G(4) and G(1), |
| G(I) + G(J) | J | 1 | 400.0 + (−11.2), |
|  | 2 * I | 8 | store sum (388.8) in G(8) |

**Table 5.1 Manipulating array G.**

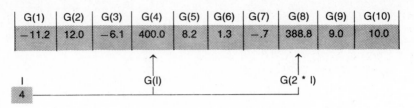

**Fig. 5.2** New array G.

**Example 5.9:**   Let ALPHA be a real array declared as

$$\text{REAL ALPHA(10)}$$

a) The statements

```
DO 30 I = 1, 10
    ALPHA(I) = REAL(10 * I)
30 CONTINUE
```

will cause the values 10.0, 20.0, 30.0, . . . , 100.0 to be placed in succession in the first, second, third, . . . , tenth elements of ALPHA, yielding

Array ALPHA

| (1) | (2) | (3) | (4) | (5) | (6) | (7) | (8) | (9) | (10) |
|------|------|------|------|------|------|------|------|------|------|
| 10.0 | 20.0 | 30.0 | 40.0 | 50.0 | 60.0 | 70.0 | 80.0 | 90.0 | 100.0 |

b) The statements

```
DO 40 I = 1, 10
    ALPHA(I) = 0.0
40 CONTINUE
```

will cause each element of ALPHA to be set to 0.0, in order, from the first (I=1) to the last (I = 10).

**Example 5.10:**   In Example 5.5, we wrote program segments to display and sum twelve bank balances. In the program shown in Fig. 5.3, each depositor's name and bank balance are first read and echo printed (loop 10). Then the sum and average of all bank balances is computed and displayed (loop 20). Finally, the names of all depositors with balances greater than the average are printed (loop 30).

The first loop reads and prints each depositor name and balance, starting with NAME(1) and BALANC(1) and ending with NAME(12) and BALANC(12). The program will read twelve data cards, each containing a character string followed by a real number (e.g., 'SMITH' 355.20). The summation of all balances could also be performed in this loop; however, we preferred to implement this as a separate step (loop 20).

In the third loop (loop 30) each value stored in the array BALANC is compared to the average balance, AVE. The program prints the name, NAME(I), of

```
C SIMPLE BANK BALANCE PROGRAM
C
C PROGRAM PARAMETERS
      INTEGER SIZE
      PARAMETER (SIZE = 12)
C
C VARIABLES
      INTEGER I
      CHARACTER * 10 NAME(SIZE)
      REAL BALANC(SIZE), SUM, AVE
C
C READ AND PRINT ALL DEPOSITOR NAMES AND BALANCES
      PRINT*, 'LIST OF ALL DEPOSITORS'
      PRINT*, '   NAME           BALANCE'
      DO 10 I = 1, SIZE
         READ*, NAME(I), BALANC(I)
         PRINT*, NAME(I), BALANC(I)
   10 CONTINUE
C
C COMPUTE AND PRINT SUM AND AVERAGE OF ALL DEPOSITS
      SUM = 0.0
      DO 20 I = 1, SIZE
         SUM = SUM + BALANC(I)
   20 CONTINUE
      PRINT*, ' '
      PRINT*, 'THE SUM OF BALANCES IS ', SUM
      AVE = SUM / REAL(SIZE)
      PRINT*, 'THE AVERAGE BALANCE IS ', AVE
C
C PRINT NAMES OF ALL DEPOSITORS WITH BALANCES OVER AVERAGE
      PRINT*, ' '
      PRINT*, 'LIST OF DEPOSITORS WITH ACCOUNTS OVER THE AVERAGE'
      DO 30 I = 1, SIZE
         IF (BALANC(I) .GT. AVE) THEN
            PRINT*, NAME(I)
         ENDIF
   30 CONTINUE
C
      STOP
      END
```

**Fig. 5.3** An illustration of simple array manipulations.

each depositor whose balance, BALANC(I), is greater than AVE. A sample run of the program is shown in Fig. 5.4.

It is important to realize that the loop control variable I in Fig. 5.3 determines which array element is manipulated during each loop repetition. The use of the loop control variable as an array subscript is very common since it allows us to easily specify the sequence in which the elements of an array are to be manipulated. Each time the loop control variable is increased, the next array element is automatically selected. Note that the same loop control variable is used in all three loops. This is not necessary, but it is permitted since I is reset to 1 when each loop is entered.

```
LIST OF ALL DEPOSITORS
     NAME              BALANCE
HENRY          6000.000000000
JOHN           3000.000000000
STEVE          1000.000000000
JACK           1234.500000000
THURGOOD       9876.500000000
JUDY           4568.500000000
FRANK          5000.250000000
DAVE           4999.750000000
BOB            8763.250000000
HARRY          3215.680000000
ED             9273.620000000
JOE            4211.110000000

THE SUM OF BALANCES IS    61143.16000000
THE AVERAGE BALANCE IS    5095.263333333

LIST OF DEPOSITORS WITH ACCOUNTS OVER THE AVERAGE
HENRY
THURGOOD
BOB
ED
```

**Fig. 5.4**   Sample run of program in Fig. 5.3.

**Exercise 5.5:**   Given the array G as shown in Fig. 5.2:

a)   What is the contents of G(2)?
b)   If I = 3, what is the contents of G(2*I−1)?
c)   What is the value of the condition

$$(G(I) \ .EQ. \ 8.2)$$

if I is equal to 3; if I is equal to 5?
d)   What will be the value of the logical variable FLAG after the statements below are executed?

```
      FLAG = .FALSE.
      DO 70 INDEX = 1, 10
          IF (G(INDEX) .EQ. 388.8) FLAG = .TRUE.
   70 CONTINUE
```

e)   What will the array G look like after the following loop is executed?

```
      DO 60 IX = 1, 10
          G(IX) = 2.0 * REAL(IX)
   60 CONTINUE
```

f)   If we are given the 4 data items shown on the next page, describe how the array G would be changed by the following statement sequence.

```
      DO 80 INX = 1, 4
          READ*, G(INX)
   80 CONTINUE
```

*data*: 12.0
      18.0
      22.0
    −9.3

## 5.5   INPUT AND OUTPUT OF ARRAYS

### 5.5.1   Reading and Printing Entire Arrays

We stated previously that an array name cannot appear without a subscript in an assignment statement or a condition. However, an array name can appear without a subscript in READ or PRINT statements. If the subscript is not included in the array reference, the compiler assumes that all of the array elements are involved in the input or output operation.

**Example 5.11:**   The statement

```
PRINT*, BALANC
```

will cause all 12 elements of the array BALANC (Fig. 5.3) to be printed on a line of output.

**Example 5.12:**   The following program contains unsubscripted references to the arrays PRES and YEAR.

```
INTEGER YEAR(5)
CHARACTER * 10 PRES(5)
READ*, PRES, YEAR
PRINT*, PRES
PRINT*, YEAR
STOP
END
```

The input list of the READ statement contains the names of two arrays of five elements each. Consequently, ten data items will be read as shown below. Note that all the values for PRES are read first, then all the values for YEAR.

*data*:   'WASHINGTON'   'ADAMS'   'JEFFERSON'   'MADISON'   'MONROE'
        1789         1797       1801        1809       1817

Array PRES

| (1) | (2) | (3) | (4) | (5) |
|---|---|---|---|---|
| WASHINGTON | ADAMS□□□□□ | JEFFERSON□ | MADISON□□□ | MONROE□□□□ |

Array YEAR

| (1) | (2) | (3) | (4) | (5) |
|---|---|---|---|---|
| 1789 | 1797 | 1801 | 1809 | 1817 |

When the name of an array appears without a subscript in a READ statement, as many data cards (or lines) as necessary will be read until all of the array elements have been filled. If you fail to provide sufficient data to fill the entire array, an INSUFFICIENT DATA diagnostic message will occur.

The two PRINT statements shown in Example 5.12 would cause the contents of each array to be printed on a single line:

```
WASHINGTON  ADAMS      JEFFERSON   MADISON    MONROE
     1789         1797       1801         1809       1817
```

The exact spacing of each output line may differ from that shown above, depending upon which FORTRAN compiler you are using.

Although these PRINT statements display all of the information stored in the arrays PRES and YEAR, the form of the printout could be improved upon. It would be more desirable to print the array contents in a table with two columns, one for the name and one for the first year in office, as shown in the following example.

**Example 5.13:** The following program segment displays the contents of the arrays PRES and YEAR (see Example 5.12) in tabular form.

```
      PRINT*, 'NAME            FIRST YEAR IN OFFICE'
      DO 10 IX = 1, 5
         PRINT*, PRES(IX), YEAR(IX)
   10 CONTINUE
```

The output list for the PRINT statement in loop 10 references a pair of array elements with subscript IX. As the value of IX goes from 1 to 5, the contents of these arrays will be printed in two columns, as shown below. Each execution of the PRINT statement causes the next president's data to be printed on a separate line.

```
      NAME              FIRST YEAR IN OFFICE
      WASHINGTON              1789
      ADAMS                   1797
      JEFFERSON               1801
      MADISON                 1809
      MONROE                  1817
```

**Exercise 5.6:** Declare an array PRIME consisting of ten elements. Prepare a single data card (or line) and READ statement for entering the first ten prime numbers into the array PRIME. Provide a statement for printing the contents of the array PRIME.

**Exercise 5.7:** Write a program segment to display the subscript and the contents of each element of the array PRIME in the tabular form on the next page.

| N | PRIME(N) |
|---|----------|
| 1 | 2 |
| 2 | 3 |
| 3 | 5 |
| 4 | 7 |
| . | . |
| . | . |
| . | . |
| 10 | 29 |

### 5.5.2  Reading Part of an Array with an Explicit DO Loop

When an array name appears without a subscript in a READ or PRINT statement, the compiler assumes that the entire array is involved in the indicated operation. In many programs, however, we may want to manipulate only a portion of an array, with the exact number of elements involved determined during each execution of the program. In this case, we should declare the size of the array to be large enough to accommodate the largest expected set of data items. Since most executions of the program will manipulate only part of the array, we will not be able to use the name of the array without a subscript in a READ or PRINT statement. Instead, we must specify that these operations should be performed on only part of the array. One obvious way to do this is through the explicit use of the DO loop as illustrated in the next example.

**Example 5.14:**   Due to classroom space limitations, the maximum size of a class at the New University is 100. The students at the University are given a series of achievement examinations, and we are asked to write a program to perform some statistical computations on the exam scores on a class-by-class basis. We will be given the size of each class and a list of the achievement exam scores for the class.

The program segment shown in Fig. 5.5 can be used to read the input data into computer memory for subsequent processing. The sample output was obtained by echo printing the input data below.

```
data:     7        (number of items)
          95  ⎫
          87  ⎪
          32  ⎪
          72  ⎬  7 data items
          65  ⎪
          81  ⎪
          70  ⎭
```

```
      INTEGER MAXSIZ
      PARAMETER (MAXSIZ = 100)
      INTEGER SCORES(MAXSIZ), N, IRD
C
C READ IN SIZE AND CHECK THAT IT IS WITHIN BOUNDS
      READ*, N
      PRINT*, 'THE CLASS SIZE IS ', N
      IF (N .LT. 1 .OR. N .GT. MAXSIZ) THEN
          PRINT*, 'N IS NOT WITHIN CLASS SIZE BOUNDS OF 1 TO ', MAXSIZ
          PRINT*, 'PROGRAM EXECUTION TERMINATED '
          STOP
      ENDIF
C
C READ AND PRINT SCORES
      PRINT*, ' '
      PRINT*, 'LIST OF SCORES'
      DO 10 IRD = 1, N
          READ*, SCORES(IRD)
          PRINT*, IRD, SCORES(IRD)
   10 CONTINUE
```

**Fig. 5.5**  Reading part of an array with a DO loop.

The DO loop (loop 10) causes the data items to be read into the array SCORES. Only one score may be on each data card (or line) since each execution of the READ statement causes a new data card (or line) to be read. Each exam score will be printed on a separate line of output since each execution of a PRINT statement causes a new output line to be displayed.

---

**Program Style**

*Parameters revisited*

The maximum class size (100 students) for the New University is a good example of an important program parameter. It is used to specify the size of the array SCORES, and to verify that the variable N falls within a meaningful range of values, as specified by the bounds of this array. Program values such as this should always be given a constant name using a PARAMETER statement. If the value of the parameter is ever changed (for example, if a new building with larger lecture rooms is built), then only the PARAMETER statement would require alteration. Neither the related array declarations nor other statements (such as the IF statement in Fig. 5.5) would need to be changed. This can save considerable time, and greatly reduce the potential for error should such a change be required.

> **Program Style**
>
> *Verifying array bounds*
>
>     The input variable N is an essential part of the program segment shown in Fig. 5.5. It is used to specify exactly what portion of the array SCORES is manipulated in loop 10. If the value of N does not fall within a range of values that is meaningful $(1 \leq N \leq 100)$, the DO loop will not execute correctly. It is, therefore, important to print the value of N and to verify that it is within range.

**Exercise 5.8:**   Write program segments (using a DO loop) to:

    a)   print the contents of the first eight elements of the array PRIME.
    b)   print the contents of the middle six elements (the third through the eighth) of PRIME.
    c)   print the contents of the last four elements of PRIME.
    d)   print the contents of the first K elements of PRIME, where K is an integer variable.

**Exercise 5.9:**   Write a loop containing a single READ statement that will enter the first seven prime numbers into the first seven elements of PRIME. How would you prepare your data for processing by the loop? *Hint:* The data card used for Exercise 5.6 will not work here. Why not?

### 5.5.3   Implied DO loop

    When using explicit DO loops for reading and printing, a separate data card (or line) is required for each execution of a read, and a separate line of output is displayed for each execution of a print (see Fig. 5.5). This can be inconvenient and wasteful. For this reason and because read and print operations are often performed on only part of an array, FORTRAN provides a simple shorthand feature, called an *implied DO loop*, which can be used to specify these operations in a concise manner.

**Example 5.15:**   The explicit DO loop shown below (left) may be rewritten using the implied DO also shown below (right).

        *Explicit DO*                            *Implied DO*

```
   DO 10 IRD = 1, N
      READ*, SCORES(IRD)      READ*, (SCORES(IRD), IRD = 1, N)
10 CONTINUE
```

    When an implied DO loop is used in a READ statement, as many data cards (or lines) as needed will be read until all of the array elements indicated by the variable list in the READ statement are filled. You may enter as many items as you wish on each card; you do not have to place each data item on a separate card as is necessary when the explicit DO is used.

**Example 5.16:** If we wished to print the contents of elements 1 through N of the array SCORES, we could use the implied DO feature.

```
PRINT*, (SCORES(IRD), IRD = 1, N)
```

This statement causes as many elements of SCORES as will fit to be printed on each line. (The exact number of items that will fit on a line is a function of the compiler you are using.)

The rules for specifying the loop control parameters (initial and final value) of an implied DO are identical to the rules that we have learned for the explicit DO. The implied loop control also functions in the same manner as the explicit loop. The form of the implied DO is summarized in the following display.

---

**Implied DO**

$$READ*, \ (input\ list,\ lcv = initial,\ final)$$
$$PRINT*, \ (output\ list,\ lcv = initial,\ final)$$

**Interpretation:** The READ (or PRINT) statement is executed. The *input list* (or *output list*) is repeated once for each integer value of *lcv* from *initial* to *final* inclusive. Thus, the statement

```
PRINT*, (A(I), B(I), I = 1, 3)
```

has the same effect as the statement

```
PRINT*, A(1), B(1), A(2), B(2), A(3), B(3)
```

The rules for specifying the implied DO loop parameters are the same as for the explicit DO (see Sec. 3.5).

---

**Example 5.17:** The program segment

```
INTEGER SCORES(10), N, I
READ*, N
PRINT*, 'THE NUMBER OF SCORES IS ', N
PRINT*, ' '
READ*, (SCORES(I), I = 1, N)
PRINT*, 'THE SCORES ARE...'
PRINT*, (SCORES(I), I = 1, N)
```

first reads a value into N and then reads N data items into the array SCORES. The value of N must be on a separate data card (or line); the scores may all be placed on one card or on as many as ten cards. The number of items printed on each line will depend upon the compiler you are using.

For the input data

```
      9
88   100   47
72   69    78   93
81   52
```

the variable N and array SCORES would be defined as shown below.

Array SCORES

| N | (1) | (2) | (3) | (4) | (5) | (6) | (7) | (8) | (9) | (10) |
|---|-----|-----|-----|-----|-----|-----|-----|-----|-----|------|
| 9 | 88 | 100 | 47 | 72 | 69 | 78 | 93 | 81 | 52 | ? |

The output displayed by the PRINT statement is shown next.

```
THE NUMBER OF SCORES IS        9

THE SCORES ARE...
        88        100        47        72        69        78
        93         81        52
```

### 5.5.4  Summary of Array Input Methods

The following example illustrates the three methods of reading data into arrays. For each illustration, the declarations

```
INTEGER SCORES(100), I
CHARACTER * 20 NAMES(100)
```

are assumed.

**Example 5.18:**   a) The statement

```
READ*, NAMES, SCORES
```

reads 100 character strings (student names) followed by 100 integer values (exam scores). It does not matter how many data cards (or lines) are used, but all 100 names must be entered before the 100 integers as illustrated below.

```
data:  'SMITH' 'JONES'  . . .   }
         . . .    'ZISKIN'      }  100 names
       90 100 . . .             }
         . . .    75            }  100 scores
```

b) The statements

```
READ*, N
READ*, (NAMES(I), SCORES(I), I = 1, N)
```

read N pairs of data items, each consisting of a character string followed by an integer value. The value of N must be read first, from a separate card (or line). Again, it does not matter how many data cards are used, so long as the names and scores are punched in pairs, as illustrated next.

```
data:  9
       'SMITH'    90    'JONES'    100 . . .
              9 pairs
```

c) The statements

```
READ*, N
DO 10 I = 1, N
    READ*, NAMES(I), SCORES(I)
10 CONTINUE
```

read a value of N and then read N cards (or lines), each containing a character string followed by an integer value. The second READ statement (inside
loop 10) reads one pair of data items (a name and a score) each time it is executed.

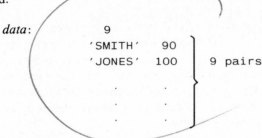

data:
```
              9
        'SMITH'    90
        'JONES'   100      9 pairs
              .     .
              .     .
              .     .
```

**Exercise 5.10:** Describe what happens when the word PRINT is substituted for
READ in Example 5.18.

**Exercise 5.11:**   *For each question, provide a brief justification for your response.*

a)  Could the data line (or card) below be used with the program segment of Ex-
    ample 5.17?

```
9    88    100    47    72    69    78    93    81    52
```

b)  Could the data above be used with the READ statement

```
            READ*, N, SCORES
```

c)  Show how the data would have to be modified if the program segment

```
            READ*, N
            DO 40 I = 1, N
                READ*, SCORES(I)
        40 CONTINUE
```

were used to read it.

**Exercise 5.12:**  In Example 5.17, the value of N should be checked after the first read
statement to verify that it lies between 1 and 10. Write an IF structure to perform the
necessary test on N, and print appropriate diagnostics if N is out of the range 1 to 10.
(The program execution should terminate if N is out of range.) Why is this test so important?

**Exercise 5.13:**  Rewrite the program segment shown in Example 5.17 using a sentinel
value (containing −99) to indicate the end of the data (rather than reading the number of
data items beforehand). *Hint:* Define a loop control variable NEXT to be used for reading
each data item. If the contents of NEXT is not equal to the sentinel value, store NEXT in
the next element of the array SCORES, and repeat. In your program segment, count the
number of items read, and print this count after all reading has been done. Be sure to
place a test in your program to ensure that it does not attempt to read more than ten
items. How would you enter your data to conform to your loop? Why can't you use a DO
loop (either explicit or implicit) to solve the problem?

## 5.5.5　The Grading Problem

The next problem makes use of both implied and explicit DO loops in manipulating array elements.

**Problem 5A:** A number of faculty members at the New University have requested a grading program that can be used to determine letter grades for their classes. The faculty members would like to be able to enter each student's name and exam score and have the program compute and print the class average (AVE), standard deviation (STDEV), and a grade frequency distribution for the number of outstanding, satisfactory, and unsatisfactory scores. For each student, the grade category will be determined as follows:

if SCORE $>$ AVE $+$ STDEV, the GRADE is outstanding
if SCORE $<$ AVE $-$ STDEV, the GRADE is unsatisfactory
otherwise, the GRADE is satisfactory.

As the program determines the grade category for each student, it should print the student name, score, and the category. Finally, the program should print the number of outstanding (OUTCNT), satisfactory (SATCNT), and unsatisfactory (UNSCNT) grades. We will assume that one student record (name and score) is placed on each data card (or line), and that there are no invalid scores—i.e., all scores lie between 0 and 100.

**Discussion:** This problem provides a complete illustration of many of the points made earlier in this section. The initial data table is shown next; the level one flow diagram is given in Figure 5.6a.

### Data Table for Problem 5A

*Program parameter*

MAXSIZ $= 100$, maximum number of scores that can be processed (integer)

| *Input variables* | *Output variables* |
|---|---|
| N: Number of students taking the exam (N $\leq$ MAXSIZ) (integer) | AVE: Average of exam scores (real) |
| NAMES: Array of student names (character * 24, size MAXSIZ) | STDEV: Standard deviation of scores (real) |
| SCORES: Array of exam scores (integer, size MAXSIZ) | OUTCNT ⎫ Counts of number SATCNT ⎬ of outstanding, UNSCNT ⎭ satisfactory and unsatisfactory scores, respectively (integer) |

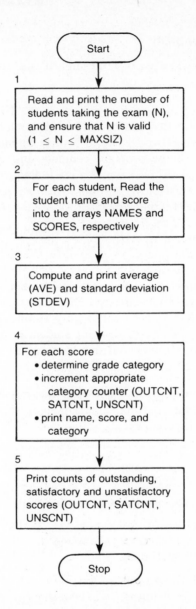

**Fig. 5.6a** Level one flow diagram for grading problem (5A).

We will use the arrays NAMES and SCORES (size MAXSIZ), to store all student data. Since the number of scores to be processed by this program is likely to vary from class-to-class, we will declare these arrays to be large enough to accommodate the largest set of data items that we might anticipate.

The refinements of Steps 1, 2 and 5 are relatively straightforward. Step 5 requires simple PRINT statements. Step 1 involves a decision step: if N is not valid, the program should print a diagnostic message and stop.

We will use an implied DO loop (loop control variable I1) for reading the student names and scores (Step 2). For each loop repetition, a new name/score pair is read into the arrays NAMES and SCORES.

The computation (Step 3) of the average (AVE) and standard deviation (STDEV) involves the formulas

$$AVE = \frac{\sum_{I=1}^{N} SCORES(I)}{N} \qquad STDEV = \sqrt{\frac{\sum_{I=1}^{N} SCORES(I)^2}{N} - AVE^2}$$

where the notation $\sum_{I=1}^{N} SCORES(I)$ means the summation of elements SCORES(1) through SCORES(N): SCORES(1) + SCORES(2) + ... + SCORES(N). These formulas require the prior computation of the sum of the data items (TOTAL $= \sum_{I=1}^{N} SCORES(I)$) and the sum of the squares (TOTSQ $= \sum_{I=1}^{N} SCORES(I)^2$) of these items. For the computation of these sums, we will use a counting loop. The refinement of Step 3 is shown in Fig. 5.6b. The data table entries for TOTAL, TOTSQ and the loop variables (I1, I) are shown below.

**Additional Data Table Entries for Problem 5A**

*Program variables*

TOTAL: Accumulated sum of all
exam scores—used to com-
pute the average (integer)

TOTSQ: Accumulated sum of the
squares of all exam scores
—used to compute the
standard deviation (integer)

I1: Loop control variable for Step
2 refinement—reading the
data (integer)

I: Loop control variable for Step 3
refinement—compute sums
(integer)

In refining Step 4 we first must compute the low and high boundary scores (LOSAT and HISAT) for the category "satisfactory." We can then use these scores to determine the grade category for each student score and increment the appropriate category counter. The refinement of Step 4 also requires a loop (loop variable I3), and is shown in Fig. 5.6c. The additional data table entries for this refinement are given next.

### Additional Data Table Entries for Problem 5A

*Program variables*

LOSAT: Low bound for satisfactory grade category (integer)

HISAT: High bound for satisfactory grade category (integer)

I3: Loop control variable for Step 4 refinement (integer)

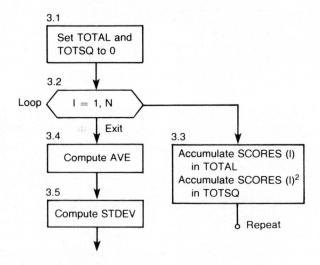

**Fig. 5.6b**   Refinement of Step 3 of Fig. 5.6a.

The program for Problem 5A is shown in Figure 5.7 and the output from a sample execution appears in Figure 5.8.

---

**Program Style**

*Multiple references to array elements*

    There are many references to the elements of array SCORES in Fig. 5.7. The exam scores are first read into the array using an implied loop and next accumulated in the sums TOTAL and TOTSQ in loop 10. Each score is then referenced again and also printed in loop 20. Note that it is not necessary to read the exam data a second time before referencing it in loop 20. Once the data has been stored in the array, it remains there and may be referenced as often as needed.

---

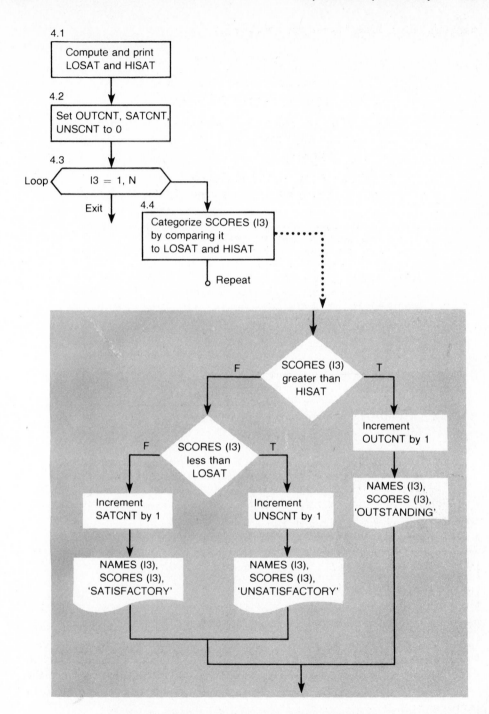

**Fig. 5.6c** Refinement of Step 4 of grading problem (5A).

```
C PROBLEM TO DETERMINE STUDENT GRADE CATEGORIES GIVEN SCORES
C
C PROGRAM PARAMETERS
      INTEGER MAXSIZ
      PARAMETER (MAXSIZ = 100)
C
C VARIABLES
      CHARACTER * 10 NAMES(MAXSIZ)
      INTEGER N, SCORES(MAXSIZ)
      INTEGER TOTAL, TOTSQ, LOSAT, HISAT
      INTEGER OUTCNT, SATCNT, UNSCNT
      INTEGER I1, I, I3
      REAL AVE, STDEV
C
C PRINT HEADING. READ AND PRINT N AND TEST THAT N IS VALID
      PRINT*, 'STUDENT GRADING PROGRAM'
      PRINT*, ' '
      READ*, N
      PRINT*, 'THE CLASS SIZE IS ', N
      IF (N .LT. 1 .OR. N .GT. MAXSIZ) THEN
         PRINT*, 'N IS NOT WITHIN CLASS SIZE BOUNDS OF 1 TO ', MAXSIZ
         PRINT*, 'PROGRAM EXECUTION TERMINATED.'
         STOP
      ENDIF
C
C READ NAME AND SCORE DATA PAIRS
      READ*, (NAMES(I1), SCORES(I1), I1 = 1, N)
C
C ACCUMULATE TOTAL SUM AND SUM OF SQUARES
      TOTAL = 0
      TOTSQ = 0
      DO 10 I = 1, N
         TOTAL = TOTAL + SCORES(I)
         TOTSQ = TOTSQ + SCORES(I) ** 2
   10 CONTINUE
C
C COMPUTE AND PRINT AVERAGE AND STANDARD DEVIATION
      AVE = REAL(TOTAL) / REAL(N)
      PRINT*, 'THE AVERAGE SCORE IS ', AVE
      STDEV = SQRT(REAL(TOTSQ) / REAL(N) - AVE ** 2)
      PRINT*, 'THE STANDARD DEVIATION IS ', STDEV
      PRINT*, ' '
C
C COMPUTE LOW AND HIGH SATISFACTORY SCORE BOUNDARIES
      LOSAT = NINT(AVE - STDEV)
      PRINT*, 'THE LOWEST POSSIBLE SATISFACTORY SCORE IS ', LOSAT
      HISAT = NINT(AVE + STDEV)
      PRINT*, 'THE HIGHEST POSSIBLE SATISFACTORY SCORE IS ', HISAT
      PRINT*, ' '
C
C DETERMINE FREQUENCY COUNTS AND PRINT STUDENT RECORDS
      PRINT*, '    NAME              SCORE    CATEGORY'
      OUTCNT = 0
```

(Continued)

```
      SATCNT = 0
      UNSCNT = 0
      DO 20 I3 = 1, N
         IF (SCORES(I3) .GT. HISAT) THEN
            OUTCNT = OUTCNT + 1
            PRINT*, NAMES(I3), SCORES(I3), ' OUTSTANDING'
         ELSE
            IF (SCORES(I3) .LT. LOSAT) THEN
               UNSCNT = UNSCNT + 1
               PRINT*, NAMES(I3), SCORES(I3), ' UNSATISFACTORY'
            ELSE
               SATCNT = SATCNT + 1
               PRINT*, NAMES(I3), SCORES(I3), ' SATISFACTORY'
            ENDIF
         ENDIF
   20 CONTINUE
C
C PRINT FREQUENCY COUNTS
      PRINT*, ' '
      PRINT*, OUTCNT, ' OUTSTANDING SCORES'
      PRINT*, SATCNT, ' SATISFACTORY SCORES'
      PRINT*, UNSCNT, ' UNSATISFACTORY SCORES'
C
      STOP
      END
```

**Fig. 5.7**  Program for Problem 5A.


```
STUDENT GRADING PROGRAM

THE CLASS SIZE IS                 9
THE AVERAGE SCORE IS   78.44444444444
THE STANDARD DEVIATION IS    17.97597848433

THE LOWEST POSSIBLE SATISFACTORY SCORE IS                60
THE HIGHEST POSSIBLE SATISFACTORY SCORE IS               96

    NAME                SCORE   CATEGORY
FRIEDMAN                   99 OUTSTANDING
STEVERINO                100 OUTSTANDING
SOLOMANSKY                 42 UNSATISFACTORY
OSHEA                      76 SATISFACTORY
BILLIE                     72 SATISFACTORY
STAFFORD                   86 SATISFACTORY
KATHY                      82 SATISFACTORY
SHELLEY                    91 SATISFACTORY
DARA                       58 UNSATISFACTORY

        2 OUTSTANDING SCORES
        5 SATISFACTORY SCORES
        2 UNSATISFACTORY SCORES
```

**Fig. 5.8**  Sample output for grading problem (5A).

---

**Program Style**

*Implied DO's for READ and explicit DO's for PRINT*

In the grading problem, data are entered into the arrays NAMES and SCORES using an implied DO loop. The input list in the statement

```
READ*, (NAMES(I1), SCORES(I1), I1 = 1, N)
```

specifies that the first two data items are to be entered into NAMES(1) and SCORES(1), respectively, the next two data items into NAMES(2) and SCORES(2), etc. Although data for several students could be placed on one card (or line), normally a separate card should be used for each student.

On the other hand, an explicit DO loop is used to print each student's name, score and category designation. This DO loop contains a decision structure that causes one of the messages 'OUTSTANDING', 'SATISFACTORY', or 'UNSATISFACTORY' to be printed on the same line as the student name and score. The explicit DO is required since we wish each student's data displayed on a separate output line. The use of the implied DO for reading and the explicit form for printing, as illustrated here, is very common.

---

The grading problem was solved with the assumption that all of the scores were valid—i.e., that they all fell between 0 and 100 inclusive. In fact, we have no guarantee that this is the case, and should have checked each score as it was read to verify that it was between 0 and 100. The modifications to the problem required by the added validation are discussed in Problem 5R at the end of the chapter.

**Exercise 5.14:** It may be desirable to use an array of counters to keep track of the number of grades in each category where COUNT(1) would represent the count of outstanding scores, COUNT(2), the count of satisfactory scores, etc. Show what modifications would be required to the data table, flow diagram and program for Problem 5A.

**Exercise 5.15:** Rewrite the program in Fig. 5.7 using a pair of sentinel values to indicate the end of student data rather than entering the number of students beforehand. *Hints:* Use a WHILE loop and read the data into two temporary cells. If these temporary cells do not contain the sentinel values, copy them into the next elements of NAMES and SCORES. Count the number of students entered in this fashion.

## 5.6  SEARCHING AN ARRAY

A very common problem in working with arrays of data items is the need to *search* an array to determine whether a particular data item, called a *key,* is in the array. We might also want to know how many times the key is present, and where in the array each copy of the key is located.

**Problem 5B:** Write a program that searches a real array A of size 30 to determine whether a given item, KEY, is present. The program should print the location in array A of each occurrence of KEY.

**Discussion:**   We must first read the array element values and KEY. We will assume enough data items are provided to fill the entire array A. In processing the array A, each element must be examined in sequence and its value compared to KEY. The value of the subscript of each array element that is equal to KEY must be printed. The initial data table for this problem is shown below, and the level one flow diagram is shown in Fig. 5.9a.

**Data Table for Problem 5B**

*Program parameter*

COUNT = 30, maximum number of data items that can be processed (integer)

| *Input variables* | *Output variables* |
|---|---|
| KEY: Item being searched for (real)<br>A: Array being searched (real, size 30) | INDEX: The value of the subscript of each element of A that is equal to KEY (integer) |

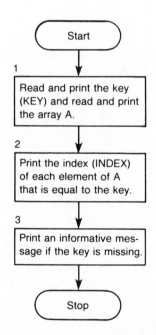

**Fig. 5.9a**   Level one flow diagram for the search problem (5B).

Refinements of Steps 2 and 3 are shown in Fig. 5.9b. The logical variable NOTFND is used as a program flag, and is initially set to .TRUE.. Each array element is compared to KEY (Step 2.3); if a match occurs, NOTFND is set to .FALSE. (Step 2.4), and the value of the array subscript, LV, is stored in IN-DEX and printed (Steps 2.5 and 2.6). After the array is completely searched, the value of NOTFND will still be true only if there were no occurrences of the key in the array A. The additional data table entries follow.

### Additional Data Table Entries for Problem 5A

*Program variables*

NOTFND: Program flag—initially set to .TRUE., changed to
.FALSE. if KEY is found
(logical)

LV: Loop control variable and array subscript (integer)

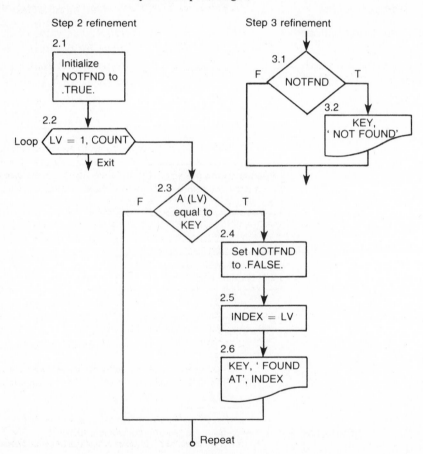

**Fig. 5.9b**  Level two refinements for Problem 5B.

The search program is provided in Fig. 5.10.

```
C SEARCH A REAL ARRAY (A) FOR A DATA ITEM (KEY)
C
C PROGRAM PARAMETERS
      INTEGER COUNT
      PARAMETER (COUNT = 30)
C
C VARIABLES
      INTEGER LV, INDEX
      REAL A(COUNT), KEY
      LOGICAL NOTFND
C
C READ DATA ITEMS
      READ*, KEY
      PRINT*, 'THE ITEM TO BE FOUND IS ', KEY
      READ*, A
      PRINT*, ' '
      PRINT*, 'LIST OF ITEMS BEING SEARCHED... '
      PRINT*, A
      PRINT*, ' '
C
C EXAMINE EACH ELEMENT OF ARRAY. IF IT MATCHES KEY, PRINT SUBSCRIPT
      NOTFND = .TRUE.
      DO 20 LV = 1, COUNT
         IF (A(LV) .EQ. KEY) THEN
            NOTFND = .FALSE.
            INDEX = LV
            PRINT*, KEY, ' FOUND IN ELEMENT ', INDEX, ' OF A'
         ENDIF
   20 CONTINUE
C
C PRINT MESSAGE IF KEY IS NOT FOUND
      IF (NOTFND) PRINT*, KEY, ' NOT FOUND IN A'
C
      STOP
      END
```

**Fig. 5.10**  Program for Problem 5B.

In this program, the entire array A is entered via a single READ statement

```
READ*, A
```

Once the array has been entered, any element value can be manipulated, as shown in decision Step 2.3, simply by using a subscripted reference to A.

**Exercise 5.16:**  Explain what is wrong with the proposed refinement of Step 2 shown in Fig. 5.11.

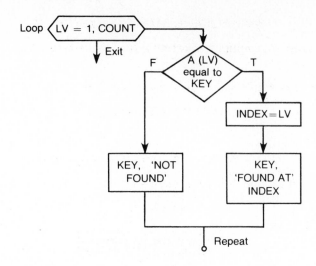

**Fig. 5.11**   Proposed refinement of Step 2 of Fig. 5.9a.

**Exercise 5.17:**   Modify the refinement for Step 2 in the search flow diagram so that only the first occurrence of KEY would be found (after which the loop would be exited). Is the program flag still needed?

**Exercise 5.18:**   Redo Problem 5B assuming that the number of elements N in A is less than 30. Treat N as an input variable. *Hint:* Your program should check that N is between 1 and COUNT inclusive; you will need an implied loop to read the N items into A. (Why?) Redo the level one flow diagram, add refinements as needed, and rewrite the program.

## 5.7   INITIALIZING ARRAYS USING THE DATA STATEMENT

Until now, we have always used a READ statement to enter data into an array. However, there are many situations in which we may wish to initialize an array to a predetermined list of values prior to program execution. In this section, we shall introduce the DATA statement as a convenient way of specifying this initialization.

**Example 5.19:**   The statement

```
DATA FIB(1), FIB(2) / 1, 1/
```

initializes the first two elements of array FIB to one. Since both constants above in this example are identical, the DATA statement could be written as

```
        DATA FIB(1), FIB(2) / 2*1 /
```

where the notation 2*1 implies two repetitions of the constant 1.


**Example 5.20:**   Let MINCAT be an array declared as

```
            INTEGER MINCAT(5)
```

Then the statement

```
        DATA MINCAT / 0, 50, 65, 80, 90 /
```

can be used to initialize MINCAT as shown below.

Array MINCAT

| (1) | (2) | (3) | (4) | (5) |
|-----|-----|-----|-----|-----|
| 0   | 50  | 65  | 80  | 90  |

The occurrence of an array name without a subscript in a DATA statement causes the entire array to be initialized. The first array element is paired with the first constant, the second array element with the second constant, etc. The same initialization could be performed using the statements

```
        MINCAT(1)  =   0
        MINCAT(2)  =  50
        MINCAT(3)  =  65
        MINCAT(4)  =  80
        MINCAT(5)  =  90
```


**Example 5.21:**   Let ZEROES be an array declared as

```
            REAL ZEROES(5)
```

Then the statement

```
        DATA ZEROES / 0.0, 0.0, 0.0, 0.0, 0.0 /
```

or

```
        DATA ZEROES / 5*0.0 /
```

can be used to initialize all of the elements of the array ZEROES to 0.0. The same initialization could be performed using a loop.

```
        DO 70 I = 1, 5
            ZEROES(I) = 0.0
    70 CONTINUE
```

The general form of the DATA statement is described in the following display.

---

### The DATA Statement

DATA *variable list* / *constant list* /

**Interpretation:** The *variable list* may contain combinations of the following:

- variable names
- array element names (array names with subscripts)*
- array names (for initializing entire arrays)
- implied loops (for initializing portions of arrays)**

The *constant list* provides the corresponding initial values for the items in the *variable list*. There may be numeric, character, or logical constants. The term constant includes named constants previously declared in a PARAMETER statement.

The type of each constant and its corresponding variable should be the same. The correspondence is established in left-to-right order. There should be as many constants as there are memory cells being initialized. The notation n*c can be used to specify n repetitions of the constant c. (n must be a positive integer constant or parameter and c must be a constant.)

Data statements must be placed after the type declarations.

*Note:* The subscripts must be either integer constants or parameters, or expressions involving integer constants and parameters.

**Note:* The loop parameters should be integer constants or parameters, or expressions involving integer constants and parameters.

---

**Example 5.22:**   Let LETTER be an array declared as

```
CHARACTER * 1 LETTER(5)
```

Any of the DATA statements

```
DATA LETTER / 'A', 'B', 'C', 'D', 'F' /

DATA (LETTER(I), I = 1, 5) / 'A', 'B', 'C', 'D', 'F' /

DATA LETTER(1), LETTER(2), LETTER(3), LETTER(4), LETTER(5)
Z / 'A', 'B', 'C', 'D', 'F' /
```

may be used to initialize the array LETTER as shown below.

Array LETTER

| (1) | (2) | (3) | (4) | (5) |
|-----|-----|-----|-----|-----|
| A   | B   | C   | D   | F   |

The statements

```
INTEGER FIRST, LAST
PARAMETER (FIRST = 1, LAST = 3)
DATA (LETTER(I), I = FIRST, LAST) / 'A', 'B', 'C' /
```

could be used to initialize the first three elements of LETTER to 'A', 'B' and 'C', respectively.

**Example 5.23:**   The statements

```
INTEGER AGE, SSNO(3)
REAL ERNING(8)
CHARACTER * 10 FIRST, LAST
DATA FIRST, LAST, SSNO, AGE, ERNING
Z / 'FRANK', 'FRIEDMAN', 219, 40, 0677, 38, 8*0.0 /
```

can be used for initialization. You should note carefully the one-to-one correspondence between the list of items and the list of constants. There are fourteen constants specified, including eight repetitions of 0.0. There are three simple variables and two arrays (SSNO and ERNING) in the variable list for a total of fourteen individual memory cells. The DATA statement has the same effect as the statements

```
          FIRST = 'FRANK'
          LAST = 'FRIEDMAN'
          SSNO(1) = 219
          SSNO(2) = 40
          SSNO(3) = 0677
          AGE = 38
          DO 50 I = 1, 8
              ERNING(I) = 0.0
       50 CONTINUE
```

DATA statements are not executable statements, and hence are not translated into machine language.

---

**Program Style**

*Guidelines for using DATA statements*

As shown in the preceding examples, an array may be initialized using a DATA statement or one or more assignment statements. Normally, the DATA statement is used when the array values are constant and will not be changed as the program executes (although this is not a requirement as it is for program parameters). For example, a table giving the number of days in each month (array DAYMON(12)) could be defined most easily using the statement

```
    DATA DAYMON / 31,28,31,30,31,30,31,31,30,31,30,31 /
```

The logical IF statement

```
        IF (MOD(YEAR, 4) .EQ. 0) DAYMON(2) = 29
```

could be used to modify the second array element for a leap year; all other element values would remain fixed.

*(Continued)*

> If the array element values are likely to change, then the initialization should be performed through one or more assignment statements. For example, an array of counters would normally be initialized using a DO loop of the form
>
> ```
>          DO 10 I = 1, N
>             COUNTR(I) = 0
>       10 CONTINUE
> ```
>
> rather than using a DATA statement. Subsequent statements of the form
>
> ```
>       COUNTR(I) = COUNTR(I) + 1
> ```
>
> would be used to update the array element values.

**Exercise 5.19:**  Each of the sequences written below contains one or more errors. Identify and correct each error. There may be more than one possible correction for each mistake.

a)
```
   REAL X
   INTEGER I
   LOGICAL Y
   CHARACTER * 4 NAME
   DATA X, I, Y, NAME / 1.0, -62.5, '.FALSE.', 'AL' /
```

b)
```
   REAL X, Y, Z, W
   INTEGER A
   DATA X, Y, Z, W, A / 5*0.0, 347 /
```

c)
```
   LOGICAL FLAG
   INTEGER SSNO1, SSNO2, SSNO3
   CHARACTER * 12 NAME1, NAME2
   DATA FLAG /1.0/
   DATA SSNO1, SSNO2, SSNO3 /219, 40, 0677/
   DATA NAME1, NAME2 / 'HERB', .TRUE. /
```

**Exercise 5.20:**  Write the declarations and DATA statements needed to carry out the following initialization of memory.

| A | B | C | D | BOO(1) | BOO(2) | BOO(3) | SWITCH |
|---|---|---|---|--------|--------|--------|--------|
| 1.0 | 1.0 | -1.0 | GIN | RUM | BEER | GIN | .TRUE. |

**Exercise 5.21:**  Let PRIME be an array declared as

```
   INTEGER PRIME(10)
```

   a)  Write a program segment to read the first 10 prime numbers into the array PRIME. Show your data.

  b)  Write a DATA statement to store the first 10 primes into the
      array PRIME.
  c)  Write a program segment to read only the first 6 primes into the
      array PRIME. Show your data.
  d)  Write a DATA statement to store only the first 6 primes into the array
      PRIME.

## 5.8  PROCESSING SELECTED ARRAY ELEMENTS

In the examples seen so far, the loop control variable of a DO loop often
serves as an array subscript as well. This technique permits us to easily reference
the elements of an array in sequential order.

There are many programming problems in which only selected elements of
an array are to be processed. In these problems, the selection process involves the
determination of an index to a specific array element. Usually, the values of one
or more input data items are involved in the determination of the index.

There are two important techniques for determining the index of the desired
array element:

• through direct computation using a formula
• by searching an array

We will illustrate the use of both techniques in the following problem.

**Problem 5C:**   The IRS has provided us with a tax table (see Table 5.2) that can
be used to determine the tax amount for all salaries up to $14,000. We wish to
use this table to write a program that computes the tax owed by all graduate students in the university.

**Discussion:**   Each line of the tax table represents a different tax bracket. For example, all salaries between $1500 and $1999.99 are in tax bracket four. The table
shows that the minimum tax for this bracket is $225. In addition, an extra tax
amount is owed if the salary is greater than the base salary, $1500. This tax is
equal to 17% of the excess salary over the base salary (e.g. for a salary of $1750,
the extra tax owed is $0.17 \times \$250$ or $42.50).

| Tax bracket | Base salary | Tax due on base | Percentage for excess over base |
|:---:|:---:|:---:|:---:|
| 1 | 0 | 0 | 14% |
| 2 | $ 500 | $ 70 | 15% |
| 3 | $ 1000 | $ 145 | 16% |
| 4 | $ 1500 | $ 225 | 17% |
| 5 | $ 2000 | $ 310 | 19% |
| 6 | $ 4000 | $ 690 | 21% |
| 7 | $ 6000 | $ 1110 | 24% |
| 8 | $ 8000 | $ 1590 | 25% |
| 9 | $ 10000 | $ 2090 | 27% |
| 10 | $ 12000 | $ 2630 | 29% |

**Table 5.2    IRS tax table for salaries under $14,000**

We shall use three *parallel arrays* to store the information in Table 5.2. As shown below, each column of the tax table is stored in a separate array; the relevant data for line *i* of the table are stored in element *i* of the arrays TAXTAB, BASTAX and PERCNT.

| Base salary | | Base tax due | | Tax percentage for excess amount earned | |
|---|---|---|---|---|---|
| | TAXTAB | | BASTAX | | PERCNT |
| 1 | .00 | 1 | .00 | 1 | 0.14 |
| 2 | 500.00 | 2 | 70.00 | 2 | 0.15 |
| 3 | 1000.00 | 3 | 145.00 | 3 | 0.16 |
| 4 | 1500.00 | 4 | 225.00 | 4 | 0.17 |
| 5 | 2000.00 | 5 | 310.00 | 5 | 0.19 |
| 6 | 4000.00 | 6 | 690.00 | 6 | 0.21 |
| 7 | 6000.00 | 7 | 1110.00 | 7 | 0.24 |
| 8 | 8000.00 | 8 | 1590.00 | 8 | 0.25 |
| 9 | 10000.00 | 9 | 2090.00 | 9 | 0.27 |
| 10 | 12000.00 | 10 | 2630.00 | 10 | 0.29 |
| 11 | 14000.00 | | | | |

Since the tax table changes infrequently, DATA statements should be used to initialize these arrays. The reason for the extra element in TAXTAB will be explained later. The data table for the problem follows; the flow diagrams are shown in Fig. 5.12a.

**Data Table for Problem 5C**

*Program parameter*

SENVAL = 0.0, sentinel value for SALARY

| *Input variables* | *Program variables* | *Output variables* |
|---|---|---|
| SALARY: The salary earned by each student (real) | TAXTAB: The array of base salaries (real, size 11) | TAX: The tax owed by each student (real) |
| | BASTAX: The array of base taxes (real, size 10) | |
| | PERCNT: The array of tax percentages (real, size 10) | |

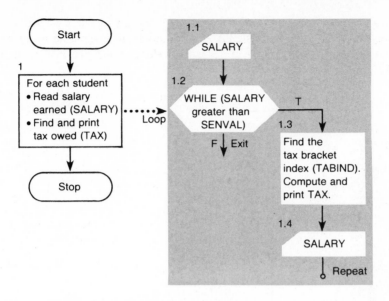

**Fig. 5.12a**   Flow diagrams for tax computation problem (5C).

In performing Step 1.3, the program must first determine whether or not SALARY is within the range of the tax table. If so, the tax owed (TAX) can be computed by finding the tax bracket (TABIND) for the current salary and then using the tax table data appropriate for that bracket. The new data table entry follows; the refinement of Step 1.3 is shown in Fig. 5.12b.

**Additional Data Table Entries for Problem 5C**

*Program variables*

TABIND: The tax bracket
   for SALARY
   (integer)

I : Loop control variable
   and subscript (integer)

DO loop 13 in Fig. 5.12b must find the tax bracket (a value from 1 to 10) for SALARY. It does this by searching for the pair of elements in array TAXTAB that "bracket" SALARY. The condition

```
(TAXTAB(I) .LE. SALARY .AND. SALARY .LT. TAXTAB(I+1))
```

is true only when SALARY falls between the adjacent array elements TAXTAB (I) and TAXTAB (I+1). The subscript, I, of the smaller element corresponds to the tax bracket index, TABIND. As shown below, the value of TABIND should be set to four for a SALARY of $1750.

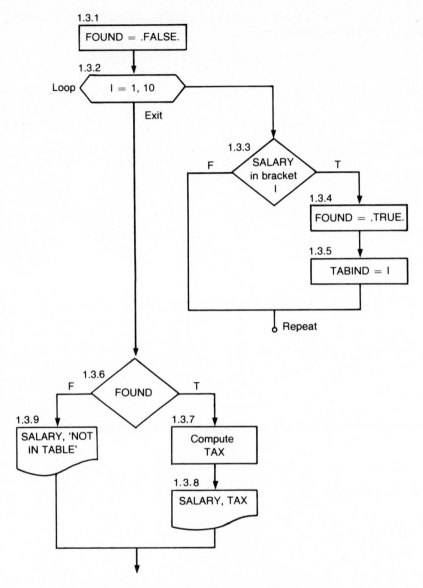

1.3.1
FOUND = .FALSE.

1.3.2
Loop ⟨ I = 1, 10 ⟩

Exit

1.3.3
SALARY
in bracket
I

F          T

1.3.4
FOUND = .TRUE.

1.3.5
TABIND = I

Repeat

1.3.6
FOUND

F          T

1.3.9
SALARY, 'NOT
IN TABLE'

1.3.7
Compute
TAX

1.3.8
SALARY, TAX

**Fig. 5.12b**   Refinement of Step 1.3 of Fig. 5.12a.

Array TAXTAB

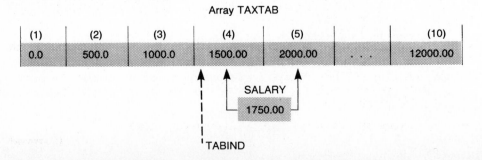

| (1) | (2) | (3) | (4) | (5) | | (10) |
|-----|-----|-----|-----|-----|-----|------|
| 0.0 | 500.0 | 1000.0 | 1500.00 | 2000.00 | . . . | 12000.00 |

SALARY

1750.00

TABIND

Once **TABIND** is defined, it can be used to compute the tax amount owed using the assignment statement

```
TAX = BASTAX(TABIND) + PERCNT(TABIND) *
Z        (SALARY - TAXTAB(TABIND))
```

In the expression above, **TABIND** is used to select the tax computation data from **TAXTAB** and the two arrays **BASTAX** and **PERCNT** defined in parallel with **TAXTAB**. This computation is performed below for a **SALARY** of $1750.

```
TAX = BASTAX(4) + PERCNT(4) * (1750.00-TAXTAB(4))
    = 225.00    + 0.17       * (1750.00-1500.00)
    = 225.00    + 42.50
    = 267.50
```

The tax computation program is given in Fig. 5.13.

```
C COMPUTE THE TAX OWED ON EACH SALARY - USE TAX TABLE
C
C PROGRAM PARAMETERS
      REAL SENVAL
      INTEGER TBLSIZ
      PARAMETER (SENVAL = 0.0, TBLSIZ = 10)
C
C VARIABLES
      INTEGER TABIND, I
      CHARACTER * 10 NAME
      REAL SALARY, TAX
      LOGICAL FOUND
C
C DEFINE TAX TABLE
      REAL TAXTAB(TBLSIZ+1), BASTAX(TBLSIZ), PERCNT(TBLSIZ)
      DATA TAXTAB /0.0, 500.0, 1000.0, 1500.0, 2000.0, 4000.0,
     Z             6000.0, 8000.0, 10000.0, 12000.0, 14000.0/
      DATA BASTAX /0.0, 70.0, 145.0, 225.0, 310.0, 690.0,
     Z             1110.0, 1590.0, 2090.0, 2630.0/
      DATA PERCNT /0.14, 0.15, 0.16, 0.17, 0.19, 0.21, 0.24,
     Z             0.25, 0.27, 0.29/
C
C FOR EACH TAXPAYER, READ NAME AND SALARY. COMPUTE TAX
      PRINT*, 'TAX SUMMARY'
      PRINT*, ' '
      PRINT*, '   NAME          SALARY            TAX'
      READ*, NAME, SALARY
      WHILE (SALARY .NE. SENVAL) DO        12 IF (SALARY .EQ. SENVAL)
C         SEARCH TABLE FOR BRACKET INDEX     Z GO TO 19
         FOUND = .FALSE.
         DO 13 I = 1, 10
            IF (SALARY.GE.TAXTAB(I) .AND. SALARY.LT.TAXTAB(I+1))THEN
               FOUND = .TRUE.
               TABIND = I
            ENDIF
   13    CONTINUE
C
C        COMPUTE AND PRINT TAX IF TABIND WAS FOUND
```

*(Continued)*

```
          IF (FOUND) THEN
             TAX = BASTAX(TABIND) + PERCNT(TABIND)
   Z               * (SALARY - TAXTAB(TABIND))
             PRINT*, NAME, SALARY, TAX
          ELSE
             PRINT*, 'FOR ', NAME, ', ', SALARY, ' IS NOT IN TAX TABLE'
          ENDIF
C         READ NEXT SALARY
          READ*, NAME, SALARY
       ENDWHILE                                GO TO 12
C                                            19 CONTINUE
       STOP
       END
```

*** PROGRAM EXECUTION OUTPUT ***

TAX SUMMARY

| NAME | SALARY | TAX |
|------|--------|-----|
| PETER | 463.0000000000 | 64.82000000000 |
| BILL | 8192.000000000 | 1638.000000000 |
| FOR ELLIOT | , 150000.0000000 | IS NOT IN TAX TABLE |
| FOR FRANK | , 14000.00000000 | IS NOT IN TAX TABLE |

**Fig. 5.13**  Program for tax computation problem (5C).

---

**Program Style**

*Testing subscripts at loop boundary values*

The program shown in Fig. 5.13 contains a potential subscript error. When loop 13 is executed for I = 10, the value of the subscript I+1 of TAXTAB will be eleven. In checking to ensure that a loop will function as desired, it is essential to focus upon the boundary values (1 and 10 in this case) for the loop control variable. For each array referenced in the loop, you should verify that the values of all subscript expressions are in range and are correct. If they are correct at the boundaries, they will most likely be correct for all values of the loop control variable.

If such a check indicates an error, then a change in the algorithm may be required. To avoid a subscript error, we added an extra element to the array TAXTAB (TAXTAB (11) = 14000.) Alternatively, we could have handled the case of a salary in excess of $12,000 separately, prior to entry to the loop, and rewritten the DO loop header as DO 13 I = 1, 9 rather than DO 13 I = 1, 10. The addition of the extra array element is the preferred solution because it is so simple to do.

---

An alternate approach to selecting the tax bracket is to directly compute the index as shown in Fig. 5.14. Direct computation is more convenient and efficient than an array search if there is a constant increment between table items.

```
IF (SALARY .LT. 0.0 .OR. SALARY .GE. TAXTAB (11)) THEN
   PRINT*, SALARY,' IS OUT OF RANGE AND IS IGNORED.'
ELSE
   IF (SALARY .LT. 2000.0) THEN
      TABIND = INT(SALARY/500.0) + 1
   ELSE
      TABIND = INT(SALARY/2000.0) + 4
   ENDIF
ENDIF
```

**Fig. 5.14**   Direct computation of tax bracket index TABIND.

In Fig. 5.14, the statement

$$\text{TABIND = INT(SALARY/500.0) + 1}$$

correctly computes the index of the tax bracket for all values of SALARY less than \$2000 (brackets one through four). There is a constant increment of \$500 for the tax table entries in this range. The statement

$$\text{TABIND = INT(SALARY/2000.0) + 4}$$

is used for all larger salaries (tax increment of \$2000).

For a salary of \$1750, the computation would proceed as follows:

$$\text{TABIND = INT(1750.0/500.0) + 1}$$
$$\text{= INT(3.5) + 1}$$
$$\text{= 3 + 1 = 4}$$

The value of TABIND should always be checked to verify that it is in range before it is used to select array elements.

**Exercise 5.22:**   Verify by hand simulation that the above formulas for TABIND are correct. Test the following values of SALARY:

$250.0, $1275.0, $2750.0, $4000.0, $11700.0, $23000.0

**Exercise 5.23:**   An exam with the grade ranges

A: 90 − 100
B: 80 − 89
C: 70 − 79
D: 60 − 69
F:  0 − 59

is given to a math class. Write a program segment that examines a student score (SCORE) and prints the student grade. Use both the array search and direct computation techniques. *Hint:* Store the lower grade boundaries in an array LOWGRD and the letter grades in an array LETTER. For the direct computation technique, process the grades 100 and 0 through 49 separately.

**Exercise 5.24:**   In Fig. 5.13 it is really not necessary to continue testing elements of TAXTAB once the bracket is found. Rewrite this code segment using a WHILE loop with a program flag that controls its repetition. The flag should be reset to cause loop exit once TABIND is defined.

## 5.9   THE GENERAL FORM OF THE ARRAY DECLARATION

In the previous sections of this chapter, we have assumed that array declarations of the form

type name(size)

designated an array whose size was given by the value of the constant *size*, and whose elements could be designated as *name*(1), *name*(2), . . . , *name*(*size*).

Thus, for example, the declaration

<code>INTEGER SCORE(6)</code>

specified an array SCORE of size 6 with elements SCORE(1), SCORE(2), . . . , SCORE(6). It was assumed that the first element of the array always had a subscript value of 1. Although this type of array is sufficient for most problems, there are some problems (mainly in statistics and numerical analysis) which can be solved more conveniently if this restriction is removed. For this reason, FORTRAN 77 permits the user to specify a lower bound other than one for an array. In this section we illustrate the declaration syntax for the general form of the array, and provide an example of its use.

---

**Array Declaration (General Form)**

type name(lowbound : highbound)

**Interpretation:** The range of the array *name* is given by the pair of integers *lowbound* and *highbound*. This means that the smallest legal subscript value is *lowbound* and the largest is *highbound*.

*Note:*  1) The individual array elements are designated as *name*(*lowbound*), *name*(*lowbound*+1), . . . , *name*(*highbound*)

2) The number of elements associated with the array *name* is given by the formula

highbound − lowbound + 1

3) Both *lowbound* and *highbound* must be integer constants or parameters, and *lowbound* may not exceed *highbound*.

---

**Example 5.24:**   The declaration

<code>REAL MIDVAL(-3: 4)</code>

defines an array with elements designated as MIDVAL(−3), MIDVAL(−2), MIDVAL(−1), MIDVAL(0), . . . , MIDVAL(4). In this declaration, lowbound = −3, and highbound = 4; hence, there are eight (4 − (−3) + 1) elements in the array.

The program segment

```
      DO 40  I = -3, 4
         MIDVAL(I) = REAL(I) / 2.0
   40 CONTINUE
```

would define values for **MIDVAL** as illustrated below.

Array MIDVAL

| (−3) | (−2) | (−1) | (0) | (1) | (2) | (3) | (4) |
|------|------|------|-----|-----|-----|-----|-----|
| −1.5 | −1.0 | −0.5 | 0.0 | 0.5 | 1.0 | 1.5 | 2.0 |

**Exercise 5.25:** Identify which of the following declarations are legal. For each legal declaration, indicate the size of the array and its subscript range.

a) INTEGER AGE(6)

b) REAL ORD(-10: 10)

c) CHARACTER DIGIT(9: 0)

d) LOGICAL FOUND(0: -100)

e) INTEGER DIGIT(0: 9)

f) LOGICAL FLAG(10: 20)

**Exercise 5.26:** Given the array declarations for Exercise 5.25, which of the following subscripted references are legal?

a) for I = 2

AGE(2*I-1)

AGE(4*I)

b) for I = 5

ORD(2*I-1)

ORD(2*I+10)

c) for K = 3

DIGIT(3*K-2)

DIGIT(4*K-2)

d) for INDEX = 2

FLAG(5*INDEX)

FLAG(10*INDEX)

**Exercise 5.27:** Produce a table of sines and cosines for angles between −180° and +180° in steps of 10°. Save the results in two arrays.

## 5.10   ROLE OF THE COMPILER IN PROCESSING ARRAYS*

We have now had a little experience in manipulating arrays of data, and should have a better understanding of the use of the FORTRAN subscript. In this section, we will briefly describe one way in which a compiler might convert a subscripted array reference into a memory-cell address.

The compiler recognizes that it may be dealing with an array reference when it encounters a symbolic name followed by a left parenthesis. The only other symbolic names that may be followed by a left parenthesis are function names. The compiler can distinguish between function and array references, since the names of all arrays must be identified at the beginning of a program through appropriate array-declaration statements.

When the compiler processes an array declaration, it associates a block of adjacent memory cells with the array name, and reserves these cells to be identi-

---

*This section may be omitted.

fied with that name. The compiler keeps track of the array size, the value of *lowbound* used in the array declaration and the address of the first cell of the block area associated with the array name. When a reference is made to a particular element of an array, the compiler uses the address of the first element of the array as the *base address* for computing the address of the indicated element. The array size may be used to check an array reference to ensure that the subscript value *falls within the range* of the array.

The calculation of the address of a memory cell indicated by a subscripted array reference is quite straightforward:

Array element address = (*base address*) + subscript value − *lowbound*

If *lowbound* is not specified, it is assumed to be one.

**Example 5.25:** Let A be a real array of size 16 (*lowbound* = 1) with a base address of 3706. Then the address of the tenth element in A, A(10), is 3706 + 10 −1 or 3715. The address of A(1) is 3706, and the address of A(16) is 3721.

## 5.11  COMMON PROGRAMMING ERRORS

There are two very common programming errors associated with arrays. One involves the failure to declare a name that is to be used to represent an array, and the other involves the use of subscripts with values exceeding the size of the array being referenced.

### 5.11.1  Failure to Declare an Array

The use of a subscript reference with a symbolic name that has not been declared as an array will cause the compiler to treat the reference as a *function call* rather than an array reference. (Only function and array names may be followed by a left parenthesis.) If this reference occurs at a point where a function call is illegal, then a syntax error will occur. Otherwise, no diagnostic will be given, and the error will not be detected until execution begins and the erroneous function call is encountered. For example, consider the two statements

```
GRADE(I) = 0
GRDSUM = GRDSUM + GRADE(I)
```

If GRADE is not declared as an array, then the reference in the first statement will be treated as a call to the function GRADE. Since function calls are illegal in the lefthand side of an assignment statement, the FORTRAN compiler will provide a diagnostic for the statement GRADE(I) = 0. This diagnostic will inform you that you have illegally used a function reference on the left side of an assignment statement. However, the function reference in the second statement is perfectly legal, and no diagnostic will be provided by the compiler. The error will be detected only when the "function" GRADE cannot be found.

If an array name is misspelled, the compiler will not be able to find its declaration. When illegal array references occur, the spelling of the array name involved should be carefully compared with the array declaration.

### 5.11.2   Out-of-Range Subscript Values

Out-of-range subscript values (subscripts that are less than the lowbound for an array or exceed the highbound) are often caused by errors in index computation (see Problem 5C), or by loops that do not terminate properly. These are not syntax errors and cannot be diagnosed during compilation. If they are not detected during program execution, such errors can result in the destruction of the contents of memory cells that are adjacent to the array whose subscript reference is in error.

For this reason most compilers provide subscript checking during program execution. If a subscript error occurs, the compiler will print a message indicating the line number of the program statement at which the error occurred, and the value of the subscript. For example, the message

```
SUBSCRIPT RANGE ERROR AT LINE NO. 28 FOR ARRAY BALANC, IX=0
```

indicates that the subscript IX in the array reference BALANC(IX) (occurring on line 28 of the program) has value 0. When such errors occur, the statements used to define the value of IX must be corrected in order to produce the proper in-range value.

### 5.12   SUMMARY

In this chapter we introduced a special *data structure* called an *array*, which is a convenient facility for naming and referencing a collection of like items. We discussed how to inform the compiler that an array of elements is to be allocated (by using the type declaration statement), and we described how to reference an individual array element by writing a parenthesized expression (called a subscript) following the array name, for example, GRADES(I). A summary of statements that manipulate arrays is provided in Table 5.3.

Two common programming techniques used to reference array elements were described. The first of these involves referencing each element in a predetermined sequence. The DO loop was shown to be a convenient structure for implementing such sequential references. We have used the explicit DO loop to read and print arrays, and to control the sequential processing (such as a search) of an array. We have used the implied form of the DO loop to read, print and to initialize (via the DATA statement) an array. These uses of the DO loop are summarized in Table 5.3.

| Statements | Meaning |
|---|---|
| *Array Declaration*<br>`REAL X(5), Y(0 : 5)` | Used to allocate storage for 5 integer data items in array X (X(1), . . . , X(5)) and six integer data items in array Y (Y(0), . . . , Y(5)). |
| *Array Initialization*<br>`DATA (X(I), I = 1, 5)`<br>`Z / 5*0 /`<br>`DATA X / 5*0 /` | Used to initialize the array X.<br>All 5 elements are initialized to 0 |
| *Array Read and Print*<br>`DO 30 I = 1, 5`<br>`   READ*, X(I)`<br>`   PRINT*, X(I)`<br>`30 CONTINUE` | Explicit loop—reads and prints one item at a time. Each item must be on a separate line (or card); each item is printed on a separate line. Can be used to read or print all or a portion of an array. |
| `READ*, (X(I), I = 1, 5)`<br>`PRINT*, (X(I), I = 1, 5)` | Implied loop—for reading and printing all or a portion of an array. Up to five data lines (or cards) may be read. As many items as will fit will be printed on a line. |
| `READ*, X`<br>`PRINT*, X` | For reading and printing an entire array. Data entry and printing rules are the same as those for an implied loop. |
| *Array Manipulation*<br>`R = X(5) - X(1)` | Assign the difference of X(5) and X(1) to R. |
| *Array Assignment*<br>`X(I) = X(I-2) + X(I-1)` | Assign to X(I) the sum of the two preceeding array element values. |
| *Array Search*<br>`DO 40 I = 1, N`<br>`   IF (X(I) .EQ. KEY) THEN`<br>`      INDEX = I`<br>`      PRINT*, 'INDEX OF KEY IS ', INDEX`<br>`   ENDIF`<br>`40 CONTINUE` | Search for KEY in X; print index of all elements of X that equal KEY. |

**Table 5.3 Summary of FORTRAN Statements for Manipulating Arrays.**

Sometimes array elements are referenced in a "random" order. This requires the determination of an index, either through a search or direct computation, that selects a single array element for processing. With this technique, the programmer must be especially careful not to reference an element that is outside the range of the array.

We illustrated one technique for referencing array elements in random order in Problem 5C. There we found the "category" (tax bracket) of a data item by comparing it with boundary values stored in an array. If there is a constant increment between the boundary values, the index of the appropriate category can also be computed directly, as was illustrated in Fig. 5.14. Once the category index is defined, it can be used to select elements from one or more parallel arrays for further processing.

The arrays discussed in this chapter are often called *linear arrays* or *lists*. These arrays are "one-dimensional," in that a single subscript is used to uniquely identify each array element. In Chapter 10, we shall examine a more complex data structure, an array with multiple dimensions.

## PROGRAMMING PROBLEMS

**5D**  Instructor X has given an exam to a large lecture class of students. The grade scale for the exam is 90–100(A), 80–89(B), 70–79(C), 60–69(D), 0–59(F). Instructor X needs a program to perform the following statistical analysis of the data:

i)   Count the number of A's, B's, C's, D's, and F's.
ii)  Determine the averages of the A, B, C, D, and F scores, computed on an individual basis—i.e., the average A score, the average B score, . . . , the average F score.
iii) Find the total number of students taking the exam.
iv)  Compute the average and standard deviation for all of the scores.

**5E**  Let A be an array consisting of 20 elements. Write a program to read a collection of up to 20 data items into A, and then find and print the subscript of the largest item in A.

**5F**  The Department of Traffic Accidents each year receives accident count reports from a number of cities and towns across the country. To summarize these reports, the Department provides a frequency-distribution printout that gives the number of cities reporting accident counts in the following ranges: 0–99, 100–199, 200–299, 300–399, 400–499, 500 or above. The Department needs a computer program to read the number of accidents for each reporting city or town and to add one to the count for the appropriate accident range. After all the data has been processed, the resulting frequency counts are to be printed.

**5G**  Write a program which, given the *taxable income* for a single taxpayer, will compute the income tax for that person. Use Schedule X shown in Fig. 5.15. Assume that "line 47," referenced in this schedule, contains the taxable income.

## Tax Rate
## Schedules

SCHEDULE X—Single Taxpayers Not Qualifying for Rates in Schedule Y or Z

Use this schedule if you checked the box on Form 1040, line 1—

| If the amount on Form 1040, line 47, is: | | Enter on Form 1040, line 16a: | |
|---|---|---|---|
| Not over $500...14% of the amount on line 47. | | | |
| Over— | But not over— | | of the amount over— |
| $500 | $1,000 | $70+15% | $500 |
| $1,000 | $1,500 | $145+16% | $1,000 |
| $1,500 | $2,000 | $225+17% | $1,500 |
| $2,000 | $4,000 | $310+19% | $2,000 |
| $4,000 | $6,000 | $690+21% | $4,000 |
| $6,000 | $8,000 | $1,110+24% | $6,000 |
| $8,000 | $10,000 | $1,590+25% | $8,000 |
| $10,000 | $12,000 | $2,090+27% | $10,000 |
| $12,000 | $14,000 | $2,630+29% | $12,000 |
| $14,000 | $16,000 | $3,210+31% | $14,000 |
| $16,000 | $18,000 | $3,830+34% | $16,000 |
| $18,000 | $20,000 | $4,510+36% | $18,000 |
| $20,000 | $22,000 | $5,230+38% | $20,000 |
| $22,000 | $26,000 | $5,990+40% | $22,000 |
| $26,000 | $32,000 | $7,590+45% | $26,000 |
| $32,000 | $38,000 | $10,290+50% | $32,000 |
| $38,000 | $44,000 | $13,290+55% | $38,000 |
| $44,000 | $50,000 | $16,590+60% | $44,000 |
| $50,000 | $60,000 | $20,190+62% | $50,000 |
| $60,000 | $70,000 | $26,390+64% | $60,000 |
| $70,000 | $80,000 | $32,790+66% | $70,000 |
| $80,000 | $90,000 | $39,390+68% | $80,000 |
| $90,000 | $100,000 | $46,190+69% | $90,000 |
| $100,000 | | $53,090+70% | $100,000 |

**Fig. 5.15**  Schedule X (from IRS Form 1040).

*Example:* If the individual's taxable income is $8192, your program should use the tax amount and percent shown in column 3 of line 7 (arrow). The tax in this case is

$$\$1590 + .25(8192. - 8000) = \$1638.$$

For each individual processed, print taxable earnings and the total tax. *Hint:* Set up three arrays, one for the base tax (column 3), one for the tax percent (column 3), and the third for the excess base (column 4). Your program must then compute the correct index to these arrays, given the taxable income.

5H   Assume for the moment that your computer has the very limited capability of being able to read and print only single decimal digits at a time; and to add together two integers consisting of one decimal digit each. Write a program to read in two ten-digit integers, add these numbers together, and print the result. Test your program on the following numbers.

$$X = 1487625$$
$$Y = \phantom{000}12783$$

$$X = 60705202$$
$$Y = 30760832$$

$$X = 1234567890$$
$$Y = 9876543210$$

*Hints:* Store the numbers X and Y in two arrays XAR, YAR, of size 10, one decimal digit per element. If the number is less than 10 digits in length, punch enough *leading zeros* (to the left of the number) to make the number 10 digits long.

Leave a space between each digit punched. (Thus, the first two numbers should be punched as

```
 0  0  0  0  0  1  2  7  8  3 ─────── Y
0  0  0  1  4  8  7  6  2  5 ─────── X
 1 2 3 4 5 6 7 8 9 10 11 12 13 14 15 16 17 18 19 20 21 22
```

Use the statements

```
            READ*, XAR
            READ*, YAR
```

to read the numbers into the arrays XAR and YAR respectively.)

You will need a loop to add together the digits in the array elements. You must start with the element with subscript value 10 and work toward the left. Do not forget to handle the carry, if there is one!

Use an integer variable, OFLOW, to indicate if a carry occurred in adding together XAR(1) and YAR(1). OFLOW is set to 1 if a carry occurs here; otherwise, OFLOW will be 0.

**5I**   Write a data table, flow diagram, and a program for the following problem. You are given a collection of scores for the last exam in your computer course. You are to compute the average of these scores, and then assign grades to each student according to the following rule.

If a student's score S is within 10 points (above or below) of the average, assign the student a grade of SATISFACTORY. If S is more than 10 points higher than the average, assign the student a grade of OUTSTANDING. If S is more than 10 points below the average, assign the student a grade of UNSATISFACTORY.

Test your program on the following data:

```
'RICHARD LUGAR'  62
'DONALD SCHAEFFER'  84
'KEVIN WHITE'  93
'JAMES RIEHLE'  74
'ABE BEAME'  70
'TOM BRADLEY'  45
'RICHARD HATCHER'  82
```

*AVG = SUM / 7*

*PG. 150*

*CHARACTER \*16 NAME*

*REAL \*7 SCORE*

The output from your program should consist of a labelled 3-column list containing the name, exam score, and grade of each student.

**5J**   Write a program to read N data items into each of two arrays X and Y of size 20. Compare each of the elements of X to the corresponding element of Y. In the corresponding element of a third array Z, store:

$$+1 \quad \text{if X is larger than Y}$$

$$0 \quad \text{if X is equal to Y}$$

$$-1 \quad \text{if X is less than Y}$$

Then print a three-column table displaying the contents of the arrays X, Y, and Z, followed by a count of the number of elements of X that exceed Y, and a count of the number of elements of X that are less than Y. Make up your own test data, with N less than 20.

**5K**   The results of a true-false exam given in a class of Computer Science students has been punched on cards. Each card contains a student identification number, and the students' answers to 10 true-false questions. The data cards are as follows:

| Student identification | Answers (1 = true; 0 = false) | | | | | | | | | |
|---|---|---|---|---|---|---|---|---|---|---|
| 0080 | 0 | 1 | 1 | 0 | 1 | 0 | 1 | 1 | 0 | 1 |
| 0340 | 0 | 1 | 0 | 1 | 0 | 1 | 1 | 1 | 0 | 0 |
| 0341 | 0 | 1 | 1 | 0 | 1 | 1 | 1 | 1 | 1 | 1 |
| 0401 | 1 | 1 | 0 | 0 | 1 | 0 | 0 | 1 | 1 | 1 |
| 0462 | 1 | 1 | 0 | 1 | 1 | 1 | 0 | 0 | 1 | 0 |
| 0463 | 1 | 1 | 1 | 1 | 1 | 1 | 1 | 1 | 1 | 1 |
| 0464 | 0 | 1 | 0 | 0 | 1 | 0 | 0 | 1 | 0 | 1 |
| 0512 | 1 | 0 | 1 | 0 | 1 | 0 | 1 | 0 | 1 | 0 |
| 0618 | 1 | 1 | 1 | 0 | 0 | 1 | 1 | 0 | 1 | 0 |
| 0619 | 0 | 0 | 0 | 0 | 0 | 0 | 0 | 0 | 0 | 0 |
| 0687 | 1 | 0 | 1 | 1 | 0 | 1 | 1 | 0 | 1 | 0 |
| 0700 | 0 | 1 | 0 | 0 | 1 | 1 | 0 | 0 | 0 | 1 |
| 0712 | 0 | 1 | 0 | 1 | 0 | 1 | 0 | 1 | 0 | 1 |
| 0837 | 1 | 0 | 1 | 0 | 1 | 1 | 0 | 1 | 0 | 1 |
| 9999 | (Sentinel card, punch 10 zeros) | | | | | | | | | |

The correct answers are

$$0 \quad 1 \quad 0 \quad 0 \quad 1 \quad 0 \quad 0 \quad 1 \quad 0 \quad 1$$

Write a program to read the data cards, one at a time, and compute and store the number of correct answers for each student in one array, and store the student ID number in the corresponding element of another array. Determine the best score, BEST. Then print a three-column table displaying the ID number, score, and grade for each student. The grade should be determined as follows: If the score is equal to BEST or BEST − 1, give an A; if it is BEST − 2 or BEST − 3, give a C. Otherwise, give an F.

**5L**   Write a program to read N data items into two arrays X and Y of size 20. Compute the products of the corresponding elements in X and Y and store the result in a third array XY, also of size 20. Print a three-column table displaying the arrays

X, Y, and XY. Then compute and print the square root of the sum of the items in XY. Make up your own data, with N less than 20.

**5M**    The results of a survey of the households in your township have been punched in cards. Each card contains data for one household, including a four-digit integer identification number, the annual income for the household, and the number of members of the household. Write a program to read the survey results into three arrays and perform the following analyses:

1. Count the number of households included in the survey, and print a three-column table displaying the data read in. (You may assume that no more than 25 households were surveyed.)
2. Calculate the average household income, and list the identification number and incomes of all households that exceed the average.
3. Determine the percentage of households having incomes below the poverty level. The poverty level income may be computed according to the formula

$$p = \$3750.00 + \$750.00 * (m - 2)$$

where m is the number of members of each household.

Test your program on the following data.

| Identification number | Annual income | Household members |
|---|---|---|
| 1041 | $12,180 | 4 |
| 1062 | 13,240 | 3 |
| 1327 | 19,800 | 2 |
| 1483 | 22,458 | 8 |
| 1900 | 17,000 | 2 |
| 2112 | 18,125 | 7 |
| 2345 | 15,623 | 2 |
| 3210 | 3,200 | 6 |
| 3600 | 6,500 | 5 |
| 3601 | 11,970 | 2 |
| 4725 | 8,900 | 3 |
| 6217 | 10,000 | 2 |
| 9280 | 6,200 | 1 |
| 9999 (Sentinel card) | 0 | 0 |

**5N**    The Major Risk Bank has a checking account program for all of its investors. Each bank member has a different checking-account number. Each month fewer than 20 bank members take advantage of the checking program. Write a program that will be used on a monthly basis to do the following.

1. Read in the account (5 digits) and past month's balance for each bank member using the checking facility (the account and past balance should be punched on one card). Store the account numbers in an array (ACCNS, size 20), and store the past balance in another array (BALANS, size 20).
2. Read in the transactions for the month. Each transaction card should contain the 5-digit account number and a transaction amount (a positive amount indicates a deposit, and a negative amount indicates a withdrawal). 
3. After all transactions have been read, print each account number and the new bank balance.

*Note:* There may be many transactions for each account read in during (1) and they need not all be together. For each transaction you will have to search the array ACCNS to find the index of that account. Then you must use this index to alter the contents of the appropriate element in the array BALANS.

**5O**   It can be shown that a number is prime if there is no smaller prime number that divides it. Consequently, in order to determine whether N is prime, it is sufficient to check only the prime numbers less than N as possible divisors (see Problem 4B). Use this information to write a program that stores the first one hundred prime numbers in an array. Have your program print the array after it is done.

**5P**   *for interactive programmers* Write a program that plays the game of HANGMAN. Read each letter of the word to be guessed into successive elements of the character array WORD. The player must guess the letters belonging to WORD. The program should terminate when either all letters have been guessed correctly (player wins), or a specified number of incorrect guesses have been made (computer wins). *Hint:* Use a character array SOLUT to keep track of the solution so far. Initialize each element of SOLUT to the symbol "*". Each time a letter in WORD is guessed, replace the corresponding "*" in array SOLUT with that letter.

**5Q**   The Grading Problem was simplified considerably by the assumption that there were no invalid scores—i.e., that all scores fell between 0 and 100. Redo this problem with added steps to validate each score as it is read.
*Hints:* a) Introduce two new program variables, INVCNT and VALCNT, to count the number of invalid and valid scores, respectively.

   b) Use two new input variables, NAMTMP and SCRTMP, to store each name and score pair as it is read. Replace the current loop (step 2) with a loop to read each input pair into NAMTMP and SCRTMP. If the score is valid, accumulate sums, copy NAMTMP to NAMES(I1) and SCRTMP to SCORES(I1), and increment VALCNT. Otherwise, increment INVCNT and print a diagnostic indicating that the score was ignored.

   c) Add a level one step between steps 2 and 3 to print the number of valid and invalid scores.

   d) For all subsequent level one steps, use VALCNT in place of N.

# ADVANCED
# CONTROL
# STRUCTURES

6

## 6.1   INTRODUCTION

In Chapter 3 we introduced four fundamental control structures to be used in computer programming. We presented flow diagram patterns, and described the FORTRAN syntax for each of these structures. We illustrated the application of these structures in the solution to a number of problems.

In this chapter we will present a more general form of the decision structure, called the multiple-alternative decision structure, and a more general form of the DO loop structure. We will also examine some of the rules for forming combinations or nests of structures. A number of examples and solved problems illustrating the nesting of structures will be provided.

## 6.2   THE MULTIPLE-ALTERNATIVE DECISION STRUCTURE

Most decision steps in the problems examined so far have involved only two alternatives. However, there are many situations in which it is necessary to choose from among several alternative courses of action.

As an example of this, consider a grading problem similar to Problem 5A in which there are five grade categories, A, B, C, D and F, instead of three, and where the range of exam scores for each grade category is allowed to vary each time the exam is administered. In solving this new problem, we shall use an array, FREQ, with five elements to count the number of exam scores in each grade category. Assuming, as in Problem 5A, that the range of all exam scores is between 0 and 100, we can define the four variables listed below to represent the grade boundaries.

LOWA:    Lowest score in the A grade category
LOWB:    Lowest score in the B grade category
LOWC:    Lowest score in the C grade category
LOWD:    Lowest score in the D grade category

The only restriction on the values of these variables is that LOWD must be less than LOWC, LOWC less than LOWB, and LOWB less than LOWA.

If the grade boundary values (LOWA, LOWB, etc.) were stored in an array (along with 0 for LOWF), we could use the direct search technique described in Section 5.8 to determine the correct grade category of an arbitrary exam score, CURGRD. Although this is the most desirable solution, it is instructive to examine a solution that uses a nest of double-alternative decision structures, as shown in Fig. 6.1.

This is a rather complicated nest of structures, which is not particularly easy to follow, much less program. The necessary decisions for this problem can be more easily written if we generalize the flow diagram pattern for the IF-THEN-

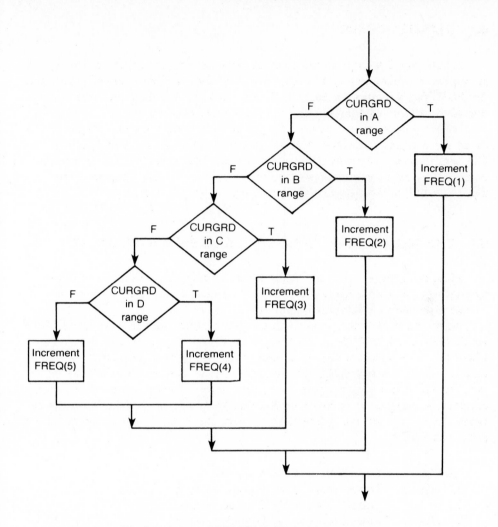

**Fig. 6.1**  A nest of IF-THEN-ELSE decision structures.

ELSE (double-alternative decision) into a *multiple-alternative decision structure,*
so that more than two alternatives may be represented in a single structure. The
flow diagram pattern for the multiple-alternative decision structure is shown in
Fig. 6.2, along with an example of the structure defined for the new grading
problem.

This flow diagram pattern implies the following program action:

a) The conditions are evaluated from top to bottom.
b) The task (Task$_i$) corresponding to the first condition (condition$_i$) to evaluate
   to true is performed, and the structure is exited immediately.
c) If no condition evaluates to true, Task$_E$ is performed.

General form

Example

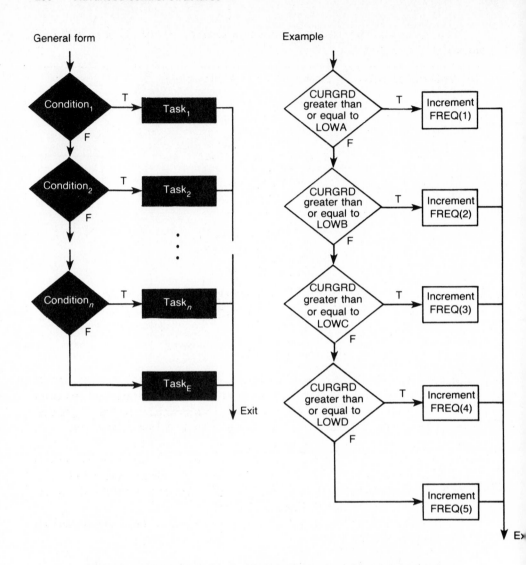

**Fig. 6.2** Multiple-alternative decision structure, general form and example.

Thus, the steps in exactly one of the tasks will be performed. More than one condition may actually be true, but only the topmost task will be executed because of the top-to-bottom order of evaluation of conditions, and the fact that the structure exit immediately follows the execution of the task corresponding to the first true condition.

The bottom task, $TASK_E$, may be omitted from this structure. In this case, if all conditions evaluate to false, none of the tasks in the multiple-alternative decision structure will be performed. The description of the subtasks in a multiple-alternative decision structure should be kept short, and refined, if necessary, in separate flow diagrams.

The syntax of the multiple-alternative decision structure (also called the block-IF) is described next.

---

**Multiple-Alternative Decision Structure (block-IF)**

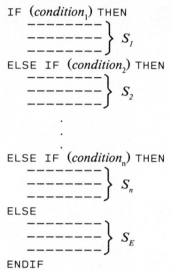

IF (*condition*₁) THEN

    $\left.\begin{array}{l} \text{------------} \\ \text{------------} \\ \text{------------} \end{array}\right\} S_1$

ELSE IF (*condition*₂) THEN

    $\left.\begin{array}{l} \text{------------} \\ \text{------------} \\ \text{------------} \end{array}\right\} S_2$

    ·
    ·
    ·

ELSE IF (*condition*ₙ) THEN

    $\left.\begin{array}{l} \text{------------} \\ \text{------------} \\ \text{------------} \end{array}\right\} S_n$

ELSE

    $\left.\begin{array}{l} \text{------------} \\ \text{------------} \\ \text{------------} \end{array}\right\} S_E$

ENDIF

**Interpretation:** *Condition*₁, *condition*₂, etc., are tested until a condition is reached that evaluates to true. If *condition*ᵢ is the first condition that evaluates to true, $S_i$ is executed; if none of these conditions evaluates to true, $S_E$ is executed. Regardless of which of the statement groups is carried out, execution next resumes with the first instruction following the ENDIF.

*Notes:* Any of the statement groups may be empty. In this case, nothing is done when the associated condition is true. If $S_E$ is empty, then nothing is done when all conditions are false. If any statement group is empty, the corresponding ELSE IF (or ELSE) line may be deleted.

The single-alternative and double-alternative forms discussed earlier are special cases of the block-IF.

---

The FORTRAN implementation of the example flow diagram shown in Fig. 6.2 follows.

```
IF (CURGRD .GE. LOWA) THEN
    FREQ(1) = FREQ(1) + 1
ELSE IF (CURGRD .GE. LOWB) THEN
    FREQ(2) = FREQ(2) + 1
ELSE IF (CURGRD .GE. LOWC) THEN
    FREQ(3) = FREQ(3) + 1
ELSE IF (CURGRD .GE. LOWD) THEN
    FREQ(4) = FREQ(4) + 1
ELSE
    FREQ(5) = FREQ(5) + 1
ENDIF
```

For a grade of A, all four of the conditions shown above would evaluate to true. However, only the first of these is tested and only FREQ(1) is incremented, as desired.

If there is no ELSE alternative ($S_E$ is empty), then the word ELSE may be deleted. However, in most instances, it is a good idea to retain the ELSE alternative, even if it consists solely of a diagnostic PRINT statement. This statement should indicate that all tests have failed for a particular data value; it may be the only indication of a program error or bad data.

**Exercise 6.1:**   Implement Step 4 in the level one flow diagram for Problem 5A using the multiple-alternative (block-IF) structure (see Fig. 5.8).

**Exercise 6.2:**   Let MINCAT be an integer array of size 5 used to store the values of LOWA, LOWB, LOWC, LOWD and LOWF. Redo the flow diagrams and FORTRAN code segment to read the lowest scores into MINCAT. Then search the array MINCAT (see Problem 5C) to determine which element of FREQ to update for a given score.

**Exercise 6.3:**   You are writing a program to print grade reports for students at the end of each semester. After computing and printing each student's grade point average (GPA) for the semester, you are supposed to use the grade point average to make the following decision:

If the GPA is 3.5 or above, print 'DEAN'S LIST';
If the GPA is above 1.0 and less than or equal to 1.99, print 'PROBATION WARNING';
If the GPA is less than or equal to 1.0, print 'YOU ARE ON PROBATION NEXT SEMESTER'

Draw a flow diagram and write the FORTRAN program segment for this decision. Use a multiple-alternative decision structure.

**Exercise 6.4:**   Replace each set of nested decision structures in Problem 4A and Example 4.19 with a multiple-alternative decision structure.

## 6.3   THE BOWLING PROBLEM

The next problem makes use of the multiple-alternative decision structure.

**Problem 6A:**   Write a program that will compute a person's ten pin bowling score for one game, given the number of balls rolled, NBALLS, and the number of pins knocked down per ball. Print the score for each frame, as well as the cumulative score at the end of each frame.

**Discussion:**   A bowling *game* consists of 10 *frames*. In ten pin bowling, a maximum of two balls may be rolled in each of the first nine frames, and two or three balls may be rolled in frame ten. Each frame is scored according to the following rules.

1. If the first ball rolled in a frame knocks down all 10 pins (called a *strike*), then the score for the frame is equal to 10 + (the total score on the next two balls

rolled). Since all 10 pins are down, no other balls are rolled in the current frame.

2. If the two balls rolled in the frame together knock down all 10 pins (called a *spare*), then the score for the frame is equal to 10 + (the score on the next ball rolled).

3. If the two balls rolled knock down fewer than 10 pins (no *mark*), then the frame score is equal to the number of pins knocked down.

The initial data table for this problem is shown below; an example of the array PINS is shown in Fig. 6.3. The array shows that 10 pins were knocked down by the first ball, seven by the second ball, etc.

**Data Table for Problem 6A**

| *Input variables* | *Output variables* |
|---|---|
| PINS: Array containing the number of pins knocked down by each ball rolled (integer, size 21) | FSCORE: The score in each frame (integer) |
| NBALLS: The number of balls actually rolled (integer) | TOTAL: The total score accumulated (integer) |

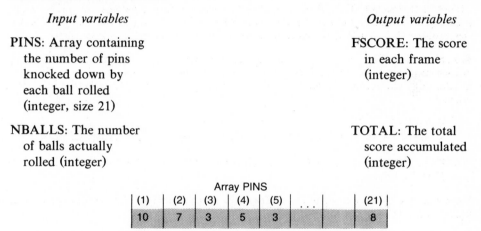

**Fig. 6.3**  Array of pin counts for each ball.

The level one flow diagram for the problem is shown on the left side of Fig. 6.4a. Steps 2 and 3 in this diagram are refined on the right side of Fig. 6.4a; Step 1 will be an implied DO loop (see Section 5.5) with final value parameter NBALLS.

In the refinement for Step 2, the loop control variable FCOUNT serves to count each frame as it is processed. In the computation of the scores (Step 2.3), the variable FIRST is used as the index to the array PINS to indicate which ball in PINS is the first ball of each frame. The use of FIRST and the computation of the score for each of the first three frames are illustrated in Table 6.1.

| Frame | FIRST | Frame score | Effect |
|---|---|---|---|
| 1 | 1 | 10 + 7 + 3 = 20 | Strike: Only one ball rolled in frame 1 |
| 2 | 2 | 7 + 3 + 5 = 15 | Spare: Two balls rolled in frame 2 |
| 3 | 4 | 5 + 3 = 8 | No Mark: Two balls rolled in frame 3 |
| 4 | 6 | . | |
|  |  | . | |
|  |  | . | |

**Table 6.1 Processing the array PINS**

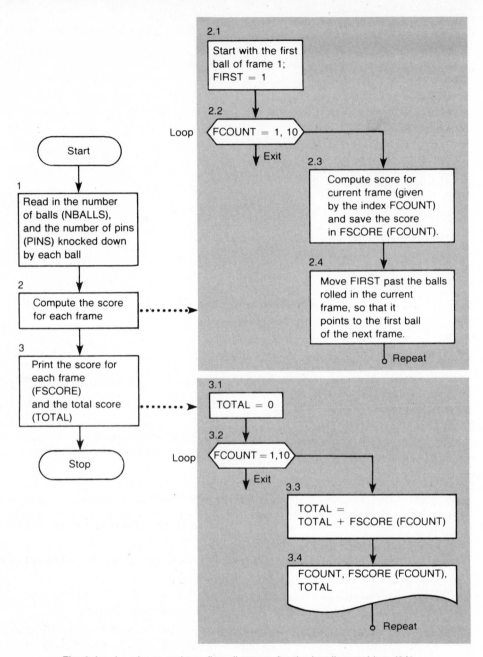

**Fig. 6.4a** Level one and two flow diagrams for the bowling problem (6A).

Since PINS(1) is 10, a strike was bowled in the first frame. The frame score (20) is determined by computing 10 + PINS(2) + PINS(3); FIRST, which is next used as the index to the first ball of the second frame, is then set to 2 (Step 2.4). In the second frame, balls 2 and 3 are needed to knock down all 10 pins (a spare). Adding in the pins knocked down by the next ball, PINS(4), gives a frame score of 15; the index FIRST is then set to 4 (the fourth ball is the first one for frame 3). Two balls are rolled in the third frame (balls 4 and 5). The frame score is 8 (PINS(4) + PINS(5)), and FIRST is set to 6.

Since FIRST contains the index to the first ball bowled in each frame, the subscript expressions FIRST, FIRST+1, and (possibly) FIRST+2 are used to select the elements of the array PINS (the number of pins knocked down by each ball) to be used in each frame score. Thus, PINS(1), PINS(2) and PINS(3) are used to compute the score for frame 1, PINS(2), PINS(3), PINS(4) for frame 2, and so on.

The refinement of the computation step is drawn in Fig. 6.4b. The test for a strike involves a test to see if PINS(FIRST) is equal to 10; the test for a spare is whether the sum PINS(FIRST) + PINS(FIRST+1) is equal to 10. FIRST should be incremented by 1 each time a strike is bowled; otherwise, it should be incremented by 2. (Why?) The additional data table entries for the Step 2.3 refinement are listed below.

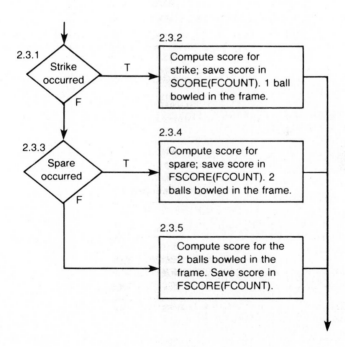

**Fig. 6.4b**  Refinement of Step 2.3 of Fig. 6.4a.

### Additional Data Table Entries for Step 2.3 Refinement:

*Program variables*

> FIRST: Index to the array
> PINS indicating the
> first ball of each frame
> (integer)

> FCOUNT: Loop control
> variable, indicates the
> number of the current
> frame (integer)

FCOUNT is also used as the loop control variable for the Step 3 refinement in which the total score (TOTAL) is accumulated (Step 3.3), and the frame number, frame score, and TOTAL are all printed (Step 3.4).

The program for the Bowling Problem is shown in Fig. 6.5. The loop control variable IX, used to enter and echo print the array PINS, should be listed in the data table as a program variable. A sample printout from this program is shown in Fig. 6.6; the input lines used to produce the output are shown below.

```
line 1:   18 (number of balls rolled)
line 2:   10  7 3  5 3  8 2  10  4 5  0 6  10  9 1  0 10 6

frame     1    2    3    4    5    6    7    8    9    10
```

```
C BOWLING PROBLEM
C
      INTEGER NBALLS, FIRST, FCOUNT, IX
      INTEGER FSCORE(10), PINS(21), TOTAL
C
C READ NUMBER OF BALLS AND NUMBER OF PINS KNOCKED DOWN BY EACH BALL
      READ*, NBALLS
      READ*, (PINS(IX), IX = 1, NBALLS)
      PRINT*, 'NUMBER OF BALLS ROLLED IS ', NBALLS
      PRINT*, 'SCORES FOR EACH BALL ARE ..'
      DO 10 IX = 1, NBALLS
         PRINT*, PINS(IX)
   10 CONTINUE
C
C INITIALIZE TO FIRST BALL OF FRAME ONE
      FIRST = 1
C
C COMPUTE SCORE FOR EACH FRAME
      DO 20 FCOUNT = 1, 10
C        TEST FOR A STRIKE OR A SPARE
         IF (PINS(FIRST) .EQ. 10) THEN
C            STRIKE - ONE BALL GETS ALL PINS: TWO BONUS BALLS
             FSCORE(FCOUNT) = 10 + PINS(FIRST+1) + PINS(FIRST+2)
             FIRST = FIRST + 1
```

*(Continued)*

```
            ELSE IF ((PINS(FIRST) + PINS(FIRST+1)) .EQ. 10) THEN
C               SPARE - TWO BALLS GET ALL PINS: ONE BONUS BALL
                FSCORE(FCOUNT) = 10 + PINS(FIRST+2)
                FIRST = FIRST + 2
            ELSE
C               NO MARK - NO STRIKE, NO SPARE, TWO BALLS ROLLED
                FSCORE(FCOUNT) = PINS(FIRST) + PINS(FIRST+1)
                FIRST = FIRST + 2
            ENDIF
   20 CONTINUE
C
C PRINT RESULTS
      PRINT*, ' '
      PRINT*, 'THE FRAME-BY-FRAME SCORES ARE ...'
      PRINT*, '              FRAME           SCORE              TOTAL'
      TOTAL = 0
      DO 30 FCOUNT = 1, 10
C        ACCUMULATE TOTAL BY ADDING IN CURRENT FRAME SCORE
         TOTAL = TOTAL + FSCORE(FCOUNT)
         PRINT*, FCOUNT, FSCORE(FCOUNT), TOTAL
   30 CONTINUE
C
      STOP
      END
```

**Fig. 6.5**  Program for Problem 6A.

```
      NUMBER OF BALLS ROLLED IS                    18
      SCORES FOR EACH BALL ARE ..
                        10
                         7
                         3
                         5
                         3
                         8
                         2
                        10
                         4
                         5
                         0
                         6
                        10
                         9
                         1
                         0
                        10
                         6
```

(Continued)

```
THE FRAME-BY-FRAME SCORES ARE ...
        FRAME              SCORE              TOTAL
          1                 20                 20
          2                 15                 35
          3                  8                 43
          4                 20                 63
          5                 19                 82
          6                  9                 91
          7                  6                 97
          8                 20                117
          9                 10                127
         10                 16                143
```

**Fig. 6.6**  Sample output for bowling problem (6A).

**Exercise 6.5:**  The program shown in Fig. 6.5 was written without the use of program parameters. Instead, several constants, such as 10 and 21 were written directly in-line in the program. Choose appropriate names for these constants, and rewrite all statements in the Bowling Program that are affected by the use of these names. Be careful; the constant 10 is used in two ways in the program.

**Exercise 6.6:**  They do things a little differently in Massachusetts where Dr. Koffman grew up. The bowling pins (called candlepins) are narrow at the top and bottom and wider in the middle. The balls are about the size of a softball. The rules for a strike and a spare are the same; however, the bowler gets to roll a third ball in each frame that is neither a strike nor spare. Modify the bowling program to score a candlepin game. (Any pins that fall on the lane are not cleared away in candlepins. This can help the bowler but should not affect your program).

## 6.4   THE GENERAL FORM OF THE DO LOOP

### 6.4.1   Examples of the General DO Loop

We have been using a limited form of the DO loop as a convenient structure for counter-controlled loops. In most applications to date, both the initial counter value and increment value were one. In general, however, it is possible to specify an increment value that is different from one by adding a third loop parameter, called the step. The step parameter may be negative as well as positive. FORTRAN permits all three parameters to be arbitrary arithmetic expressions, although we recommend using integer expressions only.

**Example 6.1:**  The program below contains two DO loops. The first DO loop has a step value of one; the loop is executed for values of $I = -3, -2, -1, 0, 1, 2, 3$. Since the step value is one, it may be omitted. The second DO loop has a negative step value $(-1)$ and is executed for values of $I = 3, 2, 1, 0, -1, -2, -3$.

```
        INTEGER MIN, MAX
        PARAMETER (MIN = -3, MAX = 3)
        INTEGER I
        DO 10 I = MIN, MAX, 1
           PRINT*, I
     10 CONTINUE
```

(Continued)

```
         DO 20 I = MAX, MIN, -1
            PRINT*, I
      20 CONTINUE
         STOP
         END
```

```
      *** PROGRAM EXECUTION OUTPUT ***
```

```
                       -3
                       -2
                       -1
                        0
                        1
                        2
                        3
                        3
                        2
                        1
                        0
                       -1
                       -2
                       -3
```

The general form of the DO loop is described in the next display.

---

## DO Loop Structure (General Form)

    DO  *sn lcv = initial, limit, step*
      ----- ⎫
      ----- ⎬  *loop body*
      ----- ⎭
    *sn* CONTINUE

**Interpretation:** The loop parameters *initial, limit* and *step* are expressions that represent the initial value, limit value and step value for the loop control variable, *lcv. Lcv* must be a variable. The *loop body* will be repeated once for each value of *lcv*, starting with *lcv* equal to the value of *initial,* and continuing until *lcv* "passes" the value of *limit.* After each loop repetition, the value of *lcv* is automatically updated by the value of *step.*

   If *step* is positive, repetition will continue as long as *lcv* is less than or equal to *limit;* the loop will be exited when *lcv* becomes greater than *limit.* If *step* is negative, repetition will continue as long as *lcv* is greater than or equal to *limit;* the loop will be exited when *lcv* becomes less than *limit.* Upon exit, *lcv* retains the last value assigned to it, i.e., the value that caused the exit.

*Notes:* The loop parameters may be arbitrary arithmetic expressions. Each expression is evaluated only once—when the loop is first entered. It is permissible for the values of variables in these expressions (but not the *lcv*) to be changed in the loop body; however, this will not affect the values of *initial, limit,* or *step,* nor will it change the number of loop repetitions.

   A *step* value of zero is not allowed. If *step* is omitted, it is assumed to be one.

Since the loop parameters may be arbitrary arithmetic expressions, they may contain constants, variables, array references, arithmetic operators and parentheses. The *step* value determines the magnitude and direction of change in the loop control variable after each loop repetition. It is possible for the loop body not to be executed at all.

- If *step* is negative, (*step* $<0$), then the loop will not be executed if *initial* is less than *limit* (e.g., DO 10 I $= -4$, 10, $-1$).
- If *step* is positive (*step* $>0$), then the loop will not be executed if *initial* is greater than *limit* (e.g., DO 15 I $= 10$, $-4$, 1).

If the loop is not executed, then the value of the loop control variable is set equal to the value of *initial*, and the first statement following the loop terminator is executed.

**Example 6.2:**    Illustrations of DO loop execution.

a)
```
        DO 10 L = 1, 5
            PRINT*, L
    10 CONTINUE
```

This loop executes once for each of the five values of L: 1,2,3,4,5; it terminates when L becomes greater than the limit 5.

| loop repetition | Values of L that are printed |
|---|---|
| first | 1 |
| second | 2 |
| third | 3 |
| fourth | 4 |
| fifth | 5 |

Repetition terminates when L becomes equal to 6.

b)
```
        SUM = 0
        DO 10 ODD = 1, 10, 2
            SUM = SUM + ODD
    10 CONTINUE
```

This loop executes once for ODD $= 1,3,5,7,9$; it terminates when ODD becomes 11.

| loop repetition | values of ODD | values of SUM |
|---|---|---|
| first | 1 | 1 |
| second | 3 | 4 |
| third | 5 | 9 |
| fourth | 7 | 16 |
| fifth | 9 | 25 |

Repetition terminates when ODD is 11.

**Example 6.3:** The program in Fig. 6.7 is used to compute the factorial (NFACT) of a positive integer N. The factorial of N (represented as N!) is defined below as the product of N and all positive integers less than N.

$$N! = N \times (N-1) \times \ldots \times 2 \times 1$$

For example

$$6! = 6 \times 5 \times 4 \times 3 \times 2 \times 1 = 720$$

as illustrated in the sample output at the bottom of Fig. 6.7.

```
C COMPUTE THE FACTORIAL (NFACT) OF N
C
      INTEGER N, I, NFACT
      READ*, N
      PRINT*, 'COMPUTE THE FACTORIAL OF ', N
      PRINT*, ' '
      PRINT*, '                 NFACT               I'
      NFACT = N
      PRINT*, NFACT
C
C FORM THE PRODUCT OF ALL INTEGERS LESS THAN OR EQUAL TO N
      DO 10 I = N-1, 1, -1
         NFACT = NFACT * I
         PRINT*, NFACT, I
   10 CONTINUE
C
C PRINT RESULTS
      PRINT*, ' '
      PRINT*, 'THE FACTORIAL OF ', N, ' IS ', NFACT
      PRINT*, 'I = ', I
C
      STOP
      END


*** PROGRAM EXECUTION OUTPUT ***

COMPUTE THE FACTORIAL OF                   6

            NFACT            I
              6
             30             5
            120             4
            360             3
            720             2
            720             1

THE FACTORIAL OF              6 IS              720
I =               0
```

**Fig. 6.7**   Factorial computation.

As shown in Fig. 6.7, NFACT is initialized to N before loop entry. During each execution of the loop body, the current value of I is incorporated into the product that is being computed by the statement

$$NFACT = NFACT * I$$

The loop body is executed for each integer value of I between $N-1$ and 1 inclusive. The value of I after loop exit is 0.

**Example 6.4:**   The following loop uses a limit expression involving three library functions; the step is 1.

```
C          DETERMINE IF N IS PRIME
           PRIME = .TRUE.
           DO 40 DIV = 2, INT(SQRT(REAL(N)))
              IF (MOD(N, DIV) .EQ. 0) THEN
                 PRIME = .FALSE.
                 PRINT*, 'DIVISOR = ', DIV
              ENDIF
    40     CONTINUE
C
C          TEST WHETHER A DIVISOR WAS FOUND
           IF (PRIME) THEN
              PRINT*, N, ' IS PRIME'
           ELSE
              PRINT*, N, ' IS NOT PRIME'
           ENDIF
```

This loop can be used to determine if a number N is prime (see Problem 4B).

The largest possible divisor of N is represented by the *limit* value $\sqrt{N}$ (e.g., if N is 55, the value of *limit* is 7). The argument of the SQRT function must be type real. The INT function converts the real square root value to an integer value. FORTRAN does not require integer-valued loop parameters; however, we strongly recommend their use.

**Example 6.5:**   In Example 3.3, we wrote a program that produced a table showing equivalent Celsius and Fahrenheit temperatures. This table covered the range of temperatures from 0°C to 100°C in increments of one degree. In general, we might prefer to use input variables for the loop parameters so that any desired table, covering a variety of temperature ranges, could be produced. To do this, we could modify the original program using TMINIT, TMLIM and TMSTEP as input variables (see Fig. 6.8).

The loop will be executed for all integer values of TMINIT, TMLIM, TMSTEP satisfying the requirements below.

$$TMSTEP > 0 \text{ and } TMLIM \geq TMINIT$$
$$\text{or } TMSTEP < 0 \text{ and } TMLIM \leq TMINIT$$

Each value of FAHREN that is computed in the loop will be rounded to the nearest integer.

```
C PROGRAM TO PRODUCE A TABLE OF CELSIUS TO FAHRENHEIT CONVERSION
C
      INTEGER CELSUS, FAHREN, TMINIT, TMLIM, TMSTEP
C
C READ LOOP PARAMETERS
      READ*, TMINIT, TMLIM, TMSTEP
      PRINT*, '            CELSIUS          FAHRENHEIT'
C
C COMPUTE TABLE ENTRIES.  PRINT DEGREES C AND ROUNDED DEGREES F
      DO 30 CELSUS = TMINIT, TMLIM, TMSTEP
         FAHREN = NINT(1.8 * REAL(CELSUS) + 32.0)
         PRINT*, CELSUS, FAHREN
   30 CONTINUE
C
      STOP
      END
```

```
*** PROGRAM EXECUTION OUTPUT ***
```

| CELSIUS | FAHRENHEIT |
|---------|------------|
| 100 | 212 |
| 95 | 203 |
| 90 | 194 |
| 85 | 185 |
| 80 | 176 |
| 75 | 167 |
| 70 | 158 |
| 65 | 149 |
| 60 | 140 |
| 55 | 131 |
| 50 | 122 |
| 45 | 113 |
| 40 | 104 |
| 35 | 95 |
| 30 | 86 |
| 25 | 77 |
| 20 | 68 |
| 15 | 59 |
| 10 | 50 |
| 5 | 41 |
| 0 | 32 |
| -5 | 23 |
| -10 | 14 |
| -15 | 5 |
| -20 | -4 |

Fig. 6.8  General Celsius to Fahrenheit conversion with sample output.

**Example 6.6:**  We have the job of writing a program to assist a registrar in updating class lists at the middle of a semester. The following program segment can be used in this program to delete the identification number of any student who has dropped out of class.

```
C SHIFT EACH VALUE ONE ELEMENT TO THE LEFT
C START WITH WHERE+1 AND WORK RIGHT TOWARD N
        DO 30 IX = WHERE + 1, N
            CLIST(IX-1) = CLIST(IX)
     30 CONTINUE
C
C DECREASE LIST SIZE
        N = N - 1
        PRINT*, 'NEW CLASS LIST'
        PRINT*, (CLIST(IX), IX = 1, N)
```

Assume the class list is represented by the first N numbers stored in the array CLIST. The variable WHERE represents the index of the number to be deleted. The program segment above accomplishes the deletion by simply shifting left the portion of the array starting at index WHERE + 1. The contents of each element is moved to the element with next smallest index (loop 30). This process is illustrated in Fig. 6.9; the number 28697 is deleted from CLIST.

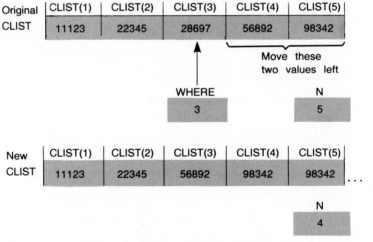

**Fig. 6.9**  Deletion of an item from a list.

If the original list was in numerical order, the new one will be, too. Following the deletion, the value of N is decreased by one, indicating that there is one less item in the list. Thus, the second copy of the last item in the list will be ignored in future manipulations. The output from the program segment above would be

```
NEW CLASS LIST
11123       22345       56892       98342
```

We now turn our attention to the problem of inserting a new number while maintaining the numerical order. We will assume that there is enough room in

CLIST to accommodate the insertion. In this case, the program must shift each value that is larger than the new number to the right in order to make room for the new number. This is shown in Fig. 6.10 for WHERE = 2, and a new insertion value (IDNUM) of 13468.

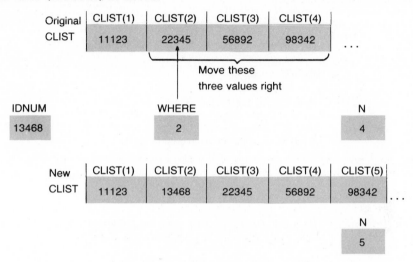

**Fig. 6.10**  Insertion of an item into a list.

At first we might try to implement the insertion using the program segment in Fig. 6.11.

```
C SHIFT EACH VALUE ONE ELEMENT TO THE RIGHT
      DO 30 IX = WHERE, N
         CLIST(IX+1) = CLIST(IX)
   30 CONTINUE
C
C INSERT IDNUM AT WHERE
      CLIST(WHERE) = IDNUM
      N = N + 1
      PRINT*, 'NEW CLASS LIST'
      PRINT*, (CLIST(IX), IX = 1, N)
```

**Fig. 6.11**  Incorrect insertion program.

Unfortunately, this insertion program segment does not work; it produces multiple copies of the item originally at position WHERE (CLIST(3) through CLIST(5) would contain the value 22345). The output printed would be:

```
NEW CLASS LIST
11123      13468      22345      22345      22345
```

To avoid this problem, we must start with the last element in the array and shift it to the right. We should then work backwards (from right to left) through the array until we shift the number at position WHERE. Finally, we can insert IDNUM. A correct program segment is shown in Fig. 6.12; the output from this segment would be:

```
NEW CLASS LIST
11123     13468     22345     56892     98342

C SHIFT EACH VALUE ONE ELEMENT TO THE RIGHT
C START WITH THE LAST VALUE AND WORK TOWARD POSITION WHERE
        DO 30 IX = N, WHERE, -1
            CLIST(IX+1) = CLIST(IX)
   30 CONTINUE
C
C INSERT IDNUM AT WHERE
        CLIST(WHERE) = IDNUM
        N = N + 1
        PRINT*, 'NEW CLASS LIST'
        PRINT*, (CLIST(IX), IX = 1, N)
```

**Fig. 6.12**  Correct insertion program.

**Exercise 6.7:**  Let A and B be the two integer arrays shown below:

| A(1) | A(2) | A(3) | A(4) | A(5) |
|------|------|------|------|------|
| 3    | 10   | −2   | 0    | 4    |

| B(1) | B(2) | B(3) | B(4) | B(5) |
|------|------|------|------|------|
| 5    | −6   | 10   | 7    | 8    |

a)  Trace the execution of the code segment

```
        OUT = 3
        LAST = 5
        DO 40 I = OUT, LAST-1
            A(I) = A(I+1)
    40 CONTINUE
```

What will be the new arrangement of data in this array? What will be the value of the loop control variable I after the loop has completed execution?

b)  Trace the execution of the following code segment. Show your work. What will be the contents of the real array C following the execution of this segment?

```
        DO 60 INDEX = LAST, 1, -1
            C(INDEX) = REAL(A(INDEX) * B(INDEX))
    60 CONTINUE
```

**Exercise 6.8:**  Write a decision structure that is executed before the code segment in Fig. 6.12 to ensure that there is enough room in CLIST for the insertion. Let MAXSIZ be the parameter representing the size of CLIST. If there is insufficient room, print an appropriate message.

**Exercise 6.9:**  As part of the program for the registrar, write program segments that will perform the following two tasks:

a)   Scan the list of identification numbers to determine the index, WHERE, of a number, NDROP, to be deleted from the class list. If NDROP is not found, print an appropriate message.

b)   Scan the ordered list of identification numbers to determine the index, WHERE, of the point of insertion of a number IDNUM. WHERE should indicate the array element that currently contains the smallest number larger than IDNUM.

### 6.4.2   Real Values as Loop Parameters*

We have cautioned against using real values as loop parameters several times. The reason may be better understood by examining the following loop:

```
DO 10 X = 0.0, 10.0, 0.1
   -------
   ------- } loop body
   -------
10 CONTINUE
```

We would expect this loop to be executed for values of X = 0.0, 0.1, 0.2, . . . , 9.8, 9.9, 10.0. However, because of the numerical error in the representation of the real number 0.1 as a binary number (see Section 4.8), the actual value of the loop control variable will be different from those listed. Since each new value of X is obtained by adding the binary representation of 0.1 to the previous value of X, the error will grow as X becomes closer to 10.0. If we do indeed desire to execute the loop for the values of X listed above, it would be better to use integer parameters and compute X as shown below.

```
DO 10 I = 0, 100, 1
   X = REAL(I) / 10.0
   -------
   ------- } loop body
   -------
10 CONTINUE
```

To verify that this is correct, we should check the loop boundary values:

| I | X |
|---|---|
| 0 | 0.0 |
| 100 | 10.0 |

As shown above, X ranges from 0.0 to 10.0. Since I increases by one after each loop repetition, X increases by 0.1 as desired.

*This section may be omitted.

## 6.5   CONTROL STRUCTURE NESTING, ENTRY AND EXIT

### 6.5.1   Introduction

Until now, we have been using structures with very little concern for any rules that might govern their use in a FORTRAN program. There are, in fact, only a few rules that we must follow. These rules concern the nesting of structures, and the entry into and transfer from these structures. More than likely, you have been following these rules ever since you began using structures. Nevertheless, you should study these rules before beginning to write more complicated computer programs.

### 6.5.2   Nested Structures

You have already studied and written programs containing nested structures and encountered little difficulty. To ensure that you have no problems in writing more complicated nests of structures, we will give a precise statement of the rule for nesting structures.

> **Structure Nesting Rule**
>
> Any structure may be nested within any other, subject to the following rule:
>
> *All nested structures must be wholly contained within a single statement group of the structure(s) in which they appear.*

To fully understand this rule, you must recall that all loops and the single-alternative decision structure contain a single *statement group.*

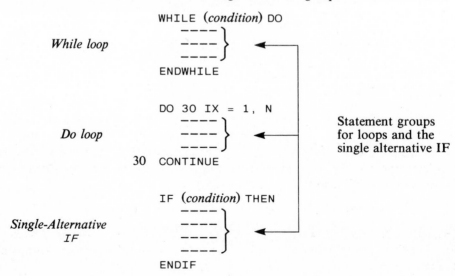

The double-alternative and multiple-alternative decision structures contain more than one statement group.

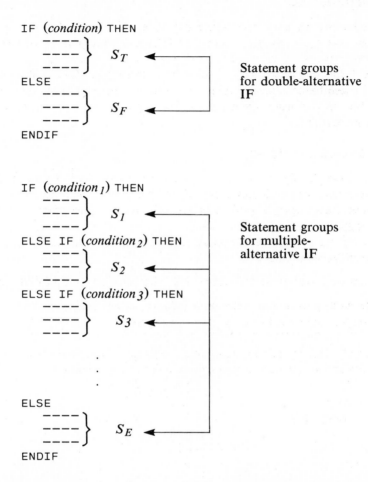

The *structure nesting rule* states that each nested structure must be wholly contained within a single statement group; it cannot overlap two or more groups. All of the programs written in the text are properly nested. (See, for example, the bowling problem (Fig. 6.5), in which a multiple-alternative decision structure is nested within a DO loop.)

**Example 6.7:** The following is an example of illegal structure nesting. The loop nested within the IF structure overlaps both the $S_T$ and $S_F$ statement groups of the IF.

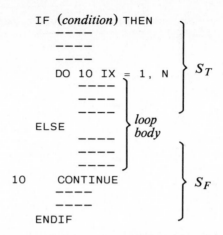

If you carefully draw your level one flow diagrams and then refine each step separately, it is impossible to draw a flow diagram that contains overlapping structures. However, if you are careless in converting your flow diagrams to FORTRAN program statements (or neglect to draw a flow diagram), you may end up with overlapping structures in your program. This will result in a compiler diagnostic since the compiler cannot translate overlapping structures.

**Exercise 6.10:**   What is wrong with the following structure nesting?

```
      DO 40 I = 1,10
          -----
          -----
          -----
      IF (condition) THEN
          -----
          -----
          -----
  40      CONTINUE
          -----
          -----
          -----
      ENDIF
```

### 6.5.3  Nested Loops

Nested loops, especially nested DO loops, are perhaps the most difficult of all nested structures to write, read, and debug. For this reason, we will examine some examples involving nested DO loops.

A flow diagram of a pair of nested loops is shown in Fig. 6.13. The refinement of Step 2.3 in the outer loop is itself a loop. This means that during each repetition, or *iteration,* of the outer loop, the inner loop must also be entered and repeated as indicated by the loop parameters.

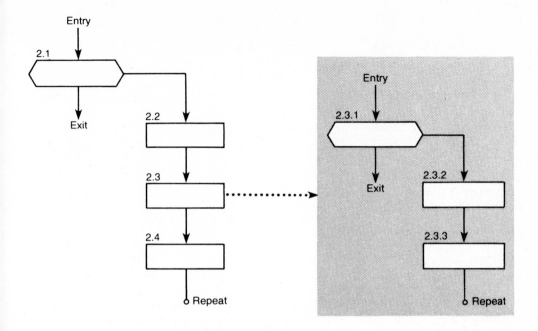

**Fig. 6.13**  Nested loops.

It is permissible to use the loop control variable of the outer loop as a parameter in the initialization, update, or test of an inner loop control variable. However, the same variable should never be used as the loop control variable of both an outer loop and an inner loop in the same nest. This would result in an illegal modification of the outer loop control variable within the loop body.

**Example 6.8:**   The following program contains a pair of nested DO loops.

```
      PRINT*, 'LOOP                      I                   J'
      DO 10 I = 1, 3
         PRINT*, 'OUTER ', I
         DO 20 J = 1, 2
            PRINT*, 'INNER ', I, J
   20    CONTINUE
   10 CONTINUE
```

For each execution of the outer loop (10), the PRINT statement in the inner loop (20) would execute twice. Thus, the PRINT statement in loop 20 would execute a total of 3 × 2, or 6 times. The output from these loops is shown below.

| LOOP  | I | J |
|-------|---|---|
| OUTER | 1 |   |
| INNER | 1 | 1 |
| INNER | 1 | 2 |
| OUTER | 2 |   |
| INNER | 2 | 1 |
| INNER | 2 | 2 |
| OUTER | 3 |   |
| INNER | 3 | 1 |
| INNER | 3 | 2 |

**Example 6.9:** The program in Fig. 6.14 plots the contents of an integer array, FREQ, in the form of a bar graph.

```
C PLOT THE ARRAY FREQ AS BAR GRAPH
C
C PROGRAM PARAMETERS
      CHARACTER * 1 STAR
      PARAMETER (STAR = '*')
C
C VARIABLES
      INTEGER FREQ(10), NBARS, NSTARS, COUNT, VALUE
C
C READ DATA INTO COUNT AND FREQ
      READ*, COUNT
      READ*, (FREQ(NBARS), NBARS = 1, COUNT)
      PRINT*, '             CLASS        FREQUENCY PLOT'
C
C PLOT EACH ELEMENT OF FREQ AS A "BAR"
      DO 10 NBARS = 1, COUNT
C        DISPLAY A PORTION OF IMAGE
         PRINT*, NBARS, ' I', (STAR, NSTARS = 1, FREQ(NBARS))
   10 CONTINUE
C
C BAR GRAPH IS COMPLETE
      PRINT*, '         I----I----I----I----I----I----I----I'
      PRINT*, '         5   10   15   20   25   30   35'
C
      STOP
      END
```

**Fig. 6.14**  Program for Example 6.9.

The program has three DO loops, including two implied DO loops. One implied loop is used for reading data into FREQ; the other is used for printing a string of asterisks and is nested within loop 10.

NBARS is the outer loop control variable and is used to cycle through the elements of the array FREQ. A bar consisting of a row of asterisks will be printed for each element of FREQ. NSTARS is the loop control variable for the inner loop of the nest (an implied DO loop). Since STAR is associated with the symbol '*', the statement

```
      PRINT*, NBARS, ' I', (STAR, NSTARS = 1, FREQ(NBARS))
```

instructs the computer to print a string of asterisks on each output line. The number of asterisks printed on a line is determined by the value of the element of

FREQ being represented, FREQ(NBARS), on that line. No asterisks are printed if this value is zero.

The bar graph printed by the plot program is shown in Fig. 6.15 for the sample values of COUNT and FREQ.

Array to be plotted

| FREQ(1) | FREQ(2) | FREQ(3) | FREQ(4) | FREQ(5) |
|---------|---------|---------|---------|---------|
| 8 | 32 | 24 | 16 | 3 |

Number of items in the array

| COUNT |
|-------|
| 5 |

Actual plot output

```
CLASS       FREQUENCY PLOT
   1  |********
   2  |********************************
   3  |************************
   4  |****************
   5  |***
      |----|----|----|----|----|----|----|
        5   10   15   20   25   30   35
```

**Fig. 6.15**  Bar graph for Example 6.9.

**Exercise 6.11:**  Write out each line of the printout for the following program.

```
      DO 10 I = 1, 2
         PRINT*, 'OUTER', I
         DO 20 J = 1, 4, 2
            PRINT*, 'INNER J', I, J
  20     CONTINUE
         DO 30 K = 2, 4, 2
            PRINT*, 'INNER K', I, K
  30     CONTINUE
  10 CONTINUE
```

## 6.5.4  Structure Entry and Exit

There is one important rule that pertains to all of the structures that have been presented in this text. This rule concerns the manner in which these structures should be entered.

---

**Structure Entry Rule**

All structures should be entered only "through the top". That is, *no statement within a structure can be executed without prior execution of the header statement of the structure.* Transfers into the middle of a structure from outside the structure are prohibited.

---

Transfers of control within a structure, or out of an inner structure, are acceptable although you should have little use for them; however, transfers from one alternative of a block-IF to another are prohibited.

**Example 6.10:**  The program segment below uses a GO TO statement to exit from DO loop 15; the GO TO statement transfers control to the first statement following the loop (label 20). This code segment is a solution to Exercise 6.9a; it finds the index (WHERE) of an item (NDROP) to be deleted from an ordered list (CLIST) of N numbers.

```
      FOUND = .FALSE.
      DO 15 IX = 1, N
         IF (CLIST(IX) .EQ. NDROP) THEN
            FOUND = .TRUE.
            WHERE = IX
            GO TO 20
         ENDIF
   15 CONTINUE
C
C TEST WHETHER NDROP WAS FOUND
   20 CONTINUE
```

The loop is exited either because **NDROP** has been found (GO TO 20 executed) or the end of the list was reached (normal exit). The list deletion should only be performed if the program flag **FOUND** is true.

**Example 6.11:**  The program segment below searches an array M for the first occurrence of a specified value, KEY. An entire list of keys is processed by the outer loop (30), and each is searched for in the inner loop (40). The index of a key is printed in loop 40 when it is found, and the GO TO statement causes the next key to be processed. If a key is not in array M, loop 40 will be exited normally (no transfer), and the message 'NOT FOUND' will be printed before the next key is processed.

```
      DO 30 J = 1, KEYCNT
C        GET NEXT KEY
         READ*, KEY
C        SEARCH FOR FIRST OCCURRENCE OF KEY
         DO 40 I = 1, N
            IF (M(I) .EQ. KEY) THEN
               PRINT*, KEY, 'FOUND AT INDEX', I
               GO TO 30
            ENDIF
   40    CONTINUE
         PRINT*, KEY, 'NOT FOUND'
   30 CONTINUE
```

The code segments shown in Examples 6.10 and 6.11 illustrate two of the most common uses of the GO TO statement in FORTRAN (aside from the

WHILE loop implementation described in Chapter 3). You should be familiar with both uses:

a) The statement GO TO 20 is used in Example 6.10 to *exit* from DO loop 15 before completing all N repetitions. Control is normally transfered to the statement immediately following the loop terminator. This statement must have a label; it is suggested that the label be attached to a CONTINUE statement. The loop control variable retains the value that it had at the time of the transfer.

b) The statement GO TO 30 is used in Example 6.11 to initiate immediate execution of the *next iteration* of DO loop 30. This is useful when conditions arise in an inner loop which call for the immediate termination of the current loop iteration, and the start of the next iteration of an outer loop. The transfer to the outer loop terminator (30 CONTINUE) causes the outer loop control variable to be incremented and tested, and if it has not passed the limit value, the next repetition of the outer loop is initiated. Note that if the GO TO statement were eliminated from this example, the message 'NOT FOUND' would be printed for all keys.

---

**Program Style**

*GO TO considered harmful*

The program segments in Examples 6.10 and 6.11 can both be written without using the GO TO statement. In general, it is preferable not to use the GO TO since it leads to program segments that are more difficult to read and debug. In fact, overuse of the GO TO is considered bad programming practice.

Use of the GO TO is justified to exit from a nest of loops while executing an inner loop or to start the next iteration of an outer loop from within an inner loop (Example 6.11). In these cases, a program written with the GO TO would be easier to understand than one without it.

---

**Exercise 6.12:**    Implement Examples 6.10 and 6.11 without using the GO TO. *Hints:* Use a WHILE loop in Example 6.10; use a program flag in Example 6.11.

## 6.6   SORTING AN ARRAY

The problem that follows is an example of the use of nested loops in sorting, or rearranging in order, the data stored in an array. Sorting programs are used in a variety of applications, and the program developed here could be easily modified to sort alphabetic data (such as last names) stored in a character array. In this example, we will sort numeric data in ascending numerical order (smallest value first); however, it would be just as easy to sort the data in descending order (largest value first).

**Problem 6B:**   Write a program to sort, in ascending order, an array of integer values.

**Discussion:**   There are many different algorithms for sorting. We will use one of the simplest of these algorithms, the *Bubble Sort.* The Bubble Sort is so named because it has the property of "bubbling" the smallest items to the top of a list. The algorithm proceeds by comparing the values of adjacent elements in the array. If the value of the first of these elements is larger than the value of the second, these values are exchanged, and then the values of the next adjacent pair of elements are compared. This process starts with the pair of elements with indices 1 and 2 and continues through to the pair of elements with indices n − 1 and n, in an array of size n. Then this sequence of comparisons (called a *pass*) is repeated, starting with the first pair of elements again, until the entire array of elements is compared without an exchange being made. At this point we know that the array is sorted.

As an example, we will trace through the sort of the integer array M as shown in Fig. 6.16. In this sequence of diagrams, diagram (1) shows the initial arrangement of the data in the array; the first pair of values are out of order and they are exchanged. The result is shown in diagram (2).

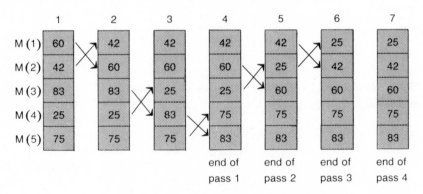

**Fig. 6.16**   Trace of Bubble Sort on a small array.

The sequence in Fig. 6.16 shows all exchanges that would be made during each pass through the adjacent pairs of array elements. After pass one, we see that the array is finally ordered except for the value 25. Subsequent passes through the array will "bubble" this value up one array element at a time until the sort is complete. In each pass through M, the elements are compared in the following order: M(1) and M(2); M(2) and M(3); M(3) and M(4); M(4) and M(5). Note that even though the array is sorted at the end of pass 3, it will take one more pass through the array without any exchanges to complete the algorithm.

Now that we have a general idea of how the algorithm works, we can write the initial data table and the flow diagrams for the Bubble Sort. The data table is shown next, and the level one flow diagram appears in Fig. 6.17a.

**Data Table for Problem 6B**

*Program parameter*

MAXSIZ = 10, maximum number of items that can be processed (integer)

| *Input variables* | *Output variables* |
|---|---|
| M: Array containing the data to be sorted (integer, size 10) | M: At the conclusion of the program, this array will contain the data sorted in ascending order (integer) |
| COUNT: Contains the number of array elements (integer) | |

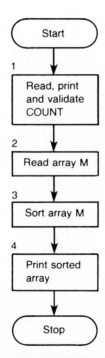

**Fig. 6.17a**   Level one flow diagram for Bubble Sort.

The refinement of Step 3 is a WHILE loop that processes array M until it becomes sorted. As shown in Fig. 6.17b, the program flag SORTED will be used to control repetition of this loop. Step 3.3 initializes SORTED to .TRUE. before each pass through array M (Step 3.4). If an exchange is made during the current pass, the array is not yet sorted; so, SORTED is reset to .FALSE. causing the WHILE loop to be repeated. When a pass is made without any exchanges, the value of SORTED will remain .TRUE. and the WHILE loop will be exited.

Since there must be at least one pass through the array, SORTED is initialized to .FALSE. before entering the WHILE loop (Step 3.1).

Step 3.4 must perform a complete pass through array M; consequently, its refinement is a DO loop. In the loop, adjacent pairs of elements of M are compared (Step 3.4.1).

As indicated in Step 3.4.2, if a pair of array elements is out of order, their values are exchanged (Step 3.4.3). A temporary storage cell, TEMP, is used to hold one of these values during the exchange. Note that INDEX, the loop control variable for the Step 3.4 refinement, always points to the first array element of any pair being compared; consequently, the limit expression for the DO loop must be COUNT − 1.

The data table additions for these refinements are shown below. Also listed is the loop control variable (I) for the implied input/output loops. The program is shown in Fig. 6.18.

**Additional Data Table Entries for Problem 6B**

*Program variables*

SORTED: Program flag—
a value of .TRUE. at the
completion of a pass in-
dicates no exchanges
were made and the ar-
ray is sorted. A value of
.FALSE. indicates at
least one exchange was
made (logical)

TEMP: Temporary storage cell
required for the exchange
(integer)

INDEX: Loop control variable
and array index (integer)

I: Loop control variable for the
implied loops (integer)

Remember that whenever nested loops are used, the inner loop is executed from start to finish for each iteration of an outer loop. In the program in Fig. 6.18, loop 34 (Step 3.4) will be executed for all values of INDEX between 1 and COUNT −1 for each repetition of the outer loop.

This kind of repetition can be quite difficult to understand, much less to program. It is, therefore, often extremely helpful to maintain a clear separation among all loops in a program, and to outline the logic of each loop separately,

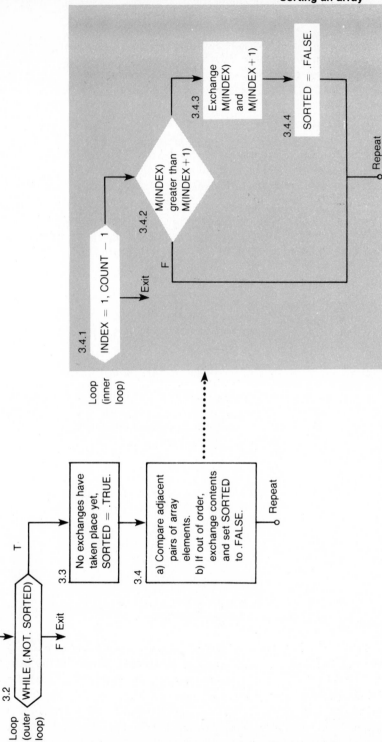

**Fig. 6.17b** Refinements of Step 3 of Fig. 6.17a.

```
C BUBBLE SORT PROGRAM
C ARRANGE A LIST OF INTEGER DATA ITEMS IN ASCENDING ORDER
C
C PROGRAM PARAMETERS
      INTEGER MAXSIZ
      PARAMETER (MAXSIZ = 10)
C
C VARIABLES
      INTEGER M(MAXSIZ), COUNT, INDEX, I, TEMP
      LOGICAL SORTED
C
C READ, PRINT AND VALIDATE COUNT
      READ*, COUNT
      PRINT*, 'THE NUMBER OF ITEMS TO BE SORTED IS ', COUNT
      IF (COUNT .LT. 1 .OR. COUNT .GT. MAXSIZ) THEN
         PRINT*, 'COUNT IS OUT OF RANGE OF 1 TO ', MAXSIZ
         PRINT*, 'PROGRAM EXECUTION TERMINATED'
         STOP
      ENDIF
C
C READ AND PRINT DATA TO BE SORTED
      READ*, (M(I), I = 1, COUNT)
      PRINT*, 'THE ORIGINAL, UNSORTED LIST IS ...'
      PRINT*, (M(I), I = 1, COUNT)
C
C OUTER LOOP -- REPEATED WHILE ARRAY IS NOT SORTED
      SORTED = .FALSE.
      WHILE (.NOT. SORTED) DO                     32 IF (SORTED)
         SORTED = .TRUE.                             Z GO TO 39
C        INNER LOOP -- COMPARE SUCCESSIVE PAIRS OF ITEMS
         DO 34 INDEX = 1, COUNT-1
            IF (M(INDEX) .GT. M(INDEX+1)) THEN
C              EXCHANGE OUT-OF-ORDER PAIRS
               TEMP = M(INDEX)
               M(INDEX) = M(INDEX+1)
               M(INDEX+1) = TEMP
               SORTED = .FALSE.
            ENDIF
   34    CONTINUE
      ENDWHILE                                       GO TO 32
C                                                  39 CONTINUE
C SORT COMPLETE. PRINT SORTED ARRAY
      PRINT*, 'THE FINAL, SORTED LIST IS ...'
      PRINT*, (M(I), I = 1, COUNT)
C
      STOP
      END


*** PROGRAM EXECUTION OUTPUT ***

THE NUMBER OF ITEMS TO BE SORTED IS        5
THE ORIGINAL, UNSORTED LIST IS ...
```

*(Continued)*

```
        60         42         83         25         75
   THE FINAL, SORTED LIST IS ...
        25         42         60         75         83
```

Fig. 6.18  Program for Problem 6B.

putting the loops together only at the final stage of writing the program. This may be accomplished simply by summarizing the activity of any loop nested within another (such as has been done for Step 3.4 in Fig. 6.17b), and then providing the details of execution of the inner loop in a separate diagram.

**Exercise 6.13:**  In Fig. 6.16, note that after pass i, the i largest values are in the correct order in array elements M(COUNT − i + 1), . . . , M(COUNT). Hence, it is only necessary to examine elements with indices less than COUNT − i + 1 during the next pass. Modify the algorithm to take advantage of this.

**Exercise 6.14:**  Modify the program in Fig. 6.18 to sort the array M in descending order (largest number first). Trace the execution of your program on the initial array shown in Fig. 6.16.

**Exercise 6.15:**  Modify the program in Fig. 6.18 so that the median or middle item of the final sorted array is printed out. If COUNT is even, the median should be the average of the two middle numbers, i.e., the average of the elements with subscripts COUNT/2 and COUNT/2 + 1. If COUNT is odd, the median is simply the value of the array element with subscript COUNT/2 + 1. A number is even if it is divisible by 2. (See Example 4.20.)

**Exercise 6.16:**  A different technique for sorting consists of searching the entire array to find the location of the smallest element, and then exchanging the smallest element with the first element. Next, elements 2 through N are searched and the next smallest element is exchanged with the second array element. This process continues until only elements N − 1 and N are left to search. Flow diagram the algorithm. *Hint:* Use a pair of nested DO loops.

**Exericse 6.17:**  A more efficient version of the bubble sort advances the smaller of each pair of elements being exchanged as far up the array as it can go. For this version, the second exchange shown in Fig. 6.16 would not be completed until the value 25 was advanced to the first element of the array. At this point the array M would be sorted. Write this version of the Bubble Sort Program. *Hint:* Replace the single alternative decision with a WHILE loop.

## 6.7  COMMON PROGRAMMING ERRORS

### 6.7.1  Structure Nesting Errors

Structure nesting errors are among the most common programming errors that are made. Such errors are more likely to occur when nested decision structures or multiple-alternative decision structures, with lengthy statement groups for each alternative, are used. Most compilers can detect structure nesting errors and will provide diagnostics informing the programmer that such errors have occurred.

To aid in obtaining the proper structure nesting, we urge you to faithfully follow the process of flow diagram refinement illustrated in the text. Refine each nested structure as a separate entity, and then carefully implement the refined flow diagram as a FORTRAN program. To retain the proper structure nesting, go back to the flow diagram when making any nontrivial changes to the algorithm. Rearranging structure components without referring to the flow diagram may introduce unexpected program logic and structure nesting errors.

The use of the GO TO statement should be restricted to implementing the WHILE loop, or initiating the next iteration or exit from a loop. For the next iteration, transfer should be to the loop terminator statement; when using the GO TO for loop exit, transfer should normally be to the first statement following the loop terminator.

### 6.7.2 Multiple-Alternative Decision Structure Errors

Care must be taken in listing the conditions to be used in a multiple-alternative decision structure. If the conditions are not *mutually exclusive* (that is, if more than one of the conditions can be true at the same time), then the condition sequence must be carefully ordered to ensure the desired results.

### 6.7.3 DO Loop Errors

The most common errors in writing DO loops involve the definition of the loop parameters in the header statement. Remember that the loop control variable must be exactly that—a variable, and not an array element, array name, constant, or expression. The loop parameters (*initial, limit, step*) may be any arithmetic expression subject to the following conditions:

- *Step* cannot have a value of 0.
- If the value of *step* is greater than 0, then *initial* must be less than or equal to *limit* in order for the loop to be executed.
- If the value of *step* is less than 0, then *initial* must be greater than or equal to *limit* in order for the loop to be executed.

We suggest that you adopt the convention of using only integer loop parameters even though FORTRAN does not require you to do so.

When writing DO loops, make sure that you do not attempt to redefine the value of the loop control variable inside the loop body. Any attempt to do this (by an assignment, read, etc.) will result in an error.

If you use expressions as your loop parameters, be sure that they are correct. Incorrect expressions will result in the wrong number of loop repetitions, or no repetitions, being performed. Too many loop repetitions may cause you to run out of input data and could result in an INSUFFICIENT DATA diagnostic message. If the loop control variable is used as an array subscript, you may get a SUBSCRIPT-OUT-OF-RANGE diagnostic when the loop parameters are incorrect. It is desirable to print the value of the loop control variable if you suspect it is not being manipulated properly.

Whenever practical, you should completely trace the execution of each loop to ensure that the number of repetitions is correct. At a minimum, you should test the "boundary values"; i.e., verify that the initial and final values of the loop control variable are correct. Furthermore, you should verify that all array references which use the loop control variable in subscript computations are within the declared range at the boundary values.

## 6.8  SUMMARY

With this chapter, we conclude the discussion of FORTRAN loop and decision control structures. A total of five structures have been presented in the text as illustrated in Table 6.2.

DECISION STRUCTURES

*Single-Alternative IF*

```
IF (INDEX .EQ. O) THEN
    PRINT*, 'KEY NOT FOUND'
ENDIF
```

*Single-Alternative IF–short version* (Use only with a single dependent statement)

```
IF (INDEX .EQ. O) PRINT*, 'KEY NOT FOUND'
```

*Double-Alternative IF*

```
IF (X .GE. Y) THEN
    LARGE = X
ELSE
    LARGE = Y
ENDIF
```

*Multiple-Alternative IF*

```
IF (X .LT. O.O) THEN
    PRINT*, 'X IS NEGATIVE'
ELSE IF (X .GT. O.O) THEN
    PRINT*, 'X IS POSITIVE'
ELSE
    PRINT*, 'X IS ZERO'
ENDIF
```

LOOP STRUCTURES

*WHILE Loop*

```
  READ*, X                           READ*, X
  WHILE (X .NE. SENVAL) DO    20 IF (X .EQ. SENVAL)
    PRINT*, X                      Z GO TO 29
    READ*, X                          PRINT*, X
  ENDWHILE                            READ*, X
                                    GO TO 20
                                29  CONTINUE
```

*DO Loop*

```
    SUM = O.O
    DO 40 I = 1, N
        SUM = SUM + X(I)
 40 CONTINUE
```

**Table 6.2 Summary of FORTRAN Control Structures**

Of these structures, only the WHILE loop is not part of the 1977 FOR-TRAN standard; however, we have illustrated its implementation using standard statements.

The multiple-alternative decision structure (block-IF) introduced in this chapter is extremely useful in describing algorithms containing decisions for which there are more than two alternatives. Such situations could be described using sequences and/or nests of single- and double-alternative decision structures, but these can be extremely difficult to organize, and virtually impossible to understand. The multiple-alternative decision structure should simplify the implementation of decision sequences.

The DO loop provides a most convenient means for representing loops in which execution is controlled by a counter. This structure permits the specification of all loop control information in the loop header statement. The programmer is thereby freed from having to separately program the counter initialization, increment, and test.

All five structures may be used in a FORTRAN program, and any of them may be nested inside another. (There will usually be a limit to the depth of nesting permitted, but this limit is normally large enough that there is little need to be concerned about it.) It is essential, however, that any structure nested inside another structure begin and end within the same statement group. Thus, a loop that begins inside one alternative of a decision structure must end within the same alternative of that structure.

All structures must be entered via the execution of the header statement. Transfers into the middle of a structure are prohibited. Exit from the middle of a loop structure and the initiation of the next iteration of an outer loop may be accomplished using a GO TO statement.

## PROGRAMMING PROBLEMS

**6C**   An instructor has just given an exam to a very large class, and has punched the grades onto cards, one grade per card. The grading scale is 90–100 (A), 80–89 (B), 70–79 (C), 60–69 (D), 0–59 (F). The instructor wants to know how many students took the exam, what the average and standard deviation were for the exam (see Problem 5B), and how many A's, B's, C's, D's, and F's there were. Write a program using a loop and a multiple-alternative structure to help the instructor obtain the desired information.

**6D**   A tax table is used to determine the tax rate for a company employee, based on weekly gross salary and number of dependents. The tax table has the form shown below. An employee's net pay can be determined by multiplying gross salary times the tax rate, and subtracting this product from the gross salary. Write a program to read in the ID number, number of dependents, and gross salary for each employee of a company, and then determine the net salary to be paid to each employee. Your program should also print out a count of the number of employees with gross salary in each of the three ranges shown. *Hint.* Use a multiple-alternative decision structure to "implement" this table. Note that the increase in rate for each column is constant (0.1 for 0–100, 0.12 for 100–200, 0.13 for $\geq 200$).

Gross salary

|  | | 0–100 | 100–200 | ≥ 200 |
|---|---|---|---|---|
| Number of | 0 | 0.2 | 0.28 | 0.38 |
| dependents | 1 | 0.1 | 0.16 | 0.25 |
|  | ≥ 2 | 0.0 | 0.04 | 0.12 |

Tax rate table

**6E**   *Continuation of the insertion and deletion problem, Example 6.6 and Exercise 6.9.* In this chapter, separate program segments were written to maintain a collection of student identification numbers as an ordered list of numbers stored in the array CLIST. Write a single program that will process both deletions and insertions. Assume that all numbers to be inserted are preceded by 'I' and that numbers to be deleted are preceded by 'D'. You will have to test each value read to see whether a deletion or an insertion is required. *Hints.* The first step in the deletion process involves a search for the student number to be deleted. If the number isn't found, or if the array CLIST is empty, print an error message and ignore the request.

If the item is found in array element CLIST (ISCH), the deletion process involves moving each element from CLIST (ISCH + 1) through CLIST (N) into elements CLIST (ISCH) through CLIST (N − 1).

Read into CLIST (1) through CLIST (17) the numbers:

| | | | | | |
|---|---|---|---|---|---|
| 502 | 923 | 1045 | 2113 | 4642 | 8192 |
| 10974 | 14673 | 21892 | 33574 | 33575 | 33576 |
| 41821 | 44444 | 58912 | 71125 | 88893 | |

Then process the insertions and deletions given below.

| | |
|---|---|
| 'I' | 16891 |
| 'D' | 33575 |
| 'I' | 43627 |
| 'I' | 121 |
| 'D' | 21212 |
| 'I' | 91741 |
| 'I' | 33575 |
| '*' | 0 (Sentinel card) |

**6F**   The equation of the form

$$(1)\ mx + b = 0$$

(where $m$ and $b$ are real numbers) is called a linear equation in one unknown, $x$. If we are given the values of both $m$ and $b$, then the value of $x$ that satisfies this equation may be computed as

$$(2)\ x = -b/m$$

Write a program to read in $N$ different sets of values for $m$ and $b$ (punched one set per card), and compute $x$. Test your program for the following five value pairs.

| $m$ | $b$ |
|---|---|
| − 12.0 | 3.0 |
| 0.0 | 18.5 |
| 100.0 | 40.0 |
| 0.0 | 0.0 |
| − 16.8 | 0.0 |

*Hint.* There are three distinct possibilities concerning the values of $x$ that satisfy the equation $mx + b = 0$.

1. As long as $m \neq 0$, the value of $x$ that satisfies the original equation (1) is given by equation (2).
2. If both $b$ and $m$ are 0, then any real number that we choose satisfies $mx + b = 0$.
3. If $m = 0$ and $b \neq 0$, then no real number $x$ satisfies this equation.

**6G**     Each year the legislature of a state rates the productivity of the faculty of each of the state-supported colleges and universities. The rating is based on reports submitted by each faculty member indicating the average number of hours worked per week during the school year. Each faculty member is ranked, and the university also receives an overall rank.

The faculty productivity rank is computed as follows:

a) faculty members averaging over 55 hours per week are considered "highly productive";
b) faculty members averaging between 35 and 55 hours a week (inclusive) are considered "satisfactory";
c) faculty members averaging fewer than 35 hours a week are considered "overpaid."

The productivity rating of each school is determined by first computing the faculty average for the school:

$$\text{Faculty average} = \frac{\Sigma \text{ hours worked per week for all faculty}}{\text{Number of faculty reporting}}$$

and then applying the faculty average to the category ranges defined in (a), (b), and (c).

Use the multiple-alternative decision structure and write a program to rank the following faculty:

| | |
|---|---|
| HERM | 63 |
| FLO | 37 |
| JAKE | 20 |
| MO | 55 |
| SOL | 72 |
| TONY | 40 |
| AL | 12 |
| ZZZZ | 0 (Sentinel card) |

Your program should print a three-column table giving the name, hours, and productivity rank of each faculty member. It should also compute and print the school's overall productivity ranking.

**6H**     Write a savings-account transaction program that will process the following input data:

| | | |
|---|---|---|
| 'ADAM' | 1054.37 | |
| 'W' | 25.00 | group 1 |
| 'D' | 243.35 | |
| 'W' | 254.55 | |
| 'EVE' | 2008.24 | group 2 |
| 'W' | 15.55 | |

|  |  |  |
|---|---|---|
| 'MARY' | 128.24 | group 3 |
| 'W' | 62.48 |  |
| 'D' | 13.42 |  |
| 'W' | 84.60 |  |
| 'SAM' | 7.77 | group 4 |
| 'JOE' | 15.27 | group 5 |
| 'W' | 16.12 |  |
| 'D' | 10.00 |  |
| 'BETH' | 12900.00 | group 6 |
| 'D' | 9270.00 |  |
| 'ZZZZ' | 0.0 | (Sentinel card) |

The first card in each group (header card) gives the name of an account and the starting balance in the account. All subsequent cards show the amount of each withdrawal (W) or deposit (D) that was made for that account. Each data card that does not contain a W or D as the first item is the header card for the next account. Print out the final balance for each of the accounts processed. If a balance becomes negative, print an appropriate message and take whatever corrective steps you deem proper. If there are no transactions for an account, print a message so indicating.

**6J** (Variation on the mortgage interest problem—Problem 4L) Use DO loops to write a program to print tables of the following form.

Home loan mortgage interest payment tables

Amount _____ Loan duration (Months) _____

| Rate (Percent) | Monthly payment | Total payment |
|---|---|---|
| 8.00 | | |
| 8.25 | | |
| 8.50 | | |
| 8.75 | | |
| 9.00 | | |
| . | | |
| . | | |
| . | | |
| 13.00 | | |

Your program should produce tables for loans of 30, 40, and 50 thousand dollars, respectively. For each of these three amounts, tables should be produced for loan durations of 240, 300, and 360 months. Thus, *nine* tables of the above form should be produced. Your program should contain three nested loops. Be careful to remove all redundant computations from inside your loops, especially from inside the innermost loop.

**6K** The equation of the form

$$ax^2 + bx + c = 0 \qquad (a, b, c \text{ real numbers, with } a \neq 0)$$

is called a quadratic equation in $x$. The *real roots* of this equation are those values of $x$ for which

$$ax^2 + bx + c$$

evaluates to zero. Thus if $a = 1$, $b = 2$, and $c = -15$, then the real roots of

$$x^2 + 2x - 15$$

are $+3$ and $-5$, since

$$(3)^2 + 2(3) - 15 = 9 + 6 - 15 = 0$$

and

$$(-5)^2 + 2(-5) - 15 = 25 - 10 - 15 = 0.$$

Quadratic equations have either 2 real and different roots, 2 real and equal roots, or *no* real roots. The determination as to which of these three conditions holds for a given equation can be made by evaluating the discriminant $d$ of the equation, where

$$d = b^2 - 4ac.$$

1. If $d > 0$, then the equation has two real and unequal roots.
2. If $d = 0$, the equation has two real and equal roots.
3. If $d < 0$, the equation has no real roots.

Write a program to compute and print the real roots of quadratic equations having the following values of $a$, $b$, and $c$.

| $a$ | $b$ | $c$ |
|-----|-----|-----|
| 1.0 | 2.0 | -15.0 |
| 1.0 | -1.25 | -9.375 |
| 2.0 | 0.0 | 1.0 |
| 5.0 | -80.0 | -900.0 |
| 1.0 | -6.0 | 9.0 |
| 0.0 | 0.0 | 0.0 |

You should punch each set of values for $a$, $b$, and $c$ on one card, and terminate the program when a value for $a$ of 0.0 is read. If the equation has no real roots for a set of $a$, $b$, and $c$, print an appropriate message, and read the next set. *Hint.* If the equation has two real and equal roots, then the root values are given by the expression

$$\text{Root } 1 = \text{Root } 2 = -b/2a.$$

If the equation has two real and unequal roots, their values may be computed as

$$\text{Root } 1 = \frac{-b + \sqrt{d}}{2a}$$

$$\text{Root } 2 = \frac{-b - \sqrt{d}}{2a}$$

**6L**    Write a program to solve the following problem:

Read in a collection of $N$ data cards, each containing one integer between 0 and 9, and count the number of consecutive pairs of each integer occurring in the data card set. (Your program should print the number of consecutive pairs of 0's, of 1's, 2's, . . . , and the number of consecutive pairs of 9's found in the data.)

**6M**    Write a program which will provide change for a dollar for any item purchased that costs less than one dollar. Print out each unit of change (quarter, dimes, nickels, or pennies) provided. Always dispense the biggest-denomination coin possible.

For example, if there are 37 cents left in change, dispense a quarter (which leaves 12 cents in change), then dispense a dime, and then 2 pennies. You may wish to use a multiple-alternative decision structure in solving this problem. However, you can also use a four-element array (to store each denominational value 25, 10, 5, and 1), and a DO loop.

**6N** *Statistical measurements—a linear-curve fit problem.* Scientists and engineers frequently perform experiments designed to provide measurements of two variables X and Y. They often compute measures of central tendency (such as the mean) and measures of dispersion (such as the standard deviation) for these variables, and then attempt to decide whether or not there is any relationship between the variables, and, if so, to express this relationship in terms of an equation. If there is a relationship between X and Y that is describable using a linear equation of the form

$$Y = aX + b$$

the data collected is said to *fit* a *linear curve*.

For example, the ACE Computing Company recently made a study relating aptitude test scores to programming productivity of new personnel. The 6 pairs of scores shown below were obtained by testing 6 randomly selected applicants and later measuring their productivity.

| Applicant | Aptitude score (*Variable* X) | Productivity (*Variable* Y) |
|-----------|-------------------------------|------------------------------|
| 1 | $x_1 = 9$ | $y_1 = 46$ |
| 2 | $x_2 = 17$ | $y_2 = 70$ |
| 3 | $x_3 = 20$ | $y_3 = 58$ |
| 4 | $x_4 = 19$ | $y_4 = 66$ |
| 5 | $x_5 = 20$ | $y_5 = 86$ |
| 6 | $x_6 = 23$ | $y_6 = 64$ |

ACE wants to find the equation of the line which they can use to predict the productivity of workers tested in the future. They are also interested in obtaining means and standard deviations for the variables X and Y. The required computations can be performed as follows:

1. Compute     $\text{SUMX} = \Sigma \, X = x_1 + x_2 + \cdots + x_6$

                 $\text{SUMY} = \Sigma \, Y = y_1 + y_2 + \cdots + y_6$

                 $\text{SUMXY} = \Sigma \, X{\cdot}Y = x_1 y_1 + y_2 y_2 + \cdots + x_6 y_6$

                 $\text{SUMXSQ} = \Sigma \, X^2 = x_1^2 + x_2^2 + \cdots + x_6^2$

                 $\text{SUMYSQ} = \Sigma \, Y^2 = y_1^2 + y_2^2 + \cdots + y_6^2$

2. Compute     $\text{MEANX} = \text{SUMX}/\text{N}$ where $\text{N} = 6$

                 $\text{MEANY} = \text{SUMY}/\text{N}$

3. Compute     $\text{STDDVX} = \sqrt{\text{SUMXSQ}/\text{N} - \text{MEANX}^2}$

                 $\text{STDDVY} = \sqrt{\text{SUMYSQ}/\text{N} - \text{MEANY}^2}$

4. Compute     $a$ and $b$ in $Y = aX + b$ using the equation

$$a = \frac{\text{SUMXY} + \text{N} * \text{MEANX} * \text{MEANY}}{\text{SUMXSQ} - \text{N} * \text{MEANX}^2}$$

$$b = \text{MEANY} - a * \text{MEANX}$$

Write a program to carry out the above computations. Test your program on the aptitude/productivity data just shown.

**6O**    Write a program to print a table of the first 100 prime numbers. To determine whether a number N is prime, it is sufficient to test all primes smaller than N as divisors. Use this fact in developing your data table and algorithm.

# SUBPROGRAMS

## 7.1   INTRODUCTION

One of the most fundamental ideas of computer programming and problem solving concerns the subdivision of large and complex problems into smaller, simpler and more manageable subproblems. Once these smaller tasks have been identified, the solution to the original problem can be specified in terms of these tasks, and the algorithms and programs for the smaller tasks can be developed separately.

We have tried to emphasize this technique of programming in all earlier examples through the use of *step-wise algorithm refinement.* In this process, each major part of a problem was identified in a level one flow diagram, and then further broken down into smaller problems during successive stages of refinement. A number of special control structures were introduced that enabled us to implement the solution to each of these subproblems in terms of clearly defined groups of FORTRAN program statements.

FORTRAN has still another feature, called a *subprogram,* which facilitates solving problems in terms of their more manageable parts. The use of subprograms allows us to write separate *program modules* to solve small problems and then call (reference) these modules in the overall solution of the original problem. Each module can be implemented and tested independently of the others.

An additional advantage is that each subprogram may be reused easily in the same program or another program to manipulate different data. To accomplish this, only the statement that calls the subprogram would have to be changed, not the subprogram itself.

In the following sections we will explain how to write (*define*) subprograms and how to *call* them in a *main program* or in other subprograms. We will also show how the main program and subprograms communicate with one another, and how data manipulated in one subprogram may be transmitted to another. Two different types of subprograms will be discussed: *function* subprograms and *subroutine* subprograms.

## 7.2   FUNCTION SUBPROGRAMS

### 7.2.1   Review of Library Functions

In Chapter 4 we described a number of special subprograms called *library functions.* These functions are usually written by computer manufacturers for use by FORTRAN programmers. They enable the programmer to easily incorporate some very common numerical computations into a program. Some of the more standard functions that were described are INT and REAL (for type conversion), SQRT (square root), ABS (absolute value), MOD (compute remainder), LOG (logarithm), and the trigonometric functions SIN and COS.

Recall from Chapter 4 that a library function is referenced simply by specifying the name of the function, followed by a list of input values (*arguments*) enclosed in parentheses. Whenever a call to a function is encountered in a program, control is transferred from the *calling program* to the function referenced.

The function manipulates the argument(s), and when the function computation is complete, the result is returned and control is transferred back to the calling program at the point of the call. This process is illustrated in the next example.

**Example 7.1:** The program segment below contains two calls to the library function SQRT.

```
          .
          .
          .
     X = 24.0
     Z = 25.0
     W = SQRT(Z) + 6.5
     Q = SQRT(X + Z)
          .
          .
          .
```

In the first call, the argument Z has a value of 25.0; the result of the function execution, 5.0, is returned and added to 6.5. The value assigned to W is 11.5. In the second call, the argument is an expression. This expression, (X + Z), must be evaluated before the function can be executed. Its value, 49.0, is *passed* to the function and the result, 7.0, is *returned* and assigned to Q. This process is summarized below.

| Call | Function argument | Argument value | Function result | Final effect of statement |
|------|-------------------|----------------|-----------------|---------------------------|
| First | Z | 25.0 | 5.0 | 11.5 stored in W |
| Second | X + Z | 49.0 | 7.0 | 7.0 stored in Q |

### 7.2.2 Defining New Functions

Often, the functions provided in the function library are not sufficient for the solution of a particular problem, and we may wish to write our own. In this section, we will see how to write or *define* a function subprogram.

**Example 7.2:** The function MYSIGN defined below determines the sign ('−' or '+') of its real argument.

```
          CHARACTER * 1 FUNCTION MYSIGN (X)
     C
     C DETERMINE THE SIGN OF A REAL VALUE
     C
          REAL X
          IF (X .LT. 0.0) THEN
             MYSIGN = '-'
          ELSE
             MYSIGN = '+'
          ENDIF
     C
          RETURN
          END
```

The first statement identifies MYSIGN as a type CHARACTER * 1 function; it returns a single character as its result. X represents the data to be manipulated; it is called a *dummy argument* of the function and is declared as type REAL. The IF-THEN-ELSE structure defines the function result by assigning a value (either '−' or '+') to the function name, MYSIGN. The RETURN statement transfers control back to the statement which called the function.

The statement

```
CHSIGN = MYSIGN(-3.8)
```

would cause the function MYSIGN to be executed; the *actual argument,* −3.8, would be substituted for the dummy argument X. The result returned would be '−', which is stored in CHSIGN. The effect of this call to MYSIGN is illustrated below.

*Main program*

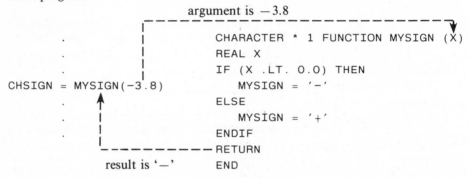

**Example 7.3:**   The function LARGER defined below finds the larger of its two integer arguments. The dummy arguments, M and N, are used in the function definition to represent the *actual arguments* that are manipulated when the function is called. The value returned by LARGER is the larger of its actual arguments (an integer value).

```
      INTEGER FUNCTION LARGER (M, N)
C
C DETERMINE THE LARGER OF TWO INTEGER VALUES
C
      INTEGER M, N
      IF (M .GT .N) THEN
         LARGER = M
      ELSE
         LARGER = N
      ENDIF
C
      RETURN
      END
```

The statement

```
      I = LARGER(I, 10)
```

imposes a lower limit of ten on the value of I. (Why?) During the execution of

LARGER, the actual argument I is substituted for the dummy argument M, and the actual argument 10 is substituted for the dummy argument N.

The general form for a function definition is described in the next display.

---

**FUNCTION Definition**

> *ftype* FUNCTION *fname* (*dummy argument list*)
> *argument definition section*
> *local declaration section*
>
> $\left.\begin{array}{l} ----\\ ----\\ ----\\ ---- \end{array}\right\}$   *function body*
>
> RETURN
> END

**Interpretation:** The FUNCTION statement specifies the function name, *fname,* and the type of the result, *ftype.* The *dummy argument list* is a list of symbolic names. The *argument definition section* contains descriptions and declarations for all the dummy arguments. Other symbolic names not appearing in the dummy argument list, but required for writing the function, should be declared separately in the *local declaration section.*

The *function body* describes the data manipulation performed by the function. The RETURN statement transfers control back to the calling statement. The END statement marks the end of the function definition.

---

There are some additional points that must be kept in mind when defining functions. These points are listed in the next display and illustrated in the rest of the section.

---

**Rules for Defining Functions**

1. The function type specifies the type of the result returned by the function (INTEGER, REAL, LOGICAL, or CHARACTER * n). If the type specification is omitted from the function statement, it is determined by the first letter typing convention (I-N, type INTEGER; otherwise, type REAL).

2. At least one statement assigning a value to the function name, *fname,* must be executed each time the function is called. Otherwise, the function result will be undefined. Usually this statement is an assignment statement of the form

$$fname = expression$$

The function name may appear only on the left side of an assignment as shown above.

3. If a dummy argument is the name of an array, both its size and type must be declared in the argument definition section of the function.

4. Only symbolic names may appear in the dummy argument list. Constants, expressions, or array elements are not permitted.

---

**Example 7.4:**  The function below can be used to calculate the tuition charge (in dollars), given the number of credit hours taken during one semester by a resident student at a public university. Full-time students taking 12 hours and over are charged a flat rate of $900.00. Students taking less than 12 hours are charged $80 per credit hour.

```
      REAL FUNCTION CHARGE (HOURS)
C
C COMPUTE TUITION CHARGE GIVEN CREDIT HOURS TAKEN
C
      REAL HOURS
      REAL COST, FULL, SEMEST
      PARAMETER (COST = 80.0, FULL = 12.0, SEMEST = 900.0)
      IF (HOURS .GE. FULL) THEN
         CHARGE = SEMEST
      ELSE
         CHARGE = COST * HOURS
      ENDIF
C
      RETURN
      END
```

In the **CHARGE** function, the dummy argument **HOURS** represents the number of credit hours taken by a student whose tuition is being computed. For example, if the statement

```
      TUITN = CHARGE(10.5)
```

were used to call **CHARGE**, **HOURS** would be replaced by the actual argument 10.5. The value returned from this call to **CHARGE** would be 10.5 × 80.00, or 840.00.

Three parameters, COST, FULL, and SEMEST, are declared in function **CHARGE**. These parameters are considered *local* to the function and would have no relevance outside the function definition. They would have to be redeclared and redefined in order to be used in any other program module or segment.

As another example, we could write the statements

```
      HOURS = 0.0
      DO 10 I = 1, N
         HOURS = HOURS + SEMHRS(I)
   10 CONTINUE
      TUITN = CHARGE(HOURS)
```

to compute the tuition charge for a student taking N courses in a semester, where N is likely to be between 1 and 6. If SEMHRS is an array containing the number of credit hours for each course, then the DO loop computes the total credit hours taken. The call to function **CHARGE** would return the tuition cost for the total credit hours taken. In this example, the actual argument is a variable with the same name (HOURS) as the dummy argument used in the function definition. This is not necessary, but it also causes no difficulties.

### 7.2.3  Using Functions

Fig. 7.1 provides an illustration of a main program that calls the function LARGER (Example 7.3) to find the largest of four numbers stored in NUM1, NUM2, NUM3, and NUM4. The function definition follows directly after the main program END statement. For a batch environment, the data separator card and all external data would follow the function END statement.

```
C FIND THE LARGEST OF FOUR NUMBERS
C
      INTEGER NUM1, NUM2, NUM3, NUM4, TEMP1, TEMP2, MAX
C
C FUNCTIONS USED
      INTEGER LARGER
C
C READ AND PRINT THE DATA
      READ*, NUM1, NUM2, NUM3, NUM4
      PRINT*, 'THE FOUR DATA ITEMS ARE ...'
      PRINT*, NUM1, NUM2, NUM3, NUM4
      PRINT*, ' '
C
C FIND AND PRINT THE LARGEST VALUE
      TEMP1 = LARGER(NUM1, NUM2)
      PRINT*, 'THE LARGER OF THE FIRST TWO ITEMS IS ', TEMP1
      TEMP2 = LARGER(NUM3, NUM4)
      PRINT*, 'THE LARGER OF THE SECOND TWO ITEMS IS ', TEMP2
      MAX = LARGER(TEMP1, TEMP2)
      PRINT*, 'THE LARGEST VALUE IS ', MAX
C
      STOP
      END


      INTEGER FUNCTION LARGER (M, N)
C
C DETERMINE THE LARGER OF TWO INTEGER VALUES
C
      INTEGER M, N
      IF (M .GT. N) THEN
         LARGER = M
      ELSE
         LARGER = N
      ENDIF
C
      RETURN
      END

*** PROGRAM EXECUTION OUTPUT ***

THE FOUR DATA ITEMS ARE ...
            10              15              20               5
```

*(Continued)*

```
THE LARGER OF THE FIRST TWO ITEMS IS                    15
THE LARGER OF THE SECOND TWO ITEMS IS                   20
THE LARGEST VALUE IS                    20
```

**Fig. 7.1**  Using the function LARGER.

Note that the type of function **LARGER** is also declared in the main program. This is necessary because the compiler must know the type of **LARGER** when translating an expression that references it. The implicit type convention is used if **LARGER** is not declared in the main program; however, we recommend that you declare all function types explicitly.

In the main program, the function **LARGER** is called three times. After each of the first two executions, the larger of its two arguments is stored in a temporary variable. Program execution terminates when the STOP statement in the main program is reached.

This sequence of statement executions is indicated below. The main program steps are shown on the left; the function steps are shown on the right.

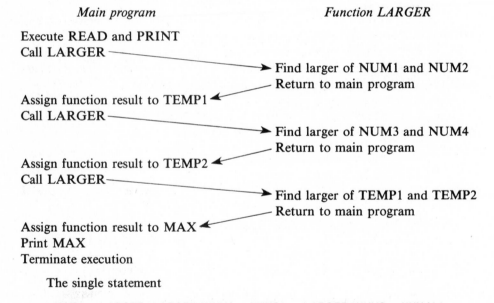

*Main program*                              *Function LARGER*

Execute **READ** and **PRINT**
Call **LARGER**
                                            Find larger of NUM1 and NUM2
                                            Return to main program
Assign function result to TEMP1
Call **LARGER**
                                            Find larger of NUM3 and NUM4
                                            Return to main program
Assign function result to TEMP2
Call **LARGER**
                                            Find larger of TEMP1 and TEMP2
                                            Return to main program
Assign function result to MAX
Print **MAX**
Terminate execution

The single statement

```
MAX = LARGER(LARGER(NUM1, NUM2), LARGER(NUM3, NUM4))
```

could also be used to find the largest value. This statement contains nested calls to the function **LARGER**. The inner calls are executed first, and then the outer call is executed. The value returned by the outer call is stored in **MAX**.

The major points illustrated in the previous examples are summarized in the following display.

---

**Rules for Using Functions**

1. The type of each function (as specified in its definition) must be declared in any module that references it; otherwise, the first letter typing convention is assumed.

> 2. To call (reference) a function, the function name and its actual arguments must be used in an expression.
>
> 3. There must be the same number of actual arguments as there are dummy arguments.
>
> 4. The actual arguments must be listed in the same order as their corresponding dummy arguments.
>
> 5. An actual argument may be a constant, variable, array name, array element, or expression; however, the type and structure of each actual argument must be the same as its corresponding dummy argument.

As indicated in the above display, each actual argument in the function call corresponds to a dummy argument used in the function definition. The value represented by the actual argument is manipulated in place of the dummy argument when the function is executed. The correspondence that is established is based on position, i.e., the first actual argument corresponds to the first dummy argument, the second actual argument corresponds to the second dummy argument, etc. We shall have more to say about this correspondence later.

According to rule 5, the type of each actual argument and its corresponding dummy argument must be the same. Further, each dummy argument that represents an array must correspond to an actual argument that is an array name. Each dummy argument that is a simple variable must correspond to an actual argument that is a variable, array element, constant, or expression.

**Exercise 7.1:** Indicate how the library function MAX0 could be used instead of LARGER to find the largest of four numbers.

**Exercise 7.2:** Write an integer function SQR(N) to compute the square of the integer argument N.

**Exercise 7.3:** Write an integer function POWER(N, K) to compute $N^K$ for any pair of integers N and K, $K \geq 0$.

**Exercise 7.4:** Write a logical function SAME(X, Y, Z) that compares the real variables X, Y and Z and returns .TRUE. if $X = Y = Z$, and returns .FALSE. otherwise.

### 7.2.4   Single Statement Functions*

Occasionally, a function computation can be specified in terms of an expression written in a single line. For example, the *single statement function* POWER defined as

```
POWER (X, N) = X ** N
```

can be used to raise a number (represented by X) to the power N.

Although it has the appearance of an executable statement, a single statement function definition is actually a declaration. Hence, it must come directly after all other declarations and before the executable statements of any program module that references it. The type of the single statement function and its arguments should also be declared before the function definition itself. The dummy

---

*This section is optional and may be omitted.

argument names (X and N for **POWER**) may be reused for other purposes in the program module.

A single statement function is called just like any other function. Given the function **POWER** as previously defined, the statement

$$Z = POWER(1.5, 10)$$

could be used to store in Z the value of 1.5 raised to the power ten.

The single statement function definition is described in the next display.

---

**Single Statement Function Definition**

*fname (dummy argument list) = expression*

**Interpretation:** *fname* is defined as a single statement function whose value is determined by evaluating the *expression*. The *expression* may reference variables, constants, and other functions as well as the names in the *dummy argument list*. When the function *fname* is called, the actual argument values are substituted for the corresponding dummy arguments, and the *expression* is evaluated.

*Note:* Fname and *expression* must be the same type, or one may be type real and the other type integer. In the latter case, the compiler will convert the expression value. Only simple variables (not arrays) may be used as dummy arguments.

---

**Example 7.5:** The program shown in Fig. 7.2 computes the average value (AVE) and the standard deviation (STDEV) for an array of N exam scores (SCORES). It uses two single statement functions, OUTBND (type LOGICAL) and SQUARE (type REAL) as well as the library function REAL. OUTBND evaluates to true when the value of N is outside the range one to one hundred; in this case, the program is terminated. The function SQUARE is used in the computation of the standard deviation. The formula used for the standard deviation is

$$STDEV = \sqrt{\frac{\sum_{I=1}^{N} SCORES(I)^2}{N} - AVE^2}$$

```
C COMPUTE AVERAGE AND STANDARD DEVIATION OF EXAM SCORES
C USING SINGLE STATEMENT FUNCTIONS
C
      INTEGER SCORES(100), N, I, SUM
      REAL AVE, STDEV, SUMSQ, X
C
C DEFINE THE SINGLE STATEMENT FUNCTIONS USED
      LOGICAL OUTBND
      REAL SQUARE
      OUTBND(N) = (N .LT. 1) .OR. (N .GT. 100)
      SQUARE(X) = X ** 2
C
```

(Continued)

```
C READ, PRINT AND TEST N -- READ AND PRINT SCORES
      READ*, N
      PRINT*, 'THE NUMBER OF ITEMS IS ', N
      IF (OUTBND(N)) THEN
         PRINT*, 'N IS OUT OF RANGE'
         STOP
      ENDIF
      READ*, (SCORES(I), I = 1, N)
      PRINT*, 'THE EXAM SCORES ARE ...'
      DO 10 I = 1, N
         PRINT*, SCORES(I)
   10 CONTINUE
C
C COMPUTE SUMS
      SUM = 0
      SUMSQ = 0
      DO 20 I = 1, N
         SUM = SUM + SCORES(I)
         SUMSQ = SUMSQ + SQUARE(REAL(SCORES(I)))
   20 CONTINUE
C
C COMPUTE AND PRINT THE AVERAGE AND STANDARD DEVIATION
      AVE = REAL(SUM) / REAL(N)
      STDEV = SQRT(SUMSQ / REAL(N) - SQUARE(AVE))
      PRINT*, ' '
      PRINT*, 'AVERAGE = ', AVE
      PRINT*, 'STANDARD DEVIATION = ', STDEV
C
      STOP
      END


*** PROGRAM EXECUTION OUTPUT ***

THE NUMBER OF ITEMS IS                    4
THE EXAM SCORES ARE ...
               10
                1
                0
                9

AVERAGE =    5.000000000000
STANDARD DEVIATION =    4.527692569069
```

**Fig. 7.2**  Computation of the standard deviation.

**Exercise 7.5:**  Write a single statement function ROUND(X, N) to round a real number (represented by X) to N decimal digits (N is a non-negative integer). *Hint:* You will have to multiply and divide by 10.0 ** N.

**Exercise 7.6:**  a) Write a single statement function CONVRT(C) which converts temperatures given in degrees Celsius (C) to temperatures in degrees Fahrenheit (F), where

$$F = 1.8 * C + 32.0$$

b) Let M be a real parameter declared as

$$\text{REAL M}$$
$$\text{PARAMETER (M = 1.8)}$$

Rewrite the single statement function CONVRT using M as the coefficient for C.

c) Using the function CONVRT defined in part b), write a FORTRAN program segment (a DO loop) to generate a table for Celsius to Fahrenheit conversion for degrees C = 0,1,2, . . . 99,100.

## 7.3   SEPARATE COMPILATION OF SUBPROGRAMS—NAME INDEPENDENCE

When preparing job decks with subprograms, the subprograms (functions or subroutines) should be placed immediately after the END statement in the main program. If there is more than one subprogram, the order of the subprograms is immaterial. Each subprogram must have its own header and END statement, and the type of a function must be declared in any module that references it.

In a batch environment (see Fig. 7.3), any external data should follow the last subprogram. The data cards should be provided in the order in which they are used regardless of whether the data are read by the main program or a subprogram.

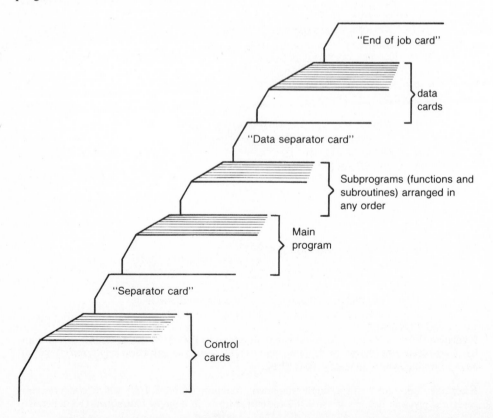

**Fig. 7.3**   Job deck for a main program and subprograms (batch environment).

The above comments also apply to the interactive environment. However, in this case, the data will be entered as requested by the program. Be sure to print prompting messages before each read operation.

When the compiler processes your program and subprograms, in either the batch or interactive environment, each module is translated separately. Thus, there is complete *name independence* between modules; that is, the names and statement labels used in one module have absolutely no relation to those used in other modules. Correspondences between the names used in different modules are established through the use of argument lists, as explained in the next section.

## 7.4 SUBPROGRAM ARGUMENTS

### 7.4.1 Arrays as Subprogram Arguments

To illustrate further the definition and use of functions, we now turn our attention to a function that manipulates an array. Figure 7.4 shows a function LARGE that finds the largest data item in an array. This function has two arguments, A and NRITMS. Argument A represents the array of data that is to be searched for the largest item, and NRITMS represents the number of items in A that are to be examined. In the argument definitions, it is permissible to use the dummy argument NRITMS (instead of an integer constant) to specify the size of the dummy array A. This point will be discussed in Section 7.4.3.

```
      REAL FUNCTION LARGE (A, NRITMS)
C
C DETERMINE THE LARGEST ITEM IN AN ARRAY OF REAL DATA ITEMS
C
C ARGUMENT DEFINITIONS --
C   INPUT ARGUMENTS
C     A - ARRAY CONTAINING THE DATA TO BE PROCESSED
C     NRITMS - NUMBER OF ITEMS IN THE ARRAY
C
      INTEGER NRITMS
      REAL A(NRITMS)
C
C LOCAL VARIABLES
      REAL CURLRG
      INTEGER I
C
C INITIALIZE CURRENT LARGEST ITEM
      CURLRG = A(1)
C LOOK FOR AN ITEM THAT IS LARGER THAN CURLRG
C REDEFINE CURLRG WHEN A NEW LARGEST ITEM IS FOUND
      DO 40 I = 1, NRITMS
         IF (A(I) .GT. CURLRG) CURLRG = A(I)
   40 CONTINUE
C
C RETURN VALUE OF THE LARGEST ITEM WHEN THE SEARCH IS COMPLETE
      LARGE = CURLRG
C
      RETURN
      END
```

**Fig. 7.4** Function LARGE.

As shown in Fig. 7.4, the dummy arguments A and NRITMS are defined
and declared first; then the local variables CURLRG and I are declared. The argument declarations are preceded by a set of comments describing the use of
each argument. The data manipulation performed by this function consists of an
assignment statement to initialize CURLRG and a DO loop that searches for the
largest element in the array represented by A. The statement

```
LARGE = CURLRG
```

defines the function result or the value to be returned.

During loop execution, the value of CURLRG represents the largest value
found so far. It would be reasonable to consider omitting CURLRG and storing
this value directly in LARGE using the statement

```
IF (A(I) .GT. LARGE) LARGE = A(I)
```

However, this statement is illegal since it uses the function name, LARGE, in a
*relational expression* (the condition). Rule 3 of the Rules for Defining Functions
(see Section 7.2.2) states that a function name may only appear on the left of an
assignment; a function name may not appear in an expression within the definition of that function.

---

**Program Style**

*Documentation of subprograms*

The function LARGE shown in Fig. 7.4 illustrates some conventions for
writing subprograms which we will use throughout the text. All subprograms
should begin with comments that
1) identify the purpose of the subprogram

and

2) describe in detail the subprogram arguments—their names and use.
This section of comments should be followed immediately by the argument
declarations.

Together, the comments and declarations provide all of the information
necessary for the correct use of the subprogram. Any programmer can very
quickly determine what a subprogram does and how to use it by examining
this front section of comments and declarations, without having to look at the
remainder of the subprogram.

---

**Exercise 7.7:**   Write a function SUM to compute the sum of a collection of two real
data items. Include all pertinent comments using Fig. 7.4 as a guide.

**Exercise 7.8:**   Write a function SUM to compute the sum of a collection of N real data
items, $N \geq 1$. Include all pertinent comments as illustrated in Fig. 7.4.

### 7.4.2  Argument List Correspondence

In order to fully understand the concept of a subprogram argument list, we
must remember that the dummy arguments listed in the definition of a subprogram serve only to describe what is to be done with each of the actual argu-

ments. The actual arguments and their correspondence with the dummy arguments are determined anew each time a subprogram is called, as illustrated in the following example.

**Example 7.6:** The program below is an example of a main program that uses the function LARGE.

```
C MAIN PROGRAM TO FIND THE LARGEST ITEM IN AN ARRAY
C
      REAL TABLE(10), BIGGST
      INTEGER N, I
C
C FUNCTIONS USED
      REAL LARGE
C
C ENTER DATA
      READ*, N, (TABLE(I), I = 1, N)
      PRINT*, 'THE NUMBER OF DATA ITEMS IS ', N
      PRINT*, 'LIST OF DATA ITEMS TO BE PROCESSED ...'
      PRINT*, (TABLE(I), I = 1, N)
C
C FIND LARGEST VALUE
      BIGGST = LARGE(TABLE, N)
      PRINT*, 'THE LARGEST ITEM IS ', BIGGST
C
      STOP
      END

      REAL FUNCTION LARGE (A, NRITMS)
C
C DETERMINE THE LARGEST ITEM IN AN ARRAY OF REAL DATA ITEMS
C
C ARGUMENT DEFINITIONS --
C   INPUT ARGUMENTS
C   A - ARRAY CONTAINING THE DATA TO BE PROCESSED
C   NRITMS - NUMBER OF ITEMS IN THE ARRAY
C
      INTEGER NRITMS
      REAL A(NRITMS)
C
C LOCAL VARIABLES
      REAL CURLRG
      INTEGER I
C
C INITIALIZE CURRENT LARGEST ITEM
      CURLRG = A(1)
C LOOK FOR AN ITEM THAT IS LARGER THAN CURLRG
C REDEFINE CURLRG WHEN A NEW LARGEST ITEM IS FOUND
      DO 40 I = 1, NRITMS
         IF (A(I) .GT. CURLRG) CURLRG = A(I)
   40 CONTINUE
C
```

(Continued)

```
C RETURN VALUE OF THE LARGEST ITEM WHEN THE SEARCH IS COMPLETE
      LARGE = CURLRG
C
      RETURN
      END
```

In this example, the function LARGE is called by the statement

```
      BIGGST = LARGE(TABLE, N)
```

The argument list correspondence defined in this call is illustrated in Fig. 7.5. As shown in this figure, the dummy argument A in the definition of LARGE represents the actual array TABLE, and NRITMS represents the actual argument N. It is essential that the number of arguments in the call of a subprogram always be the same as the number of arguments in its definition, and that corresponding arguments agree in type and *structure*. Agreement with respect to structure requires that an actual argument should be an array if the corresponding dummy argument is an array. In the above example, both TABLE and A are declared as type real arrays, and N and NRITMS are simple integer variables.

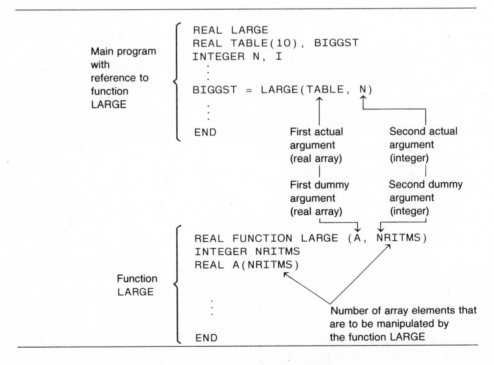

**Fig. 7.5**  Argument list correspondence.

As indicated in Fig. 7.5, the first actual argument in a call to the function LARGE corresponds to the dummy argument A which represents a type REAL array. The data in the corresponding actual array (TABLE in this case) are manipulated by the function LARGE.

**Example 7.7:** a) Given a main program with the declarations shown on the left and function LARGE with its declarations on the right

```
REAL B(30), LARGE, X     REAL FUNCTION LARGE (A, NRITMS)
INTEGER C(30), N         INTEGER NRITMS
                         REAL A(NRITMS)
```

any of the references listed below would be legal. In each case, the first actual argument is a real array (B) and the second argument is an integer variable or value. The effect of each function reference is listed alongside it. We are assuming that the arrays B and C are defined before the function call.

| *Function reference* | *Effect* |
|---|---|
| LARGE(B, 30) | The largest value in array B is determined. |
| LARGE(B, 15) | The largest value in the first fifteen elements of array B is determined. |
| LARGE(B, N) | The largest value in the first N elements of array B is determined. N must be defined before this reference $(1 \leq N \leq 30)$. |
| LARGE(B, C(1)) | The value of C(1) determines how many elements in array B will be examined $(1 \leq C(1) \leq 30)$. |

As shown in the last function reference above, the array element C(1) can be used as an actual argument. The array element value is substituted for its corresponding dummy argument. The corresponding dummy argument must be a simple integer variable (not an array name).

b) Given the program declarations in part a), each of the call statements below would be illegal for the reason indicated.

| *Function reference* | *Reason call is illegal* |
|---|---|
| LARGE(C, 30) | C is a type integer array. |
| LARGE(B, 40) | There are only thirty elements in array B. |
| LARGE(B(30), 30) | The first argument must be an array name, not an array element. Array elements can only correspond to dummy arguments that are simple variables. |
| LARGE(B, X) | X is a type REAL variable. |
| LARGE(N, B) | The argument order is incorrect. |
| LARGE(B) | One argument is missing. |
| LARGE(A, 10) | A is the dummy argument name in the definition of LARGE; however, A is not declared as an array in the main program. |

The function reference LARGE(A, 10) deserves special mention. There is no declaration for A in the main program, so the compiler assumes that it is a simple real variable (using the implied type convention). The first argument in the function definition is a type real array also named A; however, this declaration is irrelevant in the main program since the main program and the function LARGE are independent modules and are compiled separately.

**Exercise 7.9:**  Consider the following input data for the program in Example 7.6:

$$6 \quad 16.5 \quad 22.0 \quad -9.25 \quad 90.5 \quad -3.75 \quad 8.0$$

a) What values will be printed by the statement

```
PRINT*, (TABLE(I), I = 1, N)
```

in Example 7.6? Why are only six values printed by this statement when the size of the array TABLE is 10?

b) When the function LARGE is called by the assignment

```
BIGGST = LARGE(TABLE, N)
```

what is the value of CURLRG following the execution of the statement

```
CURLRG = A(1)
```

in Fig. 7.4? What is the value of CURLRG following the execution of DO loop 40? What value is returned by the function, and what statement causes this return? In the program in Example 7.6, what value is printed for BIGGST?

c) What would be returned by the statement

```
BIGGST = LARGE(TABLE, 3)
```

**Exercise 7.10:**  a) The function LARGE (see Fig. 7.4) does not test to see if the value of NRITMS is greater than 0. Rewrite the executable portion of the function to include such a test. Print an error message if NRITMS is less than or equal to 0.

b) We can rewrite the function LARGE to start the search with A(2) rather than A(1). If this change is made, what will happen when NRITMS is equal to 1? Alter the test from part a) to take care of this case too, given the DO loop header

```
DO 40 I = 2, NRITMS
```

**Exercise 7.11:**  The following sequence of FORTRAN statements can be used in a new function SMALL for determining the smallest item in a real array A containing K data items. Complete the specification of the function SMALL by writing the header statement, the appropriate dummy argument and local variable declarations, the necessary control and value return statements, and appropriate comments.

```
      CURSML = A(1)
      DO 60 I = 1, K
         IF (A(I) .LT. CURSML) CURSML = A(I)
   60 CONTINUE
```

**Exercise 7.12:**  The function COUNT computes the number of occurrences of ITEM in an array. (See Fig. 7.6.)

Let ONOFF (an integer array of size 20) and N (an integer variable) be defined as follows:

Array ONOFF

| N | 1 | 2 | 3 | 4 | 5 | 6 | 7 | 8 | 9 | 10 | 11 | 12 | 13 | 14 | 15 | 16 | 17 | 18 | 19 | 20 |
|---|---|---|---|---|---|---|---|---|---|----|----|----|----|----|----|----|----|----|----|----|
| 16 | 0 | 1 | 1 | 0 | 1 | 0 | 0 | 0 | 1 | 1 | 0 | 1 | 1 | 0 | 1 | 0 | ? | ? | ? | ? |

What value will be returned for the following references to COUNT?

a)    COUNT (ONOFF, 1, N, 1)
b)    COUNT (ONOFF, 5, N − 1,0)
c)    COUNT (ONOFF, N − 5, N, 0)
d)    COUNT (ONOFF, 12, 12, 1)

```
      INTEGER FUNCTION COUNT (LIST, FIRST, LAST, ITEM)
C
C COMPUTE THE NUMBER OF TIMES AN INTEGER ITEM APPEARS BETWEEN
C THE ELEMENTS INDEXED BY FIRST AND LAST IN AN INTEGER ARRAY
C
C ARGUMENT DEFINITIONS--
C   INPUT ARGUMENTS
C     LIST - ARRAY OF ITEMS TO BE EXAMINED
C     FIRST - SUBSCRIPT OF FIRST ELEMENT (IN LIST) TO BE CHECKED
C     LAST - SUBSCRIPT OF LAST ELEMENT (IN LIST) TO BE CHECKED
C     ITEM - DATA ITEM BEING COUNTED
C
      INTEGER FIRST, LAST, ITEM
      INTEGER LIST(LAST)
C
C LOCAL VARIABLES
      INTEGER LCV, CNTR
C
C COMPUTE COUNT
      CNTR = 0
      DO 20 LCV = FIRST, LAST
         IF (LIST(LCV) .EQ. ITEM) CNTR = CNTR + 1
   20 CONTINUE
      COUNT = CNTR
C
      RETURN
      END
```

**Fig. 7.6**  The function COUNT.

**Exercise 7.13:**  A main program contains the following declarations:

                INTEGER X(15), Y(15)
                INTEGER M, NUM, P, Q, COUNT

Explain why each of the following references to the function COUNT (Fig. 7.6) is invalid.

a)    COUNT (X, Y, P, Q)
b)    COUNT (ONOFF, X(1), X(2), X(3))
c)    COUNT (LIST, M, P, Q)
d)    COUNT (X, NUM, Y(1), Y(16))
e)    COUNT (X(15), NUM, P, Q)
f)    COUNT (X, P, Q)

## 7.4.3  Dummy Array Sizes: Adjustable Arrays

The argument declaration statement

                REAL A(NRITMS)

in function LARGE (see Fig. 7.4) identifies the dummy argument A as an *adjustable array* whose size is represented by the second dummy argument, NRITMS. Each time the subprogram is called, the actual argument corresponding to NRITMS determines the size of the dummy array A.

It is often the case that the dummy argument representing the size of an adjustable array is also used to control the array manipulation. For example, in the function LARGE, NRITMS is the limit parameter in DO loop 40, and determines how many array elements are examined each time the function is called.

Another means of representing the size of an adjustable dummy array is through the use of an asterisk

```
REAL A(*)
```

rather than through the use of another dummy argument. In this case, the size of the dummy array will always be the same as the size of the corresponding actual array in the calling program.

The use of the asterisk to declare the size of a dummy array argument is most convenient in subprograms in which the number of array elements to be manipulated is determined by the subprogram, rather than the calling program. (We will see examples of this later in this chapter.) Using the asterisk does not preclude the use of other dummy arguments for controlling execution within a subprogram, as was done in the function LARGE.

## 7.5  SUBROUTINES

### 7.5.1  Definition and Use of Subroutines

Functions are limited in that they are normally used to compute a single value. We will frequently have a need to write subprograms that return more than one value or return an array of values. We may also write subprograms that do not return any values, but instead perform some task such as printing the results of a prior computation. For these purposes, another form of subprogram, called a *subroutine,* is used.

The definition of a subroutine is similar to that of a function. However, a subroutine has no type associated with it; values to be returned are assigned to specified actual arguments (*output arguments*) rather than to a subroutine name. Unlike a function, a subroutine cannot be referenced as part of an expression; instead, a special *call statement* is used to reference it.

**Example 7.8:**  Subroutine BREAK, shown in Fig. 7.7 (bottom) finds the integral and fractional parts of a real number, represented by the dummy argument X. The integral part is assigned to dummy argument WHOLE and the fractional part to the dummy argument FRAC.

```
C MAIN PROGRAM TO TEST SUBROUTINE BREAK
C
      REAL A, B
      REAL R1, R2
      INTEGER I1, I2
C
C READ TEST DATA AND CALL SUBROUTINE BREAK
      READ*, A, B
      CALL BREAK(A, R1, I1)
      CALL BREAK(B, R2, I2)
C
C PRINT RESULTS OF CALL
      PRINT*, 'FOR A = ', A
      PRINT*, 'THE PARTS ARE ', R1, I1
      PRINT*, ' '
      PRINT*, 'FOR B = ', B
      PRINT*, 'THE PARTS ARE ', R2, I2
C
      STOP
      END

      SUBROUTINE BREAK (X, FRAC, WHOLE)
C
C BREAKS A REAL NUMBER INTO ITS FRACTIONAL AND INTEGRAL PARTS
C
C ARGUMENT DEFINITIONS --
C   INPUT ARGUMENTS
C     X - THE VALUE TO BE SPLIT
C   OUTPUT ARGUMENTS
C     FRAC - THE FRACTIONAL PART OF X
C     WHOLE - THE INTEGRAL PART OF X
C
      REAL X, FRAC
      INTEGER WHOLE
C
C COMPUTE RESULTS
      WHOLE = INT(X)
      FRAC = X - WHOLE
C
      RETURN
      END
```

```
*** PROGRAM EXECUTION OUTPUT ***

FOR A =    8.630000000000
THE PARTS ARE   .6300000000000                        8

FOR B =  -15.25000000000
THE PARTS ARE  -.2500000000000                      -15
```

**Fig. 7.7** Definition and use of subroutine BREAK.

The dummy argument X is an input argument. An input argument is used only to pass data to a subroutine. The actual argument corresponding to X is manipulated by the subroutine, but its value is not changed. As a result of the subroutine execution, the dummy output arguments FRAC and WHOLE are defined. Their new values are returned to the calling module and stored in their corresponding actual arguments. This process is illustrated in Fig. 7.8.

*Main program*

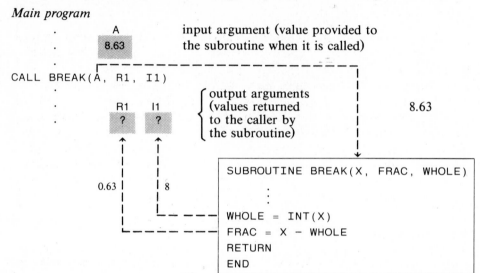

**Fig. 7.8** Passing arguments between a subroutine and its caller.

As indicated in Fig. 7.8, the dummy argument X represents the actual argument A (value is 8.63); the dummy arguments FRAC and WHOLE represent the main program variables R1 and I1, respectively (initially undefined). Consequently, the assignment statement

$$\text{WHOLE} = \text{INT}(X)$$

in BREAK assigns the value 8 to variable I1 and the assignment statement

$$\text{FRAC} = X - \text{WHOLE}$$

assigns the value 0.63 to variable R1.

The second call statement in the main program of Fig. 7.7

$$\text{CALL BREAK}(B, R2, I2)$$

would cause the integral part of variable B to be stored in variable I2 and the fractional part in variable R2 (all declared in the main program). Assuming the value $-15.25$ is read into B, the value of I2 would be $-15$ and the value of R2 would be $-0.25$.

As shown in Fig. 7.7, there is no distinction made in the subroutine definition between arguments that are used to pass data to the subroutine and arguments that are used to return results. The manner in which the argument is used determines whether it is an *input* or *output argument.* As we shall see, it is also possible to use an argument to pass data to the subroutine and return a result as well. We shall call such arguments *in-out arguments.*

---

**Program Style**

*Using drivers to test subprograms*

One important benefit of the use of subprograms is that each subprogram is an independent module that can be tested separately. Once we have confidence that all modules are correct, we can combine them to form large program systems. It is considerably easier to test and debug the complete program system if all its modules have been previously checked out.

The main program in Fig. 7.7 is an example of a *driver program.* Its sole purpose is to verify that subroutine BREAK is operating correctly. As shown in Fig. 7.7, driver programs are normally very short and consist of statements to enter test data, call the subprogram being tested, and print the results returned. We recommend that you test each subprogram that you write using similar short driver programs.

---

### 7.5.2  Additional Subroutine Examples

**Example 7.9:**  Subroutine LRGIND in Fig. 7.9 finds both the largest value in an array and its position, or index, in the array. The subroutine has two input arguments, A and NRITMS, and two output arguments, MAX and INDEX. The statements

```
INDEX = CURIND
MAX = A(INDEX)
```

assign values to the output arguments and, hence, define the values returned by the subroutine. The statement

```
CALL LRGIND(B, 10, LARGES, POS)
```

is legal only if B is a real array with 10 or more elements, LARGES is a real variable, and POS is an integer variable. This statement calls the subroutine LRGIND and causes the largest value in array elements $B(1)$ through $B(10)$ to be stored in the variable LARGES of the calling program. Similarly, the index of this value is stored in the variable POS of the calling program.

The argument correspondence specified by the above call is shown below.

| *Actual argument* | *Dummy argument* | *Use* |
|---|---|---|
| B | A | Array being searched (input) |
| 10 | NRITMS | Number of array elements searched (input) |
| LARGES | MAX | Largest value in array (output) |
| POS | INDEX | Subscript of largest value (output) |

```
      SUBROUTINE LRGIND (A, NRITMS, MAX, INDEX)
C
C DETERMINE THE LARGEST ITEM AND ITS INDEX
C FOR AN ARRAY OF REAL NUMBERS
C
C ARGUMENT DEFINITIONS --
C   INPUT ARGUMENTS
C     A - ARRAY CONTAINING THE DATA TO BE PROCESSED
C     NRITMS - NUMBER OF ITEMS IN THE ARRAY
C   OUTPUT ARGUMENTS
C     MAX - LARGEST VALUE IN THE ARRAY
C     INDEX - THE INDEX OF MAX IN THE ARRAY A
C
      INTEGER NRITMS, INDEX
      REAL A(NRITMS), MAX
C
C LOCAL VARIABLES
      INTEGER I, CURIND
C
C INITIALIZE THE INDEX OF THE CURRENT LARGEST ITEM
      CURIND = 1
C LOOK FOR AN ITEM THAT IS LARGER THAN THE LARGEST ONE SO FAR
C REDEFINE CURIND WHEN A NEW LARGEST ITEM IS FOUND
      DO 40 I = 1, NRITMS
         IF (A(I) .GT. A(CURIND)) CURIND = I
   40 CONTINUE
C
C RETURN WHEN SEARCH IS COMPLETE
      INDEX = CURIND
      MAX = A(INDEX)
C
      RETURN
      END
```

**Fig. 7.9**  Subroutine to find the largest array element and its index.

The definition and call of a subroutine are described in the following displays.

---

**Subroutine Definition**

```
      SUBROUTINE sname (dummy argument list)
      argument definition section
      local declarations section

      ----
      ---- }   subroutine body
      ----

      RETURN
      END
```

**Interpretation:** The SUBROUTINE statement specifies the subroutine name, *sname;* the *dummy argument list* is a list of symbolic names. The dummy ar-

guments are used either to pass data to the subroutine (input arguments) or
return results to the calling program (output arguments). Occasionally a dummy argument will be used for both purposes. We will refer to such arguments
as *in-out arguments*. The argument definition section contains descriptions and
declarations for all the *dummy arguments*.

Other symbolic names, not appearing in the dummy argument list, but required for writing the subroutine, should be declared in the separate *local declaration section*.

The *subroutine body* describes the data manipulation performed by the
subroutine. The RETURN statement transfers control back to the calling
statement. The END statement terminates the subroutine definition.

---

**Subroutine Call**

CALL *sname (actual argument list)*

**Interpretation:** Subroutine calls, unlike function calls, are not part of an expression. Rather, they are separate FORTRAN statements that begin with the
word CALL, followed by the name of the subroutine to be referenced, *sname,*
and the *actual argument list*. The actual arguments may be variable names, array names, or array elements. Expressions and constants may be used only to
pass data into the subroutine and must correspond to dummy input arguments.

---

**Example 7.10:** The subroutine SUMMER shown at the bottom of Fig. 7.10
forms the element-by-element sum of two real arrays (represented by A and B) in
a third real array (represented by C). The main program (preceding the subroutine) reads data into the arrays X and Y and then calls SUMMER to store the
element-by-element sum in array SUM (i.e., SUM(1) is set to X(1) + Y(1),
SUM(2) to X(2) + Y(2), etc.). The main program also prints the final results in
three columns after the subroutine execution is complete. For the test data

```
X:    10.0   15.0   20.0   25.0
Y:    10.0   20.0   30.0   40.0
```

(N = 4), the results of execution of this program are shown below.

```
THE NUMBER OF ELEMENTS TO BE PROCESSED IS 4
        X                   Y                   SUM
  10.00000000000     10.00000000000      20.00000000000
  15.00000000000     20.00000000000      35.00000000000
  20.00000000000     30.00000000000      50.00000000000
  25.00000000000     40.00000000000      65.00000000000
```

The separation of tasks or "division of labor" illustrated in this example is
quite common in programming. Usually, data entry is performed in the main
program or a separate subprogram, and the desired data manipulation (summation in this case) is performed by a subprogram. The final results are printed af-
terwards in the main program (or another subprogram).

```
C PROGRAM TO READ TWO REAL ARRAYS
C AND COMPUTE THE ELEMENT-BY-ELEMENT SUM
C
C PROGRAM PARAMETERS
      INTEGER SIZE
      PARAMETER (SIZE = 15)
C
C VARIABLES
      REAL X(SIZE), Y(SIZE), SUM(SIZE)
      INTEGER I, N
C
C ENTER DATA
      READ*, N
      PRINT*, 'THE NUMBER OF ELEMENTS TO BE PROCESSED IS ', N
      IF (N .LT. 1 .OR. N .GT. SIZE) THEN
         PRINT*, 'N IS OUT OF RANGE OF 1 TO ', SIZE
         PRINT*, 'PROGRAM EXECUTION TERMINATED'
         STOP
      ENDIF
C
C READ DATA INTO X AND Y
      READ*, (X(I), I = 1, N)
      READ*, (Y(I), I = 1, N)
C
C FORM THE ELEMENT-BY-ELEMENT SUM
      CALL SUMMER (N, X, Y, SUM)
C
C PRINT RESULTS
      PRINT*, '           X                 Y                 SUM'
      DO 10 I = 1, N
         PRINT*, X(I), Y(I), SUM(I)
   10 CONTINUE
C
      STOP
      END



      SUBROUTINE SUMMER (N, A, B, C)
C
C FORMS ELEMENT-BY-ELEMENT SUM OF ARRAYS A AND B IN THE ARRAY C
C
C ARGUMENT DEFINITIONS --
C   INPUT ARGUMENTS
C     N - SIZE OF THE ARRAYS
C     A, B - ARRAYS BEING SUMMED
C   OUTPUT ARGUMENTS
C     C - ARRAY CONTAINING THE SUMMATION RESULT
C
      INTEGER N
      REAL A(N), B(N), C(N)
C
```

*(Continued)*

```
C LOCAL VARIABLES
      INTEGER I
C
      DO 10 I = 1, N
         C(I) = A(I) + B(I)
   10 CONTINUE
C
      RETURN
      END
```

**Fig. 7.10**   Array summation program.

There is another important point illustrated in the program in Fig. 7.10. The label 10 and symbolic names I and N appear in both the main program and the subroutine. Since the main program and the subroutine are compiled separately (see Sec. 7.3), the compiler is able to handle this duplication without confusion. Any manipulation of variable I in the main program has no effect whatsoever on local variable I in the subroutine, and vice versa.

Similarly, the compiler is able to distinguish between variable N in the main program and dummy argument N in the subroutine during compilation. The only correspondence between these two items is established when the subroutine is called, with the variable N as the first argument.

**Example 7.11:**   A subroutine for sorting an array of data is provided in Fig. 7.11. This subroutine is based on the program shown in Chapter 6, Fig. 6.18. The array to be sorted is represented by the dummy argument M. M is an in-out argument since it is used to pass data (the original values) to the subroutine and to return results (the sorted values) to the calling module. The actual array to be sorted is determined when the subroutine is called.

As in the previous example, there is no need to read or enter data in the subroutine. The read operation must be performed before the SORT subroutine is called since the actual argument array must be defined before it can be sorted. The data entry task could be relegated to another subroutine, as shown in the next example.

---

**Program Style**

*Verifying subprogram arguments*

In Fig. 7.11, we introduced a very simple validation test on the dummy argument COUNT, which represents the number of array elements to be sorted. If COUNT is less than two, it makes no sense to perform the sort, and control is immediately returned from the subroutine. Similar tests could have been used in the function LARGE and the subroutine LRGIND. In fact, it is always a good idea to check any argument that is important in controlling the execution of a subprogram, especially an argument used as a loop parameter. If such an argument does not satisfy the minimal constraints required for meaningful execution of the subprogram, a diagnostic should be printed and control returned to the calling program.

```
      SUBROUTINE SORT (M, COUNT)
C
C SORT AN ARRAY OF REAL DATA IN ASCENDING ORDER
C
C ARGUMENT DEFINITIONS --
C   INPUT ARGUMENTS
C     COUNT - NUMBER OF DATA ITEMS IN M
C   IN-OUT ARGUMENTS
C     M - THE ARRAY OF DATA ITEMS, BEFORE AND AFTER SORT
C
      INTEGER COUNT
      REAL M(COUNT)
C
C LOCAL VARIABLES
      INTEGER INDEX
      LOGiCAL SORTED
C
C VALIDATE COUNT
      IF (COUNT .LT. 2) THEN
          PRINT*, '*** WARNING --- SORT NOT PERFORMED. '
          PRINT*, '    COUNT = ', COUNT, ' IS TOO SMALL.'
          RETURN
      ENDIF
C
C PERFORM THE BUBBLE SORT ON THE DATA IN M (ASCENDING ORDER)
      SORTED = .FALSE.
      WHILE (.NOT. SORTED) DO                      32 IF (.SORTED.)
          SORTED = .TRUE.                             Z GO TO 39
C         INNER LOOP -- COMPARE SUCCESSIVE PAIRS OF ITEMS IN M
          DO 30 INDEX = 1, COUNT-1
              IF (M(INDEX) .GT. M(INDEX+1)) THEN
C                 EXCHANGE OUT-OF-ORDER PAIRS
                  TEMP = M(INDEX)
                  M(INDEX) = M(INDEX+1)
                  M(INDEX+1) = TEMP
                  SORTED = .FALSE.
              ENDIF
   30     CONTINUE
      ENDWHILE                                     39 CONTINUE
C                                                     GO TO 32
      RETURN
      END
```

**Fig. 7.11**  SORT subroutine.

**Example 7.12:**  The subroutine DATAIN shown in Fig. 7.12 is used to read a collection of N data items into a real array represented by the dummy argument W. Since the number of items to be processed (N) is not known until the subroutine is already executing, an asterisk is used in declaring the size of the dummy argument array (see Section 7.4.3). MAXNUM is used as an input argument to prevent any attempt to read more data than allowed by the size of the actual array in any call to DATAIN. Note that MAXNUM could have been used in place of the asterisk in the declaration of W.

The logical argument OK is used to signal to the calling module that N was

valid and the data entry was performed. This use of a logical argument is analogous to the use of a program flag in earlier programs.

```
      SUBROUTINE DATAIN (W, N, MAXNUM, OK)
C
C READ N REAL DATA ITEMS INTO THE ARRAY W
C
C ARGUMENT DEFINITIONS --
C   INPUT ARGUMENTS
C     MAXNUM - MAXIMUM NUMBER OF ITEMS THAT CAN BE READ
C   OUTPUT ARGUMENTS
C     W - ARRAY TO CONTAIN THE DATA
C     N - NUMBER OF ITEMS TO BE READ
C     OK - INDICATES IF THE READ IS SUCCESSFUL
C
      INTEGER MAXNUM, N
      REAL W(*)
      LOGICAL OK
C
C LOCAL VARIABLES
      INTEGER I
C
C READ AND VALIDATE N.  IF N IS OK, READ THE DATA.
      READ*, N
      IF (N .LT. 1 .OR. N .GT. MAXNUM) THEN
         OK = .FALSE.
         PRINT*, '*** ARGUMENT ERROR IN SUBROUTINE DATAIN. '
         PRINT*, '    THE NUMBER OF ITEMS TO BE READ ', N
         PRINT*, '    IS OUT OF RANGE 1 TO ', MAXNUM
         PRINT*, '    NO DATA READ.'
      ELSE
         OK = .TRUE.
         READ*, (W(I), I = 1, N)
      ENDIF
C
      RETURN
      END
```

**Fig. 7.12** Subroutine DATAIN.

**Program Style**

*Using a logical argument as a program flag*

The logical argument OK is defined by the subroutine DATAIN prior to the return to the calling program. This program can then test the corresponding actual argument (also type LOGICAL) to determine what action to take next.

This technique of using a logical argument to indicate the success or failure of a subroutine is quite common in programming. It is a natural extension of the use of program flags to communicate results from one program step to another.

**Exercise 7.14:**   Given the main program variable definitions

| X(1) | X(2) | X(3) | | Y(1) | Y(2) | Y(3) | | Z(1) | Z(2) | Z(3) | | N |
|------|------|------|---|------|------|------|---|------|------|------|---|---|
| 5.0 | 2.5 | −6.1 | | 17.5 | −5.0 | 3.2 | | ? | ? | ? | | 3 |

indicate the effect of each call statement below that is legal. Describe what is wrong with each invalid call statement. The main program and subroutine SUMMER are shown in Fig. 7.10.

```
a) CALL SUMMER(X,Y,Z,N)
b) CALL SUMMER(N,X,Z,Y)
c) CALL SUMMER(N,X(*), Y(*), Z(*))
d) CALL SUMMER(N,A,B,C)
e) CALL SUMMER(N,X,Y,Z)
f) CALL SUMMER(N,X,Y,X)
```

**Exercise 7.15:**   The subroutine LRGIND in Example 7.9 returns the largest value in an array and its index. Write a function INMAX that finds the index only. How could you reference the largest value in an array using this function only?

**Exercise 7.16:**   Identify the input and output arguments and the local variables in each of the following subroutines. What do subroutines ZERO and BOUND do?

```
a) SUBROUTINE ZERO (X)
   REAL X
   IF (X .LT. 0.0) X = 0.0
   RETURN
   END

b) SUBROUTINE BOUND (M, SIZE, MAX)
   INTEGER SIZE
   REAL M(SIZE), MAX
   DO 10 I = 1, SIZE
      IF (M(I) .GT. MAX) M(I) = 0.0
10 CONTINUE
   RETURN
   END
```

**Exercise 7.17:**   Write a subroutine that will count the number of occurrences of a real item in a real array. In addition, your subroutine should indicate whether or not the data item was found by setting an output argument (type LOGICAL) to true or false. Make certain that all of your arguments are carefully defined in comment statements in the subprogram.

## 7.6   STEP-WISE PROGRAMMING

### 7.6.1   Motivation for Step-Wise Programming

Until now, the logic or flow of control in the programs and problems we have examined was relatively straightforward and easy to follow. Most programs consisted of short sequences of structures with little or no nesting. We now have the tools and skills to write more complicated programs involving several levels of nesting. Such programs can become quite cumbersome and difficult to follow unless proper procedures are used in their design and implementation.

We have seen how the step-wise process can assist in algorithm and program development, and we have used this technique in designing our algorithms by drawing level one flow diagrams and successive refinements. Unfortunately, we have not been able to carry the step-wise process through to the implementation of our programs. What we would like to do is implement a program in the same manner in which the flow diagram is designed. This involves writing a main program that closely resembles the level one flow diagram, both with respect to the order of steps and the amount of detail provided. The FORTRAN statements for those subproblems that are rather straightforward are included directly in the main program, while the solutions to the more complicated subproblems are programmed using subprograms which are referenced in the main program. Where appropriate, additional refinements should also be implemented as subprograms.

The result of this effort is a main program together with a collection of related subprogram modules that clearly reflect the separate subproblems and refinements specified by the flow diagrams. We will illustrate the step-wise approach by discussing the complete solution of a problem that is slightly more complicated than those studied thus far. In the process, we will introduce another programming tool, the *program system chart,* which we believe is useful in documenting the relationships among the various components of a problem solution.

## 7.6.2    Program System Charts

In the following problem we use the program system chart to document the *functional relationships* and *data flow* between modules.

**Problem 7A:**    Given a collection of N real numbers stored in an array, compute the range, mean (average), and median for this collection.

The initial data table is shown below; the level one flow diagram for this problem is shown in Fig. 7.13. Each box in the diagram represents a major step in the problem solution. Additional lower level subproblems may be identified within each of the Steps 2, 3 and 4. Each of these subtasks represents a refinement of a task shown at a higher level.

**Data Table for Problem 7A**

*Program parameter*
MASIZ = 100, maximum number of items that can be processed (integer)

| *Input variables* | *Output variables* |
|---|---|
| N: The number of items to be processed (integer) | RANGE: The range of the data (real) |

(*Continued*)

X: Array containing the
data to be processed
(real)

MEAN: The average of the
data (real)

MEDIAN: The median of
the data (real)

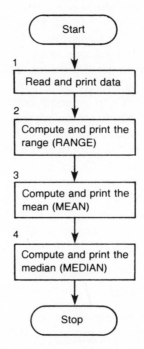

**Fig. 7.13**   Level one flow diagram for the statistics problem (7A).

The program system chart for this problem is shown in Fig. 7.14. This chart identifies the major subproblems of the original problem and illustrates the functional relationships among them. The solutions to the subproblems shown at one level in the chart can be specified in terms of the connected subproblems at the next lower level. For example, the program system chart indicates that the solution of the subproblem "compute median" may be specified in terms of the solution to the subproblems "sort data" and "compute middle value of sorted data." Similarly, in order to find the average, we must first solve the subproblem "compute sum."

Once the data table, level one flow diagram and program system chart have been completed, we can begin to add data flow information to the program system chart and to work on the lower level refinements shown in the chart. In considering the refinements, it is necessary to decide which subtasks should be implemented as subroutines or functions and which should be implemented as part of the solution of the task above it in the program system chart. In general, a subtask should be implemented using a function or subroutine unless it occurs

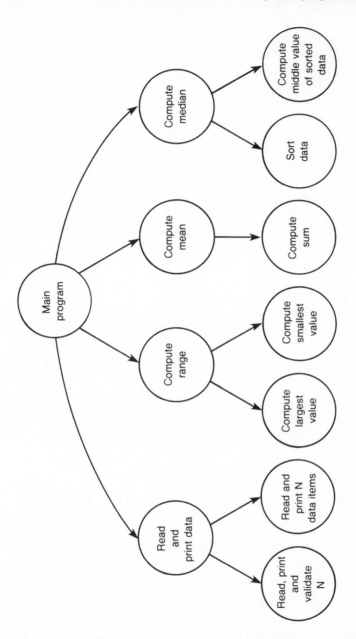

**Fig. 7.14** Program system chart for the statistics problem (7A).

only once in the program system chart and is rather trivial. The subtasks "compute range" and "compute middle value" fall in this category.

The decision as to whether to write a subroutine or a function depends upon the number of values to be returned. Functions are most convenient when a single value is to be computed. Such is the case in the subtasks for computing the largest value, the smallest value, the average, and the median. The sort task, however, rearranges an entire array of information (it does not compute a single value) and is, therefore, written as a subroutine.

Fig. 7.15 shows a program system chart (updated from Fig. 7.14) that reflects the decisions just discussed. In addition, we have added a description of the data flow between the various program modules. For example, the array X and its size N are provided as input to FNDMED; the median value, MEDIAN, is returned by FNDMED.

The next step is to add a description of each function and subroutine referenced by the main program to the initial data table for the program.

*Subprograms referenced by main program:*

DATAIN (subroutine): Used to read, print, and validate N, and to read and print N real data items in an array

*Arguments*

1. The array of items to be processed (real, output)
2. The number of items to be read (integer, output)
3. The maximum size of the array (integer, input)
4. A flag used to indicate whether or not N was valid (logical, output)

LARGE (real function): Determines the largest of a collection of real data items

*Arguments*

1. Real array containing the data to be examined (input)
2. Integer specifying the number of items in the array (input)

SMALL (real function): Determines the smallest of a collection of real data items

*Arguments*

(same as for LARGE)

FNDAVE (real function): Computes the average of a collection of real data items

*Arguments*

(same as for LARGE)

FNDMED (real function): Computes the median of a collection of real data items

*Arguments*

(same as for LARGE)

The data table now contains a complete record of the subprograms referenced by the main program, including a summary description of the arguments of each module, listed in order of appearance.

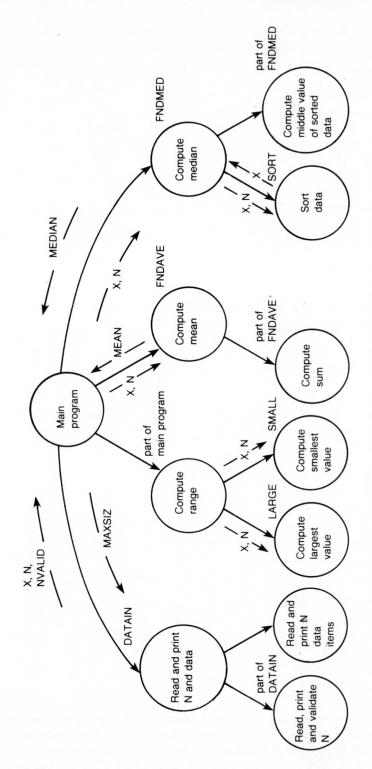

**Fig. 7.15**   Program system chart with data flow and function designation for statistics problem (7A).

Given the level one flow diagram (Fig. 7.13) and the information in this table, we can now write the entire main program for the statistics problem (see Fig. 7.16).

```
C SIMPLE STATISTICS PROBLEM - MAIN PROGRAM
C
C COMPUTE THE RANGE, MEAN AND MEDIAN
C OF A COLLECTION OF N REAL DATA ITEMS
C
C PROGRAM PARAMETERS
      INTEGER MAXSIZ
      PARAMETER (MAXSIZ = 100)
C
C VARIABLES
      REAL X(MAXSIZ)
      INTEGER N, I
      REAL RANGE, MEAN, MEDIAN
      LOGICAL NVALID
C
C FUNCTIONS USED
      REAL LARGE, SMALL, FNDAVE, FNDMED
C
C READ AND PRINT DATA
      CALL DATAIN(X, N, MAXSIZ, NVALID)
      IF (NVALID) THEN
          PRINT*, 'LIST OF DATA TO BE PROCESSED...'
          PRINT*, (X(I), I = 1, N)
          PRINT*, ' '
      ELSE
          PRINT*, N, ' IS NOT IN RANGE 1 TO ', MAXSIZ
          PRINT*, 'PROGRAM EXECUTION TERMINATED'
          STOP
      ENDIF
C
C COMPUTE THE RANGE
      RANGE = LARGE(X, N) - SMALL(X, N)
      PRINT*, 'THE RANGE IS ', RANGE
C
C COMPUTE THE MEAN
      MEAN = FNDAVE(X, N)
      PRINT*, 'THE MEAN IS ', MEAN
C
C DETERMINE THE MEDIAN
      MEDIAN = FNDMED(X, N)
      PRINT*, 'THE MEDIAN IS ', MEDIAN
C
      STOP
      END
```

**Fig. 7.16**  Main program for Problem 7A.

The executable portion of the program is written directly from the flow diagram. The four major subproblems are clearly identifiable, and the calls to the related subprograms may be written directly from the information in the data table. The only step remaining in the main program is to compute the range. As indicated earlier, this can be done easily in the main program once the largest and smallest values are determined.

The reading, verification and printing of the input data are done using the subroutine DATAIN, shown in Example 7.12. The function LARGE is shown in Fig. 7.4, and SMALL is very similar to LARGE (see Exercise 7.11).

All that remains to complete Problem 7A is to write the functions FNDMED and FNDAVE. FNDAVE is left as an exercise; we will implement FNDMED in the next section.

**Exercise 7.18:**   Write the function FNDAVE. Carefully declare the arguments and describe with comments. Make sure they follow the order specified in the function description provided in this section.

**Exercise 7.19:**   In the program in Fig. 7.16, there is no reference to the computation of the sum or to the sorting of the data items (refer back to the program system chart in Fig. 7.15). Why not?

**Exercise 7.20:**   In the program in Fig. 7.16, replace the two declarations

```
REAL X(MAXSIZ)
INTEGER N, I
```

with the declarations

```
REAL TABLE(1050)
INTEGER XCOUNT, IX
```

a) Rewrite the three statements and the three statements beginning

```
RANGE = . . .
MEAN = . . .
MEDIAN = . . .
```

given the new declarations.

b) What, if any, changes would be required to the definition of the function LARGE shown in Fig. 7.4 in order to use it in the main program with the new declarations?

### 7.6.3   Finding the Median of a Collection of Data

In this section, we will complete the statistics problem by writing the function FNDMED, which finds the median of a collection of data items. In the process, we will illustrate many of the points made so far in this chapter, and provide some additional insights concerning the use of subprograms in programming.

Although the median problem is a subproblem of the statistics problem, we will treat it as an entirely separate problem in order to illustrate the degree of independence that can be achieved when using FORTRAN subprograms.

**Problem 7B:**   Write a function that finds the median value in an array of real data items. (The median has the property that the number of array values less than the median is the same as the number of array values greater than the median.)

**Discussion:** Fig. 7.17a shows the portion of the program system chart (Fig. 7.15) that is relevant to finding the median, as well as a level one flow diagram for the problem.

As is often the case, the level one flow diagram simply reflects an ordering of the primary steps shown in the program system chart. The information involved in the solution of the problem at this level is shown in the following data table.

### Data Table for the Median Function (FNDMED)

| *Input arguments* | *Output arguments* |
|---|---|
| | (None) |

X: Represents the array
   that contains the data
   (real)

N: Represents the number
   of items in the array
   (integer)

The next step in the solution of the problem is to decide how we will deal with Steps 1 and 2 in the level one flow diagram. Since sorting a collection of data is a frequent requirement in many problems and since sorting is a somewhat complicated task, we will perform the sort in a separate subroutine. Once the data have been sorted, finding the middle value is rather easy (see Fig. 7.17b) so we will not separate this task from the function **FNDMED**. (We have decided to average the two "middle" values if there are an even number of array elements.)

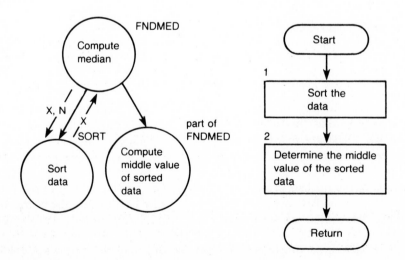

**Fig. 7.17a** Program system chart and level one flow diagram for the median problem.

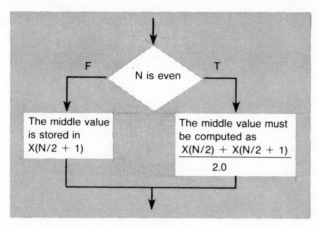

**Fig. 7.17b**  Step 2 refinement—find the middle value in a collection of sorted data items.

We can now make some additions to the data table for FNDMED. These additions reflect the decisions concerning the handling of Steps 1 and 2 in the level one flow diagram. The function FNDMED is shown in Fig. 7.18. The SORT subroutine appeared earlier in Fig. 7.11.

**Additional Data Table Entries for FNDMED**

*Program variables*

INDEX: Used to store
the value of $N/2 + 1$
to simplify the sub-
script expression
used in references
to X (integer)

*Subprograms referenced:*

SORT (subroutine): Sorts a real array in ascending order.

| *Arguments* | *Definition* |
|---|---|
| 1 | Array containing the data to be sorted (in-out, real) |
| 2 | Number of items in the array (input, integer) |

---

**Program Style**

*Documenting subprogram interfaces*

The data table for the function FNDMED is described in two sections: first the arguments and then the local variables. It is always a good idea to maintain a clear separation between the arguments and the local variables of a

*(Continued)*

subprogram. The argument section provides information about the *interface* between the subprogram and calling programs; the local variable section contains the declaration of variables and arrays that are internal to the subprogram, and not at all related to the activity of the calling program. As is very often the case for subprograms, there are no input or output local variables for the function FNDMED.

The data table for the median function contains a third section describing the subroutine SORT and its arguments. It is essential that the actual arguments used in the call statement

$$\text{CALL SORT}(X, N)$$

agree as to order and definition with the dummy arguments appearing in the subroutine header

$$\text{SUBROUTINE SORT (M, COUNT)}$$

To ensure agreement between actual arguments and dummy arguments, it is helpful to add an ordered list of argument definitions to the data table. This list should be carefully followed when writing a call statement to a subprogram. Note that similar lists of argument definitions were provided with the data table for the main program (see Section 7.6.2).

```
      REAL FUNCTION FNDMED (X, N)
C
C FIND THE MEDIAN OF AN ARRAY OF REAL DATA ITEMS
C
C ARGUMENT DEFINITIONS --
C   INPUT ARGUMENTS
C     X - ARRAY CONTAINING THE DATA ITEMS
C     N - THE NUMBER OF ITEMS IN THE ARRAY
C
      INTEGER N
      REAL X(N)
C
C LOCAL VARIABLES
      INTEGER INDEX
C
C SORT THE ARRAY
      CALL SORT(X, N)
C
C COMPUTE THE MEDIAN
      INDEX = N / 2 + 1
      IF (MOD(N,2) .EQ. 0) THEN
C         THE MEDIAN IS THE AVERAGE OF THE TWO MIDDLE ITEMS
          FNDMED = (X(INDEX-1) + X(INDEX)) / 2.0
      ELSE
C         THE MEDIAN IS THE MIDDLE ITEM
          FNDMED = X(INDEX)
      ENDIF
C
      RETURN
      END
```

**Fig. 7.18**  Function FNDMED for Problem 7B.

**Exercise 7.21:** Write a subroutine TOTALN to compute the sum (SUM) of N real numbers. Provide a complete set of comments as illustrated for the subroutine SORT. Compare this subroutine to the function written in Exercise 7.8, and list each of the important differences.

**Exercise 7.22:** Write complete data tables for DATAIN (Fig. 7.12), SORT (Fig. 7.11), and LARGE (Fig. 7.4).

**Exercise 7.23:** Rewrite the search program in Chapter 5 (Problem 5B, Section 5.6) as a subroutine. Provide a modified data table as well.

**Exercise 7.24:** If we examine the program system chart (Fig. 7.15) for the statistics problem, we can see that the subroutine SORT does not enter the picture until the third level, where sorting is required in finding the median of the data items. Yet the sort could have been quite helpful in the computation of the range. Since sorting is needed anyway, we might just as well have sorted the data in the array X in the main program immediately preceding the step to compute the range (see Fig. 7.15). Once the data have been sorted in ascending order, the range can be computed simply as

$$RANGE = X(N) - X(1)$$

and the functions LARGE and SMALL are no longer needed.

Rewrite the program system chart and the main program for the statistics problem, with the sort done immediately before the RANGE computation.

**Exercise 7.25:** Consider the following function for finding the largest of two real numbers represented by the dummy arguments P1 and P2.

```
REAL FUNCTION MAXVAL(P1, P2)
REAL P1, P2
REAL TEMP
IF (P1 .GT. P2) THEN
    TEMP = P1
ELSE
    TEMP = P2
ENDIF
MAXVAL = TEMP
RETURN
END
```

Let the real array X, and the real variables A, TEMP, and Y be defined in a program that calls MAXVAL, as shown below.

| A | TEMP | Y | X(1) | X(2) | X(3) | X(4) | X(5) |
|---|---|---|---|---|---|---|---|
| 16.0 | 8.2 | −6.0 | 4.0 | 1.0 | −2.0 | 0.0 | .5 |

a) What value would be stored in Y as a result of the execution of the statement

$$Y = MAXVAL(A, TEMP)$$

b) What value would be stored in A as a result of the execution of the statement

$$A = MAXVAL(X(3), Y)?$$

c) What value would be stored in X(3) as a result of the execution of the following statements?

```
X(1) = MAXVAL(X(4), X(5))
X(3) = MAXVAL(X(1), X(2))
```

d) What would be stored in X(5) as a result of the execution of the statement

```
X(5) = TEMP + MAXVAL(X(5) + X(2), 2.0 * X(3))
```

## 7.7  SUBPROGRAM GENERALITY AND TESTING

### 7.7.1  Introduction

At the beginning of this chapter, we presented a number of ways in which subprograms could be helpful in solving problems on the computer. We discussed the name independence feature of the subprogram and indicated why name independence was such a vital part of the subprogram concept. Indeed, the utility of the subprogram derives primarily from the name independence feature. It is this feature that allows the design, implementation, and testing of the subprograms of a large program system to be carried out individually. Independence also makes it possible for some of these subprograms to be used as the building blocks of a number of different program systems, thereby saving a duplication of effort.

The design and testing of subprograms are topics worthy of considerable study in their own right and entire books have been written on these subjects. We cannot examine these topics in any detail in this text. Rather, we confine our attention to one particular design goal (that of subprogram generality). We then present a few suggestions as to how to reduce the possibility of serious subprogram errors and to help in quickly detecting those errors that do occur.

### 7.7.2  Subprogram Generality

An important goal in designing subprograms is generality. Attempts should always be made to define the arguments of a subprogram so as to enable it to process a *logically complete set* of potential input values. The exact nature of what constitutes a logically complete set can be deduced only through a careful analysis of the given problem and its possible extensions.

**Example 7.13:**   Given a problem of rounding off employee net pay computations to the nearest two decimal places, we might initially consider writing a function for rounding off positive real numbers to the nearest two places. However, with just a little additional thought, we would see that we can easily generalize this function to round off any real number (positive or negative) to the nearest $n$ decimal places, where $n$ may be any non-negative integer. This latter function, while not useful in its fullest generality in the solution of the immediate problem, is certainly far more adaptable to changes in its input than the former. (It is even possible that such a generalized function already exists and that we could use it for the special case just defined, rather than writing a new one.)

### 7.7.3  Testing a Program System

As the number of modules and statements in a program system grows, the possibility of error also increases. However, if each module is kept to a manageable size, then the possibility of error will increase much more slowly.

Whenever possible, it is best to test each system module independently before putting the entire package together. As suggested earlier, this can be done by writing a short *driver program* consisting of necessary declarations, initialization of input arguments and a call to the subprogram being tested. The driver program should also print the results returned by the subprogram being tested. A little time spent testing each subprogram independently in this manner should significantly reduce the total time required to debug the entire program system. Examples of driver programs are shown in Example 7.6 (the driver for the function LARGE) and Fig. 7.7 (the driver for the subroutine BREAK).

If a program module being tested calls another, it is often helpful to initially substitute a dummy procedure, or *stub,* for the called subprogram. The body of the stub should consist only of a PRINT statement indicating that the stub was entered. After the subprogram is written and tested, it can be inserted in place of its stub.

After all the modules are tested independently, the entire program system should be debugged and tested. Some suggestions for preventing and detecting errors at this stage follow.

1. Accurate, written descriptions of all arguments and local variables of a subprogram should be maintained. These descriptions should be included as comments in the subprogram definition.
2. When debugging a program system, a trace of execution should be provided. This is usually accomplished by printing the name of each subprogram as it is entered.
3. At least in the debugging stage, the values of all variables input to a subprogram should be printed upon entry to the subprogram. Carefully chosen portions of input arrays might also be printed. Values critical to the control of execution in a subprogram should always be printed as a matter of course if they fall outside the range of meaningful values.
4. It is often helpful while debugging to print the values of all output arguments immediately after returning from a subprogram.

## 7.8   COMMON BLOCKS*

### 7.8.1   Introduction to Common Blocks

FORTRAN provides another facility besides argument lists for communicating data among subprograms. This facility, called the *common block,* is useful in programming systems containing several subprograms that must reference a common *base of information.* In such cases, argument lists can become quite long, introducing the possibility of numerous errors in calling the subprograms involved.

The common declaration can be used to specify that an area of memory called a common block is to be set aside so that it can be referenced by two or more program modules. Each of the modules requiring access to the common area must have a description of the information that is stored there that includes

---

*This section may be omitted.

the name of each common variable, its type, and size (if an array).

As an illustration of the use of common blocks, we shall redo the statistics problem using a common block for data communication. As we shall discuss in the program style box that follows, using common blocks is often not recommended; it is done here only to introduce the common block and not as a preferred programming practice.

**Example 7.14:** In the statistics problem, an array of data and its size were passed to each subprogram. If the declarations

```
COMMON X(MAXSIZ), N
REAL X
INTEGER N
```

were inserted in the main program (Fig. 7.16) in place of the current declarations for X, it would be possible to communicate all data through the common block and write each subprogram without arguments.

The revised declaration sections for the main program, function FNDMED and subroutine SORT are shown in Fig. 7.19. Notice that it is permissible to represent items in a common block by different names in each subprogram. The declarations

```
COMMON M(MAXSIZ), COUNT
REAL M
INTEGER COUNT
```

in subroutine SORT identify the first MAXSIZ data items in the common block as elements of a type REAL array named M; the last item in the common block is referenced by the type integer variable, COUNT (see Fig. 7.20). In the main program and in function FNDMED, however, the array is known by the name X, and the last item by the name N. The correspondence is based not upon names, but upon the relative position of each item in the block.

```
C SIMPLE STATISTICS PROBLEM - MAIN PROGRAM
C
C COMPUTE THE RANGE, MEAN AND MEDIAN
C OF A COLLECTION OF N REAL DATA ITEMS
C
C PROGRAM PARAMETERS
      INTEGER MAXSIZ
      PARAMETER (MAXSIZ = 100)
C
C COMMON BLOCKS USED
      COMMON X(MAXSIZ), N
      REAL X
      INTEGER N
C
C PROGRAM VARIABLES
             .
             .
             .
```

(Continued)

```
      REAL FUNCTION FNDMED( )
C
C FIND THE MEDIAN OF AN ARRAY OF REAL DATA ITEMS
C
C ARGUMENT DEFINITIONS
C     (NONE)
C
C PROGRAM PARAMETERS
      INTEGER MAXSIZ
      PARAMETER (MAXSIZ = 100)
C
C COMMON BLOCKS USED
      COMMON X(MAXSIZ), N
      REAL X
      INTEGER N
C
C LOCAL VARIABLES
            .
            .
            .

      SUBROUTINE SORT
C
C SORT AN ARRAY OF REAL DATA IN ASCENDING ORDER
C
C ARGUMENT DEFINITIONS
C     (NONE)
C
C PROGRAM PARAMETERS
      INTEGER MAXSIZ
      PARAMETER (MAXSIZ = 100)
C
C COMMON BLOCKS USED
      COMMON M(MAXSIZ), COUNT
      REAL M
      INTEGER COUNT
C
C LOCAL VARIABLES
```

**Fig. 7.19** Common block declaration for the statistics problem.

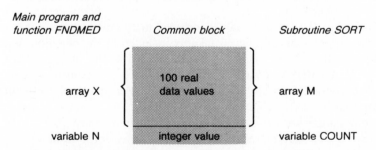

*Main program and function FNDMED*      *Common block*      *Subroutine SORT*

array X       100 real data values       array M

variable N       integer value       variable COUNT

**Fig. 7.20** Naming the elements of a common block.

The parameter MAXSIZ must be declared in all modules that use the common block; its declaration should precede the common declaration. As shown, the type of each common data item must be declared as well. The size of an array should be declared either in the common declaration or in its type declaration, but not in both places; we recommend the former.

Since all data are now communicated through the common block, arguments are not needed. The statement

```
MEDIAN = FNDMED ( )
```

would be used in the main program to call the new function FNDMED; the statement

```
CALL SORT
```

in function FNDMED would be used to call subroutine SORT. Note that the parentheses are still required in a function definition and call even though the function has no arguments.

---

**Program Style**

*Abuse of common blocks*

Unfortunately, many programmers overuse and abuse common blocks. Once they discover that common blocks exist, they tend to place all data in common rather than bother with passing arguments to subprograms. This approach has some serious shortcomings.

First, it obscures the data flow between program modules. When arguments are listed, the variables being used to pass and receive data are clearly specified in the subprogram call. However, there is no indication of which variables in a calling program are likely to be modified when data are communicated through a common block. This leads to program systems that are difficult to read, understand and maintain.

The second problem is that as common blocks grow, it becomes easier to make an error in the common declaration. A correspondence between common data and names is established by position; if the common declaration is incorrect, then the wrong correspondence will be established. This error is usually not detected by the compiler; it manifests itself insidiously and is often extremely difficult to find.

Further, as the list of variables included in common grows, the possibility of a local variable name being mistakenly included in the common declaration increases. As this "local variable" is manipulated in the subprogram, the corresponding data item in common may be modified erroneously as a *side effect* of the subprogram's execution. Side effects are also extremely difficult to detect and correct.

The potential for some of these errors can be reduced (but by no means eliminated) if you carefully separate common items, local variables, and arguments when writing your program declarations, as shown in Fig. 7.19.

---

**Exercise 7.26:**   Complete function FNDMED using a common declaration.

### 7.8.2 Named Common Blocks

As mentioned in the previous program style box, overuse of common blocks can result in program systems that are very difficult to debug because the data flow between modules is obscured. One feature of FORTRAN that can help alleviate this problem is the *named common block*. Named common blocks can be used to partition the common data into sections. We can selectively provide one or more of these blocks in a subprogram, or group of subprograms, as needed. If each named common declaration is kept to a manageable size, there should be less chance for a side effect error.

**Example 7.15:**   The declarations

```
COMMON /INBLK/ NA, NB, A(25), B(25)
INTEGER NA, NB
REAL A, B
```

and

```
COMMON /OUTBLK/ NC, C(50)
INTEGER NC
REAL C
```

describe two different named common blocks, INBLK (of size 52) and OUTBLK (of size 51). An example of INBLK is shown in Fig. 7.21.

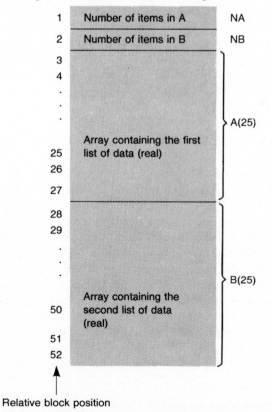

**Fig. 7.21**   The common block INBLK.

### 7.8.3  Additional Comments Concerning Common Blocks

The form of the common declaration is shown below.

---

**Common Declaration**

$$\text{COMMON } / \textit{name} / \textit{ list}$$

or

$$\text{COMMON } \textit{ list}$$

**Interpretation:** The defined common block is treated as one large array of consecutive memory cells containing data in the order indicated by the variables and arrays appearing in *list*. The statement

$$\text{COMMON } / \textit{name} / \textit{ list}$$

defines a common block called *name*. The declaration

$$\text{COMMON } \textit{ list}$$

defines a common block with a blank name (a *blank common block*). The common declarations must precede the executable statements.
*Note:* Character data may not be placed in a common block containing data of any other type (REAL, INTEGER, or LOGICAL).

---

In any program system, information may be communicated both through the use of argument lists and common blocks. Each module that contains a common declaration can reference any or all of the data items stored in the declared block, using the names appearing in the declaration. In this way, the data stored in the block is *shared* among all modules containing the block declaration. Information that is to be shared among a large number of modules in a program system is often placed in a common block. This eliminates the need to repeatedly pass this shared data through argument lists.

---

**Program Style**

*Type declarations for common data*

Each common statement used in a module should be followed by one or more type declarations defining the type of each of the items listed in the common statement. If arrays are included in a common block, it is a good practice to define the array sizes in the common statement and not in the type declarations. This can help to provide a clearer indication of the organization and structure of the common block. All common declarations should precede any executable statements. We recommend placing them at the very beginning of a main program, and immediately following the argument descriptions in a subprogram.

---

**Exercise 7.27:**

a) What values are printed as a result of the execution of the following program and sub-routine?

```
COMMON A, B              SUBROUTINE JUMBLE (X)
REAL A, B                REAL X
REAL C                   COMMON A, B
CALL JUMBLE(C)           REAL A, B
PRINT*, A, B, C          A = 1.0
STOP                     B = 2.0
END                      X = 4.0
                         RETURN
                         END
```

b) What values will be printed as a result of the execution of the following program and subroutines?

```
COMMON /WHAT/ NEXT(5)       SUBROUTINE DEFINE (ARRAY, SIZE)
INTEGER NEXT                INTEGER SIZE, ARRAY(SIZE)
INTEGER I                   INTEGER I
CALL DEFINE(NEXT, 5)        DO 40 I = 1, SIZE
CALL EXCH(1,4)                 ARRAY(I) = 2 * I - 1
I = 2                    40 CONTINUE
CALL EXCH(I, I+1)           RETURN
PRINT*, NEXT                END
STOP
END                         SUBROUTINE EXCH (S1, S2)
                            INTEGER S1, S2
                            COMMON /WHAT/ NEXT(5)
                            INTEGER NEXT
                            INTEGER TEMP
                            TEMP = NEXT(S1)
                            NEXT(S1) = NEXT(S2)
                            NEXT(S2) = TEMP
                            RETURN
                            END
```

What is the relationship between the variable I in the main program and the variable I in subroutine DEFINE? Why is it unnecessary to describe the common block WHAT in subroutine DEFINE?

## 7.9   THE SAVE STATEMENT*

There is an important restriction to be remembered when using variables or arrays declared local to a subprogram. The values stored in these variables are lost when control is returned to the calling program. The problem caused by this restriction can be illustrated via the following example.

---

*This section may be omitted.

**Example 7.16:**   Consider the following subroutine that counts the number of illegal data items that have been read by a program. The subroutine sets a flag once the number of data errors exceeds the value MAXCNT.

```
      SUBROUTINE ERRCNT (TOOBIG)
C
C COUNT ILLEGAL DATA ITEMS
C
C ARGUMENT DEFINITIONS --
C   OUTPUT ARGUMENTS
C     TOOBIG - A VALUE OF TRUE INDICATES THAT
C              MORE THAN MAXCNT ERRORS WERE FOUND
C
      LOGICAL TOOBIG
      INTEGER COUNT, MAXCNT
      DATA COUNT /0/, MAXCNT /12/
      TOOBIG = .FALSE.
      COUNT = COUNT + 1
      IF (COUNT .GT. MAXCNT) TOOBIG = .TRUE.
C
      RETURN
      END
```

There is no rule in FORTAN to guarantee that the values of the local variables COUNT and MAXCNT will be saved following the execution of the RETURN statement; thus, this subroutine may not always produce the desired results.

Fortunately, FORTRAN provides a feature, the SAVE statement, that can be used to save all, or a portion of, the local data in a subprogram. The form of this statement is illustrated in the following display.

---

**The SAVE Statement**

SAVE *list*

**Interpretation:** The *list* may contain the names of local variables or arrays whose values are to be saved following the execution of a subprogram RETURN. If the *list* is omitted, the statement is assumed to apply to all local variables within the unit in which it appears. The names of subprograms, dummy arguments, and the names of individual items in a common block must not appear in the SAVE statement.

---

The SAVE statement could be used to solve the potential problem of loss of data in Example 7.16. Inserting the statement

SAVE COUNT, MAXCNT

immediately following the INTEGER declaration in the subroutine, guarantees that the values of COUNT and MAXCNT will be retained from one call to the next of the subroutine ERRCNT.

A similar problem of data retention across subprogram calls arises in the case of variables and arrays declared in a named common block. The following rules apply:

- The contents of variables or arrays declared in the blank common block are always saved automatically.
- The contents of variables or arrays declared in a named common block are saved upon return from a subprogram only if that block is declared in the main program or in the subprogram that has called the current one.

The SAVE statement may be used to save the entire contents of a named common block. When a common block name (enclosed in slashes) appears in a SAVE statement, all items in the block will be saved. If a common block name appears in a SAVE statement in one subprogram, it must also appear in a SAVE statement in every subprogram which uses that block.

## 7.10    THE ROLE OF THE COMPILER IN PROCESSING SUBPROGRAMS*

### 7.10.1    Introduction

You have probably already written some programs that call the square root library function SQRT. Yet you didn't write this function; and, in fact, you know very little about the function aside from its name, the type of its input and output data, and that it somehow computes the square root of a positive real number. How, then, does the computer locate the SQRT function when it is called? How does the function find the argument (its input) and how does it know where to return the result? The answers to these questions can be found by examining the role of the compiler in processing subprograms. We will illustrate this role through the use of an example involving the library function SQRT. The role of the compiler is similar for user-defined functions and for the processing of subroutines. We will point out any differences in this role as we proceed with the example.

### 7.10.2    The Subprogram Linking Mechanism

Consider the program shown next.

```
REAL X, Y
READ*, X
PRINT*, 'X = ', X
IF (X .LT. 0.0) THEN
    PRINT*, 'X IS NEGATIVE, EXECUTION TERMINATED.'
ELSE
    Y = SQRT(X)
    PRINT*, 'THE SQUARE ROOT OF X IS ', Y
ENDIF
STOP
END
```

This program contains a reference to the library function SQRT with the ar-

---

*This section may be omitted.

gument X. Once the program has been translated, it must be loaded into the computer memory for execution. Furthermore, before the function SQRT can be executed, it must also be loaded into memory. Library functions (and user-defined functions and subroutines) are usually stored as machine language programs on a high-speed auxiliary memory device (such as a disk or drum) and can be loaded from this device into the computer memory whenever they are needed. It is the responsibility of the compiler to ensure that whenever SQRT is loaded into memory, its location is made known to the calling program. The compiler must also provide a mechanism that can be used to determine where to transfer control after the SQRT function has finished its task. Exactly how the compiler performs these tasks varies. However, once the appropriate communication mechanisms are set up, control can be transferred from the calling program to SQRT and back again, as shown in Fig. 7.22. The dashed arrows (C1 and R1) represent the data communication; the solid arrows (C2 and R2) represent the control transfers as described next.

C1: The address of the next instruction to be executed in the calling program is saved in a special memory cell. This address is called the *return address* and will be used to return control back to the calling program after the function is executed.

C2: Control is transferred to the address in memory that is associated with the first executable instruction in the SQRT function.

The instructions in the function SQRT are then carried out. There are two instructions in the function that have special significance when they are executed:

R1: An assignment statement of the form SQRT = *expression* causes the indicated *expression* to be evaluated and its value saved in a special memory cell.

R2: The statement RETURN causes a transfer of control back to the address which was saved at step C1.

When R2 is carried out, the calling program can continue execution at the point where it left off when the function was called. As part of this continued execution, the value returned by the function can be manipulated.

The entire mechanism just described, including steps C1, C2, R1, and R2, is set up by the compiler during translation and is carried out during the execution of the calling program.

### 7.10.3   Establishing the Correspondence between Arguments

In the preceding section we described how the compiler constructs the *transfer-of-control* between subprograms. In addition, we explained the manner in which a function result is returned to a calling program. It is also the role of the compiler to establish the correspondence between the arguments in the subprogram call (actual arguments) and those in the subprogram definition (dummy arguments).

In translating a call to a subprogram, the compiler makes certain that the addresses of all of the arguments in the argument list are saved in an *argument*

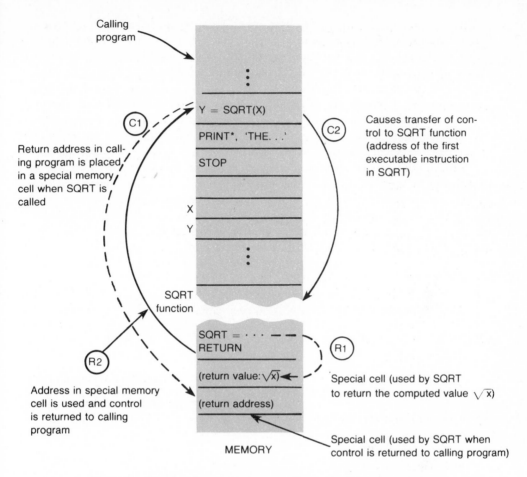

**Fig. 7.22**   Transfer of control between calling and called subprograms.

*address list.* Table 7.1 describes the entries in this table for the various kinds of subprogram arguments that are allowed in FORTRAN.

| Argument type | Address stored in argument address list |
|---|---|
| Constant | Address of the constant in the calling program |
| Variable | Address of the variable in the calling program |
| Expression | The expression is evaluated, and its value stored in a temporary memory cell. The address of this temporary memory cell is placed in the argument address list. |
| Array | Address of the first element of the array in the calling program |
| Array element | Address of the particular array element specified in the calling program |

**Table 7.1 Processing the Argument Address List**

The argument address list is used in different ways by different compilers. Some compilers translate a subprogram so that all references to dummy arguments in the subprogram will be replaced by references to the addresses of the actual calling arguments when the subprogram is called.

Other compilers use the argument addresses to obtain copies of the constants, expression values, and the contents of any variables or array elements that are used as actual arguments. These copies are saved in temporary memory cells that are manipulated when the subprogram is executed. Changes in the contents of the temporary cells must then be recorded in the memory cells occupied by the actual arguments at the completion of execution of the subprogram. (If an entire array is specified as an actual argument, it is usually not copied. Instead, the array elements themselves are manipulated by the subprogram.)

Regardless of the details of how the argument list is used, the net effect is the same. To illustrate this, we use a small subprogram EXCH with three arguments, LIST, FIRST, LAST. The purpose of this subprogram is to exchange the contents of two elements of an array represented by LIST. The subscripts of the elements to be exchanged are represented by FIRST and LAST. LAST is also used in the subroutine to specify the array size.

The subroutine EXCH and a portion of a calling program are shown below.

```
         REAL TABLE(100)              SUBROUTINE EXCH (LIST,
         INTEGER I1, I2             Z  FIRST, LAST)
                                      INTEGER FIRST, LAST
                   .                  REAL LIST(LAST)
                   .                  REAL TEMP
                   .                  TEMP = LIST(FIRST)
         CALL EXCH(TABLE, I1 + I2, 18)   LIST(FIRST) = LIST(LAST)
           {next instruction}         LIST(LAST) = TEMP
                   .                  RETURN
                   .                  END
                   .
         END
```

In this example, the statement

```
         CALL EXCH(TABLE, I1 + I2, 18)
```

is used to call the subroutine. Upon execution of this statement, the steps listed below and shown in Fig. 7.23 are carried out.

1. The address of the first element of TABLE is stored in the argument address list (see arrow A in Fig. 7.23).
2. The expression I1 + I2 is evaluated and stored in the temporary location TL. The address of TL is stored in the argument address list (arrow B ).
3. The address of the constant 18 is stored in the argument address list (arrow C ).
4. The return address is saved (arrow D ) as is the address of the first entry of the argument address list (arrow E ) and control is transferred to subroutine EXCH.

During the execution of EXCH, the values of the expression I1 + I2 and the constant 18 are used as subscripts (FIRST and LAST, respectively) to select the pair of elements of TABLE that are to be exchanged. The actual arguments will be manipulated using the addresses in the argument address list (unless copies of these arguments have been saved in temporary memory cells). When RE-TURN is executed, the return address is used to transfer control back to the calling program. (If the argument copy technique is used, the data in the temporary cells are copied back into the actual arguments using the addresses given in the argument address list.)

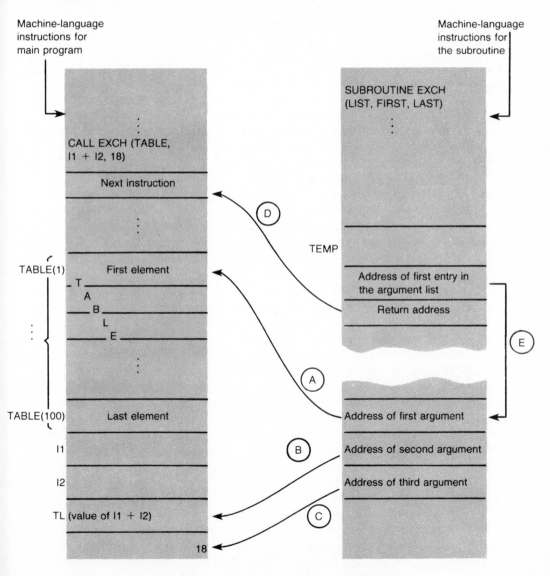

**Fig. 7.23**   Correspondence between arguments.

## 7.11    COMMON PROGRAMMING ERRORS

The most frequent and elusive errors that are made in using subprograms involve the specification of the argument list in a subprogram call and the listing of variable names in common declarations. Some compilers may be able to detect argument list errors, especially those errors that result in too few or too many arguments, or that arise from confusion in the type and structure correspondence between arguments. These compilers can also detect some length ambiguity in common block declarations.

In most cases, however, the compiler provides little help in the detection of these errors; you must, therefore, take steps to provide your own means of error detection. Some of these steps are listed in Section 7.7.3 on Testing a Program System.

## 7.12    SUMMARY

Two types of independent, separately compilable subprograms, the function and the subroutine, have been described. We discussed how to reference and define these subprograms and showed how data may be communicated among subprograms using argument lists and common blocks.

We also introduced the program system chart as a tool for representing the control and data flow relationships between a program and subprograms.

The importance of the name-independence feature of the FORTRAN subprogram was discussed in some detail. You should remember that the relationships among names used in different subprograms are determined solely by the position of the names used in argument lists (and in common declarations) and not by the names themselves.

A number of benefits of using subprograms were presented. Subprograms can be used to identify a sequence of statements that is needed in more than one place in a program. It is convenient to be able to write such a statement sequence only once and to reference it as often as it is needed. We can call a subprogram a number of times in the same program with different arguments.

A subprogram can also be referenced in programs written by people other than the subprogram author. This is often done and is, in fact, one of the most useful features of subprograms. The only information about a subprogram that a user needs to know is its name, a description of what it does (but not how it does it), and a complete description of the data that is communicated between the subprogram and the calling program. This description can be provided through a brief definition of the subprogram argument list. Once this subprogram documentation is provided, the subprogram may be used by the subprogram author or by other programmers whenever needed.

The independence of subprograms facilitates sectioning complex problems into smaller parts and designing the algorithms and writing and debugging the subprograms for these parts separately. These subprograms can then be put together to solve the original problem. Often, when more than one programmer is assigned to a project, this sectioning facility enables the project manager to assign different sections to different programmers. Each programmer can design and implement assigned sections with little knowledge of what the other programmers are doing. All that is needed is a general description of what will be done in the other program sections and the names and argument lists of the subprograms to be written for these sections.

Regardless of the size or complexity of a problem or the number of programmers involved in its solution, this sectioning technique is a most important concept in programming. The FORTRAN function and subroutine provide the capability of carrying this technique all the way through from the design stage to the final program implementation of an algorithm. By using this technique, we can divide a problem into a collection of logically meaningful sections and concentrate on the design, coding, and debugging of each section separately. This kind of complete problem modularization usually results in programs and program systems that are easier to implement and debug, easier to understand, and easier to modify.

The use of subroutines evolves most naturally when the step-wise development of algorithms is practiced. Each subproblem identified during algorithm refinement can be conveniently implemented in the form of a subprogram. A summary of subprogram features is provided in Table 7.2.

| Statement type and use | Examples |
|---|---|
| Function header statement: Identifies function name, the type of result, and the dummy arguments | REAL FUNCTION MAXTWO (A, B)<br>CHARACTER * 1 FUNCTION MYSIGN (X) |
| RETURN statement: Returns control from a subprogram | RETURN |
| Subroutine header statement: Identifies subroutine name and dummy arguments | SUBROUTINE SORT (M, NRITMS) |
| Subroutine call statement: Calls subroutine and specifies actual arguments | CALL SORT(INDATA, N) |
| Common declaration: Identifies common data items and the name of a named common block | COMMON X(100), N<br>COMMON /INBLOK/ X(100), Y(100), N |
| SAVE statement: Specifies which local variables will have their values retained after exit from the subprogram | SAVE ERRCNT, ERRFLG, /INBLOK/ |

**Table 7.2 New FORTRAN Features in Chapter 7**

## PROGRAMMING PROBLEMS

**7C**    Two positive integers I and J are considered to be *relatively prime* if there exists no integer greater than 1 that divides them both. Write a logical function **RELPRM** which has two parameters, I and J, and returns a value of true if and only if I and J are relatively prime. Otherwise, RELPRM should return a value of false.

**7D**    The *greatest common divisor*, GCD, of two positive integers I and J is an integer N with the property that N divides both I and J (with 0 remainder), and N is the largest integer dividing both I and J. An algorithm for determining N was devised by the famous mathematician Euclid; a flow diagram description of that algorithm, suitable for direct translation into FORTRAN, is provided next.

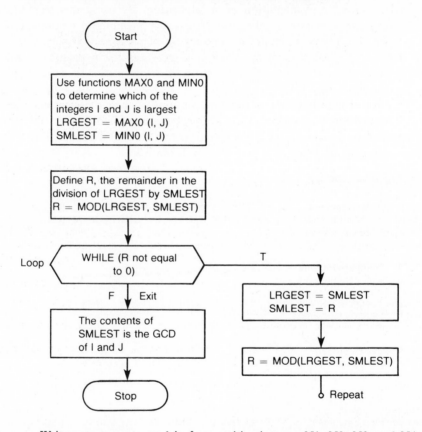

Write a program to read in four positive integers N1, N2, N3, and N4 and find the GCD of all four numbers. *Hint*: The GCD of the four integers is the largest N that divides all four of them. Implement the above algorithm as an integer function, and call it as many times as needed to solve the problem.

Note that GCD (N1, N2, N3, N4) = GCD (GCD(N1, N2), GCD(N3, N4)). Print N1, N2, N3, and N4, and the resulting GCD.

**7E**    Write a FORTRAN program that will read in a string of up to eighty 0's and 1's, and print out the number of 0's, the number of 1's, and the total number of digits in the string. Use the function COUNT given in Exercise 7.12.

**7F** Write a subroutine, MERGE, that will merge together the contents of two sorted (ascending order) real arrays A and B, storing the result (still in ascending order) in the real array C. *Hint:*

| | A(1) | A(2) | A(3) | A(4) | A(5) |
|---|---|---|---|---|---|
| A | −10.5 | −1.8 | 3.5 | 6.3 | 7.2 |

| | B(1) | B(2) | B(3) |
|---|---|---|---|
| B | −1.8 | 3.1 | 6.3 |

| | C(1) | C(2) | C(3) | C(4) | C(5) | C(6) |
|---|---|---|---|---|---|---|
| C | −10.5 | −1.8 | 3.1 | 3.5 | 6.3 | 7.2 |

When one of the input arrays has been exhausted, do not forget to copy the remaining data in the other array into the array C. If you use common blocks, refer to the named common blocks INBLK and OUTBLK shown in Example 7.15. Test your subroutine using a representative set of unsorted data in the arrays A and B. Sort both arrays before calling your MERGE subroutine.

**7G** Write a program system to process a set of exam scores. Each student's score for the exam is keypunched on a data card along with the student's last name.

  a) Determine and print the class average for the exam.
  b) Find the median grade.
  c) Scale each student's grade so that the class average will become 75. For example, if the actual class average is 63, add 12 to each student's grade.
  d) Assign a letter grade to each student based on the scaled grade: 90−100 (A), 80−89 (B), 70−79 (C), 60−69 (D), 0−59 (F).
  e) Print out each student's name in alphabetical order followed by the scaled grade and the letter grade.
  f) Count the number of grades in each letter grade category.
  g) Print a bar chart showing the distribution of exam scores (see Example 6.9).

**7H** Given the lengths $a$, $b$, $c$ of the sides of a triangle, write a function to compute the area, $A$, of the triangle; the formula for computing $A$ is given by

$$A = \sqrt{s(s-a)(s-b)(s-c)}$$

where $s$ is the semi-perimeter of the trangle:

$$s = \frac{a+b+c}{2}$$

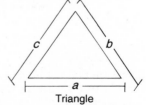

Triangle

Write a program to read in values for $a$, $b$, and $c$, and call your function to compute $A$. Your program should print $A$, and $a$, $b$, and $c$.

**7I** Write a type real function ROUND, which, given a real number X and an integer P, will return the value of X rounded to the nearest P decimal places. (*Example:* If

X is 403.7863 . . . , then ROUND (X, 2) will return the value 403.79.) Make certain that your function works for negative as well as positive values of X.

**7J**    Write an integer function FACT(N) that will compute the factorial, $n!$, of any small positive integer, $n$.

**7K**    The expression for computing $C(n, r)$, the number of combinations of $n$ items taken $r$ at a time, is

$$C(n, r) = \frac{n!}{r! \, (n - r)!} \, .$$

Assuming that we already have available a function FACT(N) for computing $n!$ (see Problem 7J), write function CNR(N, R) for computing $C(n, r)$. Write a program which will call CNR for $n = 4$, $r = 1$; $n = 5$, $r = 3$; $n = 7$, $r = 7$; and $n = 6$, $r = 2$.

**7L**    Assume the existence of a main program containing a call to a subroutine SEARCH:

```
CALLSEARCH (BUFFER, N, KEY, FOUND, INDEX)
```

Write a subroutine SEARCH to compare each of the N elements in the array BUFFER to the data item in KEY. If a match is found, SEARCH is to set FOUND to true and define INDEX to be the index of the element in the array BUFFER in which the key was located. If no key is found, FOUND is to be set false and INDEX is not to be altered.

**7M**    Do Problem 6E using subprograms. Provide a program system chart, data tables and flow diagrams before writing the required program system modules.

**7N**    The electric company charges according to the following rate schedule:

8 cents a kilowatt-hour (kwh) for electricity used up to the first 300 kwh;
6 cents a kwh for the next 300 kwh (up to 600 kwh);
5 cents a kwh for the next 400 kwh (up to 1000 kwh);
3 cents a kwh for all electricity used over 1000 kwh.

Write a function to compute the total charge for each customer. Write a program to call this function using the following data:

| Customer number | Kilowatt-hours used |
|---|---|
| 123 | 725 |
| 205 | 115 |
| 464 | 600 |
| 596 | 327 |
| 601 | 915 |
| 613 | 1011 |
| 722 | 47 |

The calling program should print a three-column table listing the customer number, hours used, and the charge for each customer. It should also compute and print the number of customers, total hours used, and total charges.

**7O** A throw of two dice may produce anywhere from a two (snake-eyes) to a twelve (box-cars). Write a program system to produce the table shown below.

| Roll value | Number of ways of getting this roll | Probability of getting this roll | Probability of a roll greater than or equal to this one |
|---|---|---|---|
| 2 | 1 | .028 | 1.000 |
| 3 | 2 | .056 | .972 |
| ⋮ | ⋮ | ⋮ | ⋮ |
| 11 | 2 | .056 | .084 |
| 12 | 1 | .028 | .028 |

For any roll value, X, the probability of getting that roll is

$$P(roll=X) = tally(X)/36$$

where tally (X) is the number of ways of getting X. Also, the probability of getting a roll greater than or equal to X is

$$P(roll \geq X) = P(roll=X)+P(roll=X+1)+. . . P(roll=12)$$

Thus

$$P(roll=10) = tally(10)/36.0 = 3.0/36.0 = .083$$

and

$$P(roll \geq 10) = .083 + .056 + .028 = .167$$

*Hints*: Store the number of ways of getting a roll value, X, the probability of each roll value, and the probability of a roll value greater than or equal to X, in three arrays NRWAYS, PX, and PGEX, each of size 12 (do not use the first elements of these arrays). Your main program should call a subroutine TALLY to compute NRWAYS for each roll (TALLY should be called just once). Given the data in NRWAYS, the probabilities of each roll can be determined, and then the data in the last column can be computed. All computations should be rounded to three decimal places.

For each X, P(roll = X) should be computed using a function. This function should be called 11 times.

Define an array ROLLS(36) of two digit integers representing all possible outcomes of a dice roll, r. Note that for any roll (represented by the two-digit integer *r*) the actual numeric value of the roll can be computed as

$$VALUE = MOD(r, 10) + r/10$$

For example, if *r* is 36, then the actual value of the roll is

$$VALUE = MOD(36,10) + 36/10 = 6 + 3 = 9$$

**7P** Each week the employees of a local manufacturing company turn in time cards containing the following information:

1) an identification number (a five-digit integer),
2) hourly pay rate (a real number),

3)  time worked Monday, Tuesday, Wednesday, Thursday and Friday (each a four-digit integer of the form HHMM, where HH is hours and MM is minutes).

For example, last week's time cards contained the following data:

| Employee number | Hourly rate | Time worked (hours, minutes) | | | | |
|---|---|---|---|---|---|---|
| | | Monday | Tuesday | Wednesday | Thursday | Friday |
| 16025 | 4.00 | 0800 | 0730 | 0800 | 0800 | 0420 |
| 19122 | 4.50 | 0615 | 0800 | 0800 | 0800 | 0800 |
| 21061 | 4.25 | 0805 | 0800 | 0735 | 0515 | 0735 |
| 45387 | 3.50 | 1015 | 1030 | 0800 | 0945 | 0800 |
| 50177 | 6.15 | 0800 | 0415 | 0800 | 0545 | 0600 |
| 61111 | 5.00 | 0930 | 0800 | 0800 | 1025 | 0905 |
| 88128 | 4.50 | 0800 | 0900 | 0800 | 0800 | 0700 |

Write a program system that will read the above data and compute for each employee the total hours worked (in hours and minutes), the total hours worked (to the nearest quarter-hour), and the gross salary. Your system should print the data shown above with the total hours (both figures) and gross pay for each employee. You should assume that overtime is paid at 1½ times the normal hourly rate, and that it is computed on a weekly basis (only on the total hours in excess of 40.00), rather than on a daily basis. Your program system should contain the following subprograms:

a)  A function for computing the sum (in hours and minutes) of two four-digit integers of the form HHMM (*Example:* 0745 + 0335 = 1120);
b)  A function for converting hours and minutes (represented as a four-digit integer) into hours, rounded to the nearest quarter hour (*Example:* 1120 = 11.25);
c)  A function for computing gross salary given total hours and hourly rate;
d)  A function for rounding gross salary accurate to two decimal places (see Problem 7I).

Test your program using the time cards shown above.

**7Q**    Redo the grading problem (Sec. 5.5) using subroutines. Also, add a step to count the number of invalid scores and print an appropriate message at the end if there are any.

**7R**    A mail order house with the physical facilities for stocking up to 20 items decides that it wants to maintain inventory control records on a small computer. For each stock item, the following data are to be stored on the computer:

1)  the stock number (a five-digit integer);
2)  a count of the number of items on hand;
3)  the total year-to-date sales count;
4)  the price;
5)  the date (month and day) of the last order for restocking an item (a four-digit integer of the form MMDD);
6)  the number of items ordered.

Both items (5) and (6) will be zero if there is no outstanding order for an item.
    Design and implement a program system to keep track of the data listed in (1)

through (6). You will need six arrays, each of size 20. Your system should contain subprograms to perform the following tasks:

a) change the price of an item (given the item stock number and the new price);
b) add a new item to the inventory list (given the item number, the price, and the initial stock on hand);
c) enter information about the date and size of a restock order;
d) reset items (5) and (6) above to zero and update the amount on hand when a restock order is received;
e) increase the total sales and decrease the count on hand each time a purchase order is received (if the order cannot be filled, print a message to that effect and reset the counts);
f) search for the array element that contains a given stock number.

The following information should be initially stored in memory (using data initialization (DATA) statements). This information should be printed at the start of execution of your program system.

| Stock numbers | On-hand count | Price |
|---|---|---|
| 02421 | 12 | 100.00 |
| 00801 | 24 | 32.49 |
| 63921 | 50 | 4.99 |
| 47447 | 100 | 6.99 |
| 47448 | 48 | 2.25 |
| 19012 | 42 | 18.18 |
| 86932 | 3 | 67.20 |

A set of typical transactions for this inventory system is given below.

**Price Changes**

| Trans no. | Card ID | Stock no. | New price |
|---|---|---|---|
| 2 | 'PRIC' | 19012 | 18.99 |
| 9 | 'PRIC' | 89632 | 73.90 |

**Add Items**

| Trans no. | Card ID | Stock no. | Price | On-hand |
|---|---|---|---|---|
| 4 | 'ADIT' | 47447 | 14.27 | 36 |
| 5 | 'ADIT' | 56676 | .15 | 1500 |

**New Orders**

| Trans no. | Card ID | Stock no. | Date | Volume |
|---|---|---|---|---|
| 3 | 'NUOR' | 00801 | 1201 | 18 |
| 8 | 'NUOR' | 47446 | 1116 | 15 |

**Orders Received**

| Trans no. | Card ID | Stock no. | Volume |
|---|---|---|---|
| 6 | 'ORIN' | 00801 | 18 |

**Purchase Orders**

| Trans no. | Card ID | Stock no. | Number wanted |
|---|---|---|---|
| 11 | 'PRCH' | 00801 | 30 |
| 12 | 'PRCH' | 12345 | 1 |
| 7 | 'PRCH' | 56676 | 150 |
| 10 | 'PRCH' | 86932 | 4 |

Each transaction should be punched using two cards: a header card, and a data card. The appearance of these cards is shown below.

Header card: Transaction number
Transaction ID
Stock number

Data card:    Stock number
Transaction data

For example, transaction four should be punched as

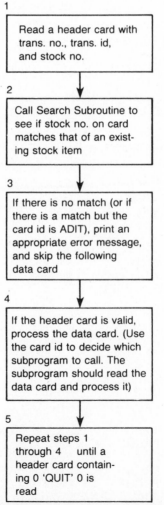

Your main program should process the transactions, one at a time, as shown below.

1

Read a header card with trans. no., trans. id, and stock no.

2

Call Search Subroutine to see if stock no. on card matches that of an existing stock item

3

If there is no match (or if there is a match but the card id is ADIT), print an appropriate error message, and skip the following data card

4

If the header card is valid, process the data card. (Use the card id to decide which subprogram to call. The subprogram should read the data card and process it)

5

Repeat steps 1 through 4    until a header card containing 0 'QUIT' 0 is read

Each subprogram should print an appropriate informative message for each transaction, indicating whether or not the transaction was processed, and giving other pertinent information about changes in the stored data that were affected by the processing of the data card.

After the quit card is read, all inventory data should be printed in tabular form.

# FORMATS AND FILES

8

## 8.1   INTRODUCTION

In Section 4.7, we provided a brief introduction to formatted output. You should review that section if you have not been using format specifications in your programming.

In this chapter, we will continue to study the use of formats for specifying the appearance of an output line. We will also discuss how format features can be used for data entry and we will introduce the FORMAT statement.

In addition, the use of data files will be discussed. These files enable data generated by one program to be reused by another. FORTRAN supports two types of data files: sequential and direct access. The properties and relative advantages of each will be covered.

## 8.2   FORMAT SPECIFICATION USING STRINGS

### 8.2.1   A Review of Format Specification

In Section 4.7 we wrote PRINT statements of the form

```
          PRINT '(A, A, 3X, F7.2, 2X, I2)',
     Z           ' ', NAME, GROSS, DEPEND
```

where the string following PRINT is the format specification describing how the output list (on the second line) is to be printed. The format specification consists of a list of edit descriptors. Two of the descriptors are used to specify horizontal spacing (3X, 2X) between output items; each of the remaining descriptors specifies how a variable or value should be printed. These descriptors are paired from left to right, with an item in the output list. The first letter of each data descriptor (A, F, I) is determined by the type of data to be printed (character, real, integer). You should review Tables 4.2 and 4.3 in Section 4.7 if the use of any of these edit descriptors is unclear.

### 8.2.2   Storing Format Specifications in Character Variables

One difficulty with the formatted PRINT statements that we have been using is their length. It is possible to remove the format specification from the PRINT statement. One way to accomplish this is to store the format specification (a string) in a character variable. If FORM1 is a character variable with capacity for at least 24 characters, then the statement

```
          FORM1 = '(A, A, 3X, F7.2, 2X, I2)'
```

assigns the format specification in our previous PRINT statement to FORM1. We can then rewrite this PRINT statement as

```
          PRINT FORM1, ' ', NAME, GROSS, DEPEND
```

Besides shortening the PRINT statement, another advantage of this approach is that it enables us to reuse the format specification easily in more than one formatted PRINT statement. All that is required is that the character variable (FORM1 in this case) be defined before the first formatted PRINT statement that references it is reached. This character variable replaces the string in the PRINT statement; the variable must be followed by a comma and then by the output list for each PRINT statement that uses it. Remember that the format specification replaces the asterisk that we used with list directed input and output, so be sure that the asterisk is not included in your formatted PRINT statements. Also, make sure there is no comma just after the word PRINT.

An additional advantage of storing a format specification in a character variable is that it becomes much easier to modify the format. We may simply change the contents of the character variable by using a READ or assignment statement. The net effect will be to modify the form of the program output.

**Example 8.1:** The statements

```
READ*, FORM1
PRINT FORM1, GROSS, NET, DEPEND
```

provide a very flexible way of specifying the appearance of an output line. The string read into the character variable FORM1 determines how the values of GROSS, NET and DEPEND will be printed. This string must be a valid format specification including the surrounding parentheses and enclosed in apostrophes.

**Example 8.2:** If OUTFRM is a character array, the DO loop

```
DO 10 I = 1, N
    PRINT OUTFRM(I), X(I), Y(I)
10 CONTINUE
```

will cause N lines of output to be printed. The form or appearance of line I is determined by the value of array element OUTFRM(I). Each element of array OUTFRM must contain a format specification. The first N elements of arrays X and Y will be printed.

**Exercise 8.1:** Describe the output generated by each PRINT statement below.

```
CHARACTER * 10 FORM1, FORM2
FORM1 = '(A, A)'
FORM2 = ' HI'
PRINT FORM1, FORM1, FORM2
PRINT FORM1, FORM2, FORM1
PRINT*, FORM2, FORM1, FORM1
```

**Exercise 8.2:** Which of the strings below are valid data items for the statement

```
READ*, FORM1
```

in Example 8.1?

```
a) (F7.2, F7.2, I3)
b) '(F7.2, F7.2, I3)'
c) '1X, F7.2, 2X, F7.2, 2X, I3 '
d) '(1X, F7.2, 2X, F7.2, 2X, I3)'
```

## 8.3   FORMAT STATEMENTS

### 8.3.1   FORMAT Statements and Numbers

There is yet another method of specifying formats in FORTRAN and that involves the use of a labelled FORMAT statement. In fact, prior to 1977, this was the only standard method available for specifying formats.

**Example 8.3:**   The statement

```
47 FORMAT (A, A, 3X, F7.2, 2X, I2)
```

is a special FORTRAN statement called a **FORMAT** statement. FORMAT statements may be used to provide format specifications in a totally separate labelled statement. This statement may then be referenced by number (47 in this case) in any PRINT statement using the format. For example,

```
PRINT 47, ' ', NAME, GROSS, DEPEND
```

The statement label, 47, is used as a *format number* identifying the number of the FORMAT statement to be used to specify the appearance of the line of output to be produced. The format specification appears in the FORMAT statement without apostrophes.

The PRINT and FORMAT statements illustrated above produce the same output as the PRINT statement

```
PRINT '(A, A, 3X, F7.2, 2X, I2)',
      ' ', NAME, GROSS, DEPEND
```

shown earlier.

The general form of a FORMAT statement is described in the next display.

---

**The FORMAT Statement**

*lab* FORMAT  (*list of edit descriptors*)

**Interpretation:** The *list of edit descriptors* provides descriptive information about the input/output operation to be performed. The FORMAT statement may be placed anywhere in the program. The link between an input/output statement and the format pertaining to that statement is provided by the statement label *lab*.

---

The FORMAT statement is a nonexecutable statement and may be placed anywhere in the program. Each FORMAT statement must always have a unique label in its label field. (Recall that a label consists of one through five digits and is placed in the first five columns of a FORTRAN statement.)

More than one PRINT statement may reference the same FORMAT statement. Generally, the FORMAT statement follows right after the first statement that references it. Make sure there is no comma before the label in a formatted PRINT.

**Exercise 8.3:**   What would be printed by the following statements?

```
      PRINT 33, 'THIS IS A TITLE'
   33 FORMAT (10X, A)
      PRINT*, 33, 'THIS IS A TITLE'
```

## 8.3.2   Strings in Formats

In earlier versions of FORTRAN, the FORMAT statement was the only standard means for specifying formatted output. The use of string values in output lists was not permitted; consequently, all strings had to be included in the FORMAT statement.

**Example 8.4:**   The two sets of statements below produce identical output.

```
      PRINT 27, '1P = ', P, ' Y = ', Y
   27 FORMAT (A, F7.2, A, F7.2)

      PRINT 37, P, Y
   37 FORMAT ('1P = ', F7.2, ' Y = ', F7.2)
```

In the top pair of statements, the strings in the output list are associated with the Edit descriptors in format 27. In the bottom pair, these strings are included in the format specification (format 37) and are no longer part of the output list. In either case, the 1 in the string '1P = ' will not be printed; it is used for carriage control and causes an immediate skip to the top of the next output page. Remember, the carriage control character controls the high-speed line printer; it is generally not needed when working at a terminal.

Whenever a string is encountered in a format specification, that string is inserted in the output line with the apostrophes removed. Its position in the output line is determined by its position in the format specification.

Strings can also be included in a format specification that is itself a string. However, remember that each apostrophe (single quote) within a string must be represented by a pair of apostrophes. The PRINT statements

```
      PRINT '(A, F7.2, A, F7.2)', '1P = ', P, 'Y = ', Y
      PRINT '(''1P = '', F7.2, ''Y = '', F7.2)', P, Y
```

produce the same output as those shown above.

It makes no difference which of the methods above you use to specify strings that annotate output. However, make sure there is an A descriptor in the format specification for each string in the output list.

**Example 8.5:**  If NAME contains 'FRANK', HOURS = 42.5, RATE = 3.50 and AGE = 37, then the statements

```
        PRINT 18, NAME
    18  FORMAT ('1MY NAME IS ', A)
        PRINT 29, AGE
    29  FORMAT (' I''M ONLY ', I3, ' YEARS OLD')
        GROSS = HOURS * RATE
        PRINT 34, HOURS, RATE
    34  FORMAT ('OHOURS = ', F6.2, ' RATE = ', F6.2)
        PRINT 38, GROSS
    38  FORMAT ('OMY GROSS PAY IS ', F7.2)
```

would produce the output

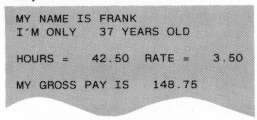

```
MY NAME IS FRANK
I'M ONLY    37 YEARS OLD

HOURS =    42.50  RATE =    3.50

MY GROSS PAY IS   148.75
```

**Exercise 8.4:**  Rewrite each PRINT statement below so that the strings are removed from the output list and placed in the format specification. Also, write a separate FORMAT and PRINT statement corresponding to each PRINT statement.

```
a)  PRINT '(A, 10X, A, 10X, A)',
    Z       ' NAME', 'SALARY', 'DEPENDENTS'

b)  PRINT '(A, A)', ' EMPLOYEE = ', NAME

c)  PRINT '(A, F6.0, 3X, A, I2)',
    Z       ' SALARY = ', GROSS, ' DEPENDENTS = ', DEPEND
```

**Exercise 8.5:**  Describe with a picture (as in Example 8.5), the output produced by the following statements. (Let NAME = 'REGGIE', BA = .300, HR = 41 and TEAM = 'YANKEES'; NAME and TEAM are type CHARACTER * 10.)

```
a)          PRINT 20, NAME, TEAM
        20  FORMAT ('1', A, ' PLAYS FOR THE ', A)

b)          PRINT 25, HR, BA
        25  FORMAT ('OLAST YEAR HE HIT ', I2,
            Z   'HOME RUNS, AND BATTED ', F4.3)

c)          PRINT 474, 'ONAME', 'TEAM', 'HOME RUNS', 'AVERAGE'
       474  FORMAT (A10, A10, A10, A10)
            PRINT 475, NAME, TEAM, HR, BA
       475  FORMAT (1X, A, A, I10, F10.3)
```

### 8.3.3  The Slash Descriptor

Often, we may want to use a single format to describe the appearance of more than one line of output. The character slash, /, may be used to separate the description of one output line from the next.

A string of consecutive slashes has the effect of causing blank lines to appear on a page. The number of blank lines will be one less than the number of consecutive slashes if the slashes are in the middle of the format specification. The number of blank lines will be equal to the number of slashes if the slashes appear at the very beginning or very end of the format specification. Commas are not required to separate consecutive slashes nor to separate slashes from other format descriptors. A large number of consecutive slashes may be indicated using the descriptor form n(/), where n is a positive integer constant indicating the number of slashes to be used.

**Example 8.6:**  The statements

```
      PRINT 18, X, Y
   18 FORMAT (1X, F8.2 ///// 1X, F8.2)
      PRINT 19, X, Y
   19 FORMAT (1X, F8.2, 5(/), 1X, F8.2)
```

will cause four blank lines to appear between the values of X and Y. The first slash in format 18 terminates the output line containing the value of X; the next four slashes terminate blank output lines.

Each format specification begins with the edit descriptor 1X which represents a blank space. This blank would be used for carriage control.

**Example 8.7:**  The Raisem Higher Home Loan Association maintains lists of home loan mortgage interest payments. A sample page of these lists is shown below.

```
RAISEM HIGHER HOME LOAN ASSOCIATION            07-04-81

HOME LOAN MORTGAGE INTEREST PAYMENT TABLES

AMOUNT = $30000.00      LOAN DURATION (MONTHS) = 300

RATE (PERCENT)     MONTHLY PAYMENT      TOTAL PAYMENT
     10.50              XXX.XX            XXXXXX.XX
     10.75              XXX.XX            XXXXXX.XX
       .                  .                   .
       .                  .                   .
       .                  .                   .
     13.50              262.11            78632.70
       .                  .                   .
       .                  .                   .
       .                  .                   .
     16.00              XXX.XX            XXXXXX.XX
```

The person who wrote the program to print this table used six variables.

DATE:      The date the list was made (a character string of length 8)
AMOUNT: The amount of the loan (real)
MONTHS: Period of time (in months) for loan repayment (integer)
RATE:      Interest rate (percent) applied to the loan (real)
MPAYMT: Monthly payment required from borrower (real)
TPAYMT:   Total amount to be paid over entire loan period (real)

The program contains the three **PRINT** statements shown below.

a)
```
    PRINT 26, DATE, AMOUNT, MONTHS
 26 FORMAT ('1RAISEM HIGHER HOME LOAN ASSOCIATION', 9X, A8/
  A 'OHOME LOAN MORTGAGE INTEREST PAYMENT TABLES'/
  B 'OAMOUNT = $', F8.2, 6X, 'LOAN DURATION (MONTHS) = ', I3)
```

b)
```
    PRINT 27
 27 FORMAT ('ORATE (PERCENT)', 4X, 'MONTHLY PAYMENT', 4X,
  A 'TOTAL PAYMENT')
```

c)
```
    PRINT 28, RATE, MPAYMT, TPAYMT
 28·FORMAT (6X, F5.2, 13X, F6.2, 10X, F9.2)
```

Together these three statements produce the sample page of output just shown. Statement (a) prints the three lines of page-heading information that appears at the top of the page. The values of the variables DATE, AMOUNT, and MONTHS are included as part of the heading. Statement (b) prints the column heading labels, and statement (c) is used inside a loop in the program to print the numbers appearing in each output line, the values of RATE, MPAYMT, TPAYMT.

You should convince yourself that formats 26, 27, and 28 do indeed produce the output shown in Example 8.7. To aid you in this effort, we provide a detailed description of the formation of the three output lines defined in format 26.

| | Edit descriptor | Corresponding output list item | Meaning |
|---|---|---|---|
| | '1RAISEM. . . ASSOCIATION' | None | Indicates that the string '1RAISEM. . .ASSOCIATION', is to be entered into the line being formed. (The "1" is not printed; it is used for carriage control.) |
| Line 1 of format 26 | 9X | None | Indicates that a field of 9 blanks is to be entered into the line being formed. |
| | A8 | DATE | Indicates that a field of width 8 (8 print positions) is to be used to print the character string stored in DATE. |
| | / | None | Indicates the end of the output line being formed. |

(*Continued*)

| Edit descriptor | Corresponding output list item | Meaning |
|---|---|---|
| 'OHOME. . . TABLES' | None | Indicates that the string 'OHOME. . . TABLES', is to be entered into the line being formed. (The "0" is not printed; it is used for carriage control.) |
| / | None | Indicates the end of an output line. |
| 'OAMOUNT = $' | None | Indicates that the string 'OAMOUNT = $', is to be entered into the line being formed. (The "0" is not printed; it is used for carriage control.) |
| F8.2 | AMOUNT | Indicates that a field of width 8 is to be used to print the real number stored in AMOUNT (two digits will appear to the right of the decimal point). |
| 6X | None | Skip 6 spaces. |
| 'LOAN. . . MONTHS) = ' | None | Indicates that the string 'LOAN. . . MONTHS) = ', is to be entered into the line being formed. |
| I3 | MONTHS | Indicates that a field of width 3 is to be used to print the integer stored in MONTHS. |

Continuation line A (brackets first two rows); Continuation line B (brackets remaining rows).

## 8.4   INPUT FORMATS

### 8.4.1   Partitioning a Data Card into Fields

The function of an input format is analogous to that of an output format. Output formats provide a line-by-line description of the external appearance of data that are to be printed. Input formats, on the other hand, are normally used to describe the card-by-card appearance of information that is to be read into the computer from punched cards (or line-by-line appearance of terminal input).

The format specification in a formatted READ statement partitions the data card into fields. It may cause some columns to be skipped and enables strings to be keypunched without enclosing quotes.

As was the case with list-directed input, each execution of a READ statement causes at least one new data card to be read. The data items in these cards are placed into the memory cells designated by the input list.

**Example 8.8:**  If the data 07-04-81 are punched in columns 1–8 of a card, then the statement

```
      READ 62, MONTH, DAY, YEAR
   62 FORMAT (I2, 1X, I2, 1X, I2)
```

will have the effect shown below.

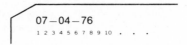

MONTH    DAY    YEAR

| 7 | 4 | 81 |

Format 62 partitions the data card into five fields are shown below. Three of these fields contain integer data that are stored in memory and two fields (columns 3 and 6) are skipped.

```
07 – 04 – 76
1 2 3 4 5 6 7 8 9 10 . . .
```

**Example 8.9:** The list-directed READ statement

   a) `READ*, FIRST, LAST, IDEMPL, HOURS, RATE, OTHRS`

```
'JOHN'  'CAGE'  37458  35.0 6.75  0.0
1 2 3 4 5 6 7 8 9 10 11 12 13 14 15 16 17 18 19 20 21 22 23 24 25 26 27 28 29 30 31 32 33 34 35 36 37 38 39 40 41 42 43 44 45 46 47 48 49      • • •            80
```

will cause the card above to be read, and its contents to be stored in memory, as shown below.

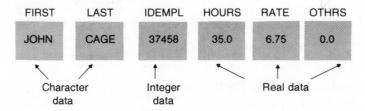

FIRST    LAST    IDEMPL    HOURS    RATE    OTHRS

| JOHN | CAGE | 37458 | 35.0 | 6.75 | 0.0 |

     Character      Integer        Real data
       data          data

Given a similar card without apostrophes, the statements

   b) `READ 25, FIRST, LAST, EMPLNO, HOURS, RATE, OTHRS`
     `25 FORMAT (1X, A4, 3X, A4, 2X, I5, 1X, F5.1, 1X, F4.2, 1X, F4.1)`

will have the same effect. Format 25 partitions the card into the fields shown below.

```
JOHN   CAGE   37458   35.0  6.75   0.0
1 2 3 4 5 6 7 8 9 10 11 12 13 14 15 16 17 18 19 20 21 22 23 24 25 26 27 28 29 30 31 32 33 34 35 36 37 38 39 40 41 42 43 44 45 46 47 48 49      • • •            80
```

In order to see why the statements (b) have the same effect as the statement in (a), it is necessary to understand the meaning of the edit descriptors listed in format 25. They are described in Table 8.1.

| Edit descriptor | Corresponding input list item | Meaning |
|---|---|---|
| ⎵1X | None | Skip the first column in the card. |
| A4 | FIRST | Indicates that the next four columns in the card (2–5 in this case) are to be treated as a field containing a string (JOHN) which is to be stored in the variable FIRST. |
| 3X | None | Skip the next three columns in the card (columns 6, 7, and 8). |
| A4 | LAST | The next four columns in the card (9–12) are to be treated as a field containing a string (CAGE) which is to be stored in the variable LAST. |
| 2X | None | Skip the next two columns in the card (columns 13, 14). |
| I5 | EMPLNO | The next five card columns (15–19) are to be treated as containing a 5-digit integer (37458) to be stored in variable EMPLNO. |
| 1X | None | Skip the next column in the card (column 20). |
| F5.1 | HOURS | The next five card columns (21–25) are to be treated as containing a real number (35.0) with one digit (0) to the right of the decimal point. This number is to be stored in variable HOURS. |
| 1X | None | Skip the next column in the card (column 26). |
| F4.2 | RATE | The next four columns (27–30) are to be treated as containing a real number (6.75) with two digits (75) to the right of the decimal point. This number is to be stored in variable RATE. |
| 1X | None | Skip the next column in the card (column 31). |
| F4.1 | OTHRS | The next four columns (32–35) are to be treated as containing a real number (0.0), with one digit to the right of the decimal point. This number is to be stored in the variable OTHRS. |

**Table 8.1 Meaning of Edit Descriptors in FORMAT 25.**

The next display summarizes the rules that must be followed when using formatted input.

## Rules for Formatted Input

1. The FORMAT statement determines which card columns are to be skipped and which are to be read. It also describes the type and width (number of columns) of the information contained in each field and the number of decimal places in each type real data item.

2. As is the case with output formats, all edit descriptors in an input format except the slash should be separated from one another using a comma.

3. The data descriptors (A, F, or I) must be compatible with the

corresponding variable type (character, real, or integer). Also, the type of data in the data field must be compatible with the data descriptor (i.e., digits and a sign for I; digits, a sign and a decimal point for F).

    4. The field width ($w$) information for each data descriptor immediately follows the type indication where $w$ is an integer constant.

A$w$:  The A indicates that a string of characters is to be read into a memory cell; the $w$ indicates the total width (or number of characters) contained in the string. If the $w$ is omitted, the field width is the same as the declared length of the corresponding input variable. Any legal FORTRAN character may appear in the string. If the length, $w$, of the data string is less than the declared length of the variable, then the string will be stored with blank padding on the right. If the length, $w$, of the data string is greater than the length of the variable, then only the leftmost $w$ characters of the string will be stored.

I$w$:  I indicates that a type integer value is to be read. The $w$ indicates the width (the number of decimal digits) contained in the integer.

F$w$.$d$:  F indicates that a type real value is to be read. The $w$ indicates the total width (including the sign and decimal point); the $d$ indicates the number of decimal places in the real number that are assumed to be to the right of the decimal point if no explicit decimal point appears.

    5. The edit descriptor $n$X indicates that a field of $n$ columns is to be skipped. These columns need not contain blanks; they will be skipped regardless of what they contain.

    6. It is not necessary that the data occupying a character, integer, or real data field fill the entire field. However, the following guidelines should be followed:

a) Integer values should be punched in the rightmost portion of the field;

b) Real values may be punched anywhere in the field as long as the decimal point is included. If the decimal point is not punched, the number should be punched in the rightmost portion of the field, and the decimal point will be assumed to be placed as indicated by the $d$ parameter of the format descriptor. If the decimal point is punched, then it overrides the $d$ parameter in the F$w$.$d$ edit descriptor.

    7. A single FORMAT statement may be used to describe the layout of more than one card. The edit descriptor slash, /, can be used to mark the end of the description of one card, and the start of the description of a new card.

Rule 6 above is concerned with the position of real and integer values in a data field. These values should generally be punched right-justified since some systems may read any blanks in an integer or real data field as zeros. An I or F field with all blanks will be read as the number 0. Real values may be punched with or without a decimal point.

**Example 8.10:**   Given the READ statement and data card

```
     REAL 672 X, Y, Z
 672 FORMAT (F4.1, F4.2, F4.2)
```

the real variables X, Y and Z will be defined as

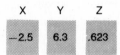

In the first field, the decimal point is assumed to be between the 2 and the 5, according to the format descriptor F4.1. In the last two fields, the keypunched decimal point overrides the format descriptor.

**Example 8.11:**   The format

```
 562 FORMAT (A8 / F10.4 / I6, I4)
```

describes the appearance of three cards. The first card is described as containing one field that is to be treated as a character string of width 8. The second card is described as having a field of width 10 containing a real number. If the decimal point is not punched in this number, it will be assumed to be located to the left of the fourth decimal digit (counting from the right). The third card is described as containing two integer fields, the first of width 6 and the second of width 4.

The general form of the edit descriptors used in input formats is summarized in Table 8.2.

| Edit<br>descriptor | Meaning |
|---|---|
| $nX$ | Indicates that a field of $n$ columns is to be skipped. |
| $Iw$ | Indicates that a field of width $w$ ($w$ columns) is to be treated as an integer. |
| $Aw$ | Indicates that a field of width $w$ is to be treated as a character string. The $w$ may be omitted. |
| $Fw.d$ | Indicates that a field of width $w$ is to be treated as a real number with $d$ digits to the right of the decimal point. If a decimal point is punched, it overrides the $d$ specification. |

**Table 8.2 Edit descriptors for formatted input.**

It is often desirable to echo print data being read. In such cases, it is tempting to try to use the same FORMAT statement for both the READ and the PRINT statements. This temptation should be avoided because there is no guarantee that the first character read from the card will be suitable for use as a carriage control character.

**Exercise 8.6:**   You are given a card with the following format:

*CHARACTER*11 so's IF NO ' QUOTES, USE A FORMAT*

| Card columns | Contents | Sample data |
|---|---|---|
| 1–11 | Social Security Number | 552–63–0179 |
| 12–31 | Last name | BROWN |
| 32–44 | First name | JERRY |
| 45 | Middle initial | L |
| 46–48 | Blanks | |
| 49–50 | Age | 38 |
| 51–54 | Blanks | |
| 55–59 | Total years of education | 23 |
| 60–63 | Blanks | |
| 64–66 | Occupation code | 12 |
| 67–80 | Blanks | |

Assign variable names to the data in each of the nonblank fields. Give these names appropriate types, and write a READ statement and appropriate format for reading such a card. Draw a picture of the card, and show how the sample data would be arranged in each field (left-adjusted, right-adjusted, and so on).

**Exercise 8.7:**   Design a card layout for an account records program. For each account, the following information must appear on a single card.

| *Information* | *Form* |
|---|---|
| Account number | a 6-digit integer |
| Name of firm | character string (maximum width of 25) |
| Previous account balance | a real number between −9999.99 and 9999.99 |
| Charges for current month | a real number between   0.0 and  999.99 |
| Credits for current month | a real number between   0.0 and  999.99 |
| Total amount due | a real number between −999.99 and  999.99 |

You should ensure that there is at least one space between each of the six data items listed. Write the appropriate variable declarations, READ statement, and FORMAT statement for reading in the card that you designed.

**Exercise 8.8:**   Suppose you decided that you did not want to have to keypunch the decimal points in the four real values in Exercise 8.7 and that you didn't want to bother punching zero entries (if any) for these four values. Will your FORTRAN statements for Exercise 8.7 still work? Why? If not, change them.

**Exercise 8.9:**   You are given the following data declarations and READ statement

```
REAL   ALPHA, BETA
INTEGER GAMMA, EPS
CHARACTER * 4  DELTA
        .
        .
        .
READ 30, ALPHA, GAMMA, DELTA, BETA, EPS
```

Write format 30 so that the information in the three cards shown below (on the left) will be read in as indicated below (on the right).

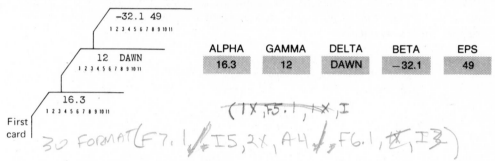

**Exercise 8.10:** How would you punch the data shown in Exercise 8.9 if format 30 appeared as shown below?

```
30 FORMAT (F3.1, I2, A4, F4.1, 3X, I3)
```

## 8.4.2   The E Format Descriptor *

Another method for reading real numbers is to use the E format descriptor (rather than F). A number to be read under E format must be punched in exponential notation, using the letter E. The E descriptor Ew.d contains a total width specification, w, and a count of the number of decimal digits to the right of the decimal point, d. In this case, however, the width specification must not only include the sign and the decimal point (if present), but also the letter E and the sign and the value of the integer exponent. Thus in the E format descriptor, w should normally be larger than d + 5.

For example, the number

$$-6.245E-6$$

could be read using a format descriptor of E9.3. Here again, if the decimal point is punched in the card, it overrides the d parameter of the E descriptor.

The E format descriptor may also be used for printing real data. It is often convenient to use the E descriptor when the magnitude of a real number is not known, or is so large that the use of the F format descriptor is impractical. Real data printed using the E format descriptor will usually appear in the form

$$\pm 0.X_1 X_2 \ldots X_d E \pm \text{exp}$$

where the $X_i$ are the d most significant digits of the number (after rounding) and exp is the base 10 exponent. When the E descriptor is used for output, w must include the zero (if present), the two signs, the decimal point, the E, the width of the exponent as well as the d digits to the right of the decimal point. Therefore, w should always be greater than d + 7 when Ew.d is used in output (the exponent should not exceed two digits in width).

---

*This section may be omitted.

**Example 8.12:**    The READ statement

```
        READ 37, INDEX, A, B, C
     37 FORMAT (I5, E10.3, E10.3, E10.3)
```

will cause the information in the following card

```
   36 6.107E+02   3.993E+4  -92.6E-3
   1 2 3 4 5 6 7 8 9 10 11 12 13 14 15 16 17 18 19 20 21 22 23 24 25 26 27 28 29 30 31 32 33 34 35 36 37 38 39 40 41 42 43 44
```

to be placed in the named variables as shown below:

| INDEX | A | B | C |
|-------|-----|---------|--------|
| 36 | 610.7 | 39930.0 | -.0926 |

## 8.5  USING FORMATS WITH ARRAYS

### 8.5.1  Repeating Format Descriptors

When using formats with arrays, there must be an edit descriptor for each array element read or printed.

**Example 8.13:**    Given the declaration

```
        INTEGER K(4)
```

and the PRINT statement

```
        PRINT 10, K
```

FORMAT statement 10 must have four edit descriptors beginning with the letter I. They can be provided individually as in

```
     10 FORMAT ( ' ', I3, 2X, I3, 2X, I3, 2X, I3, 2X)
```

or by specifying the *repetition* of one or more edit descriptors, as in

```
     10 FORMAT (' ', 4(I3, 2X))
```

In format 10 above, the notation 4(I3, 2X) indicates that the *descriptor group* (I3, 2X) is to be repeated four times.

**Example 8.14:**    Let W be a real array consisting of 200 elements. We wish to print the elements of W in 10 columns across a line. We can do this using the following PRINT and FORMAT statements:

```
        PRINT 62, W
     62 FORMAT (20(' ', 10F12.3/))
```

The descriptors that are repeated (' ', 10F12.3/) describe a single line consisting of ten real numbers, each of width 12, with three digits to the right of the decimal point. The slash / is used to mark the end of the line. Without the slash, it would be assumed that all 200 real elements of W are to be printed in F12.3 format on the same line. It is not likely that a printer exists that can accommodate such a long line—of 2,400 characters. The repeat count 20 will cause the line description (' ', 10F12.3/) to be repeated 20 times during the printing of the array W. Each time, a blank will be used for carriage control resulting in single line spacing.

In all of our previous examples, the number of *data descriptors* (beginning with I, F, or A) appearing in a format matched the number of items to be read or printed. If there are not enough data descriptors in a format specification to satisfy the input or output list, then the format specification will be automatically reused. Thus, input or output continues until all items specified in the list have been processed, regardless of the number of data descriptors in the format. This feature can be extremely useful when the exact number of array elements to be processed depends upon a value that is determined during program execution.

**Example 8.15:** Again let W be a real array of 200 elements, and let N be an integer variable used to indicate the number of data items to be stored in W. If the value of N is punched on one data card (in I3 format), and the N values to be read into W are punched on successive data cards (8 items per card in F10.3 format), then the statements

```
      READ '(I3)', N
      READ 50, (W(I), I = 1, N)
   50 FORMAT (8F10.3)
```

can be used to read in N and the items to be stored in W. Format 50 will work regardless of the size of N (as long as N is in the declared range of W). Each time the format is exhausted during the execution of the READ 50 statement, a new card will be read and the format will be repeated automatically. The format still describes a single card layout, but it will be repeated as often as necessary, until all N elements of W have been filled with data.

The statements

```
      PRINT 60
   60 FORMAT ('1THE DATA IN W ARE')
      PRINT 70, (W(I), I = 1, N)
   70 FORMAT (' ', 12F10.3)
```

can be used to print a short heading at the top of a page, and then print the data in the array W, 12 items per line. Format 70 describes a single line of output, but the format will be repeated as often as necessary until all N elements in W have been printed. Each time the format is repeated, a new line is started, and a blank is used for carriage control.

The input or output list in a READ or PRINT statement completely determines the number of data items to be processed regardless of the number of data descriptors in the associated FORMAT statement. During processing, each item in the input or output list is matched with the corresponding data descriptor. If there are more data descriptors than items in the input or output list, any extra data descriptors are ignored (although leftover space descriptors, including strings, are processed up to the next data descriptor). If there are more items than data descriptors, the FORMAT statement is repeated as many times as necessary until the list of input or output items is satisfied.

**Exercise 8.11:** How would array W be printed if format 62 in Example 8.14 were rewritten as

```
62 FORMAT (' ', 10F12.3)
```

## 8.5.2  Reading Character Arrays

A formatted read can be very useful in reading textual data. If we wish to store a paragraph or page of text in memory, we would normally do this by typing the information on consecutive data cards or data lines. Each 80-column card could then be entered into a separate element of an array as shown below.

```
      INTEGER MAXLIN, NUMLIN, CURLIN
      PARAMETER (MAXLIN = 1000)
      CHARACTER * 80 TEXT(MAXLIN)
      READ*, NUMLIN
      DO 10 CURLIN = 1, NUMLIN
         READ 17, TEXT(CURLIN)
   17    FORMAT (A80)
   10 CONTINUE
```

Each element of array TEXT can be used to store a string of 80 characters. Format 17 above specifices that each card contains a single character data field of length 80. The input variable NUMLIN determines how many cards will be read (NUMLIN must be between 1 and MAXLIN).

Often, we wish to examine individual characters of a string. For example, we might want to know how many times the letter E or the word THE occurs in a string. One way to do this would be to read each character of a data card or line into a separate element of a character array using a formatted READ, as shown in the following example. We will study other techniques of manipulating character strings in the next chapter.

**Example 8.16:** The following statements can be used to read the contents of an 80-character card or line into an 80-element array.

```
      CHARACTER * 1 BUFFER(80)
      READ 27, BUFFER
   27 FORMAT (80A1)
```

Format 27 specifies that the data card contains 80 individual character fields of width one. Consequently, each card column will be stored separately in an element of the array BUFFER. We can then process any character on the data card by specifying its corresponding array subscript (e.g., BUFFER(1) represents the first column).

The statement

```
READ '(80A)', BUFFER
```

would have the same effect as the **READ** 27 statement. Each card contains 80 fields; the width of each field must be the same as the length of an element of **BUFFER** (length one).

**Exercise 8.12:**  What would happen if formats 17 and 27 were interchanged in the two code segments shown in this section?

**Exercise 8.13:**  Describe the information read by the following statements for the card images shown. For b) and c), remember that multiple format descriptors are required when an array name (without a subscript) appears in the input list. (You may assume that the variables involved in each operation have been declared consistent with the type of the data that is punched.)

a)
```
     READ 527, COLOR, ID, COST
 527 FORMAT (3X, A4, 5X, I5, 3X, F6.2)
```

```
    BLUE      37288     672.25
    1 2 3 4 5 6 7 8 9 10 11 12 13 14 15 16 17 18 19 20 21 22 23 24 25 26 27 28 29 30 31 32 33
```

b)
```
     READ 65, NAME, FLIGHT, AIRLIN, DATE
  65 FORMAT (A3, 3X, I3, A6, I2, 1X, I2, 1X, I2)
```

Assume **DATE** is an array of size 3.

```
    DEBBIE698UNITED06/13/75
    1 2 3 4 5 6 7 8 9 10 11 12 13 14 15 16 17 18 19 20 21 22 23 24 25 26 27 28
```

c)
```
      READ 2231, NAME, AB, RUNS, HITS, RBI, AVE
 2231 FORMAT (A4, A4, I3, I3, I3, I3, F5.3)
```

Assume **NAME** is an array of size 2.

```
    ELIEKOFF62 41262140 420.312
    1 2 3 4 5 6 7 8 9 10 11 12 13 14 15 16 17 18 19 20 21 22 23 24 25 26 27 28 29 30 31 32 33 34 35 36 37 38
```

**Exercise 8.14:**

a) Let WCF be an integer array of size 12 containing values ranging from $-130°$ F to $+50°$F, and TEMP be an integer variable whose values range from $-50°$ to $+50°$. Write the **PRINT** and **FORMAT** statement to output the contents of TEMP and WCF in one row:

```
TEMP WCF(1) WCF(2) . . . WCF(12)
```

The value of TEMP should be separated from WCF(1) by at least five blanks, and the contents of the elements of WCF should be separated from one another by at least two blanks.

b)   Suppose you wished to put your PRINT statement from part (a) into a loop in which TEMP ranges from $-50$ to $+50$ in increments of 5, and the contents of WCF are recomputed for each of these 21 values of TEMP. Would any changes be required in either your PRINT or your FORMAT statement? What would be the result of the execution of such a loop containing your PRINT statement?

c)   Write one PRINT and one FORMAT statement to display the following heading:

```
WIND CHILL FACTOR TABLE (DEGREES F)
TEMPERATURE              WIND VELOCITY (MILES PER HOUR)
READING (DEG F)          5      10      15  . . .   50
```

**Exercise 8.15:**   Write the PRINT and FORMAT statements needed to produce the output described below. Start each new line described by your format with the descriptor 1X. This will indicate a blank for line control for these lines.

a)   Let X be a real array of 20 elements each containing positive real numbers ranging in value from 0 to 99999.99. Print the contents of X, accurate to two decimal places, 4 elements per line.

b)   Do the same as for part (a), but print the contents of the variable N (containing an integer ranging in value from 1 to 20) on one line, and then print the contents of the first N elements of the array X, four per line.

c)   Let QUEUE be a 1000-element array of real numbers whose range of values is not easily determined but is known to be very large. Print the contents of QUEUE, six elements per line, accurate to six decimal places.

d)   Let ROOM and TEMP be 120-element arrays. ROOM contains the numbers of the rooms in a 9-story building (these range from 101 through 961). TEMP contains the temperatures of these rooms on a given day, accurate to one decimal place. Print two parallel columns of output, one containing all room numbers, and the other containing the temperature of each room.


## 8.6   SEQUENTIAL FILES

### 8.6.1   Introduction to Files

In all our programming so far, we have provided a separate group of data items for each program. These data items "belonged" to that program and could not readily be used by another. Each data item was read in as the program executed, and could not be read more than once. Furthermore, all output was printed and could not be processed at some later time.

In many computer applications, it is important to be able to share data among programs, enabling data generated by one program to be processed by another. Also, it is often desirable to read data from more than one *input data file* or combine these data to form a new *output file*. For example, if a bank had a file of accounts and a file of daily transactions, we might want to update the file of accounts based on today's transactions. This new file of accounts could become an input file to be processed later with tomorrow's transactions.

A *file* is a collection of data that is physically located in secondary storage rather than the main computer memory (see Section 1.2.4). Consequently, files can be extremely large. Normally, only one component of a file, called a *record*, will be stored in main memory and processed at a given time.

There are two kinds of files in FORTRAN: sequential files and direct access files. A *sequential file* has the property that its records must always be processed serially, starting with the first. At any point, the next record to be processed is the record following the last one processed. On the other hand, it is possible to reference any record in a *direct access file* at any time; hence, we say that the access order is *random*. To facilitate random access to a direct access file, all records must be of uniform length. There is no requirement of this type for a sequential file.

In many batch systems, the deck of data cards for a program is actually stored in secondary storage as a file; each data card is a record of this input file. Similarly, each line printed by our program is a record of the program output file. Both are sequential files as they are processed in serial order (e.g., the first five cards must be read before the sixth card). In the remainder of this chapter, we shall study both file types, starting with sequential files.

### 8.6.2  Creating a Sequential File

In order to create a file for later use, we must read or generate data under program control and write that data onto a file. In the next example, we illustrate a number of new FORTRAN statements needed to accomplish this task.

**Example 8.17:**   The program in Fig. 8.1 creates a file of the odd numbers from 1 to 999. Each odd number becomes an individual record of the newly created file named ODDNUM, as shown below.

file ODDNUM

1 | 3 | 5  . . .                | 997 | 999

The OPEN statement is used to *connect* a file so that it can be read or written. The OPEN statement in Fig. 8.1 specifies a number of properties of the file including its unit number (UNIT = 3), its name (FILE = 'ODDNUM'), the fact that it is written using formatted output (FORM = 'FORMATTED'), that it is a sequential file (ACCESS = 'SEQUENTIAL'), and that the file is newly created (STATUS = 'NEW') by this program.

The file properties listed above are all quite straightforward given our knowledge so far. Files are either sequential (ACCESS = 'SEQUENTIAL') or direct access (ACCESS = 'DIRECT'). A file may be processed using list-directed or formatted input/output (FORM = 'FORMATTED'); otherwise the file is considered unformatted (FORM = 'UNFORMATTED').

The file specifier STATUS = 'NEW' is used when a new file is being creat-

```
C CREATE A SEQUENTIAL FILE OF ODD NUMBERS
C
C PROGRAM PARAMETERS
      INTEGER ODDMAX
      PARAMETER (ODDMAX = 999)
C
C VARIABLES
      INTEGER NEXODD
C
C OPEN FILE ODDNUM FOR OUTPUT
      OPEN (UNIT = 3, FILE = 'ODDNUM', FORM = 'FORMATTED',
     Z       ACCESS = 'SEQUENTIAL', STATUS = 'NEW')
C
C WRITE EACH ODD NUMBER TO FILE ODDNUM
      DO 10 NEXODD = 1, ODDMAX, 2
          WRITE (UNIT = 3, FMT = 19) NEXODD
   19     FORMAT (I3)
   10 CONTINUE
C
C TERMINATE AND CLOSE FILE
      ENDFILE (UNIT = 3)
      CLOSE (UNIT = 3)
C
      STOP
      END
```

**Fig. 8.1**  Creating a file of odd numbers.

ed; the file specifier STATUS = 'OLD' is used when a previously created file is being read. If we do not wish to retain a file that is being created, the file specifier STATUS = 'SCRATCH' should be used.

The file name is used to locate and identify each file that is in secondary storage. On some systems the file name must follow a prescribed form. Files with a STATUS parameter of 'SCRATCH' should not be given names since they are not retained after the execution of the program.

The unit number is an integer value associated with the file. There are usually restrictions unique to each computer system as to what integers may be specified as unit numbers. On some systems, the integers 5 and 6 are reserved for the primary input and output devices being used by the computer system (card reader and line printer for a batch program; the terminal for an interactive program). The unit number must appear in all subsequent program statements that manipulate the file.

In Fig. 8.1, the statement

```
      WRITE (UNIT = 3, FMT = 19) NEXODD
```

is used to write the output list (NEXODD) to the file designated by unit number 3 using FORMAT statement 19. Note that a carriage control character is not specified in format 19; carriage control is needed only when writing to the unit that represents the line printer. The statement

```
      ENDFILE (UNIT = 3)
```

is used to mark the end of a file by appending a special end-of-file record to the newly created file. This should be done whenever a new file is created or written. The statement

$$\text{CLOSE (UNIT = 3)}$$

disconnects the file associated with unit number 3 from the system.

    The OPEN, ENDFILE and CLOSE statements are described in the displays that follow.

---

## OPEN Statement

OPEN (speclist)

**Interpretation:** The OPEN statement connects a file to the program. *Speclist* is a list of file specifiers separated by commas. The file specifiers describe the properties of the file being connected.

*Required specifiers*
UNIT = *unum*: *unum* is the unit number (an integer expression).

*Optional specifiers*
FILE = *name*:   *name* is a string giving the file name (not required for 'SCRATCH' files).

ACCESS = *acc*:   The value of *acc* must be 'SEQUENTIAL' or 'DIRECT'. If omitted, 'SEQUENTIAL' is assumed.

FORM = *frm*:     The value of *frm* must be 'FORMATTED' or 'UNFOR-MATTED'. If omitted, 'FORMATTED' is assumed for sequential files; 'UNFORMATTED' for direct access.

STATUS = *stat*:   The value of *stat* must be 'NEW', 'OLD', 'SCRATCH', or 'UNKNOWN'. If omitted, 'UNKNOWN' is assumed.

*Other specifiers*
RECL = *len*:     *Len* is an integer expression indicating the length of each record in a direct access file (used only when ACCESS = 'DIRECT').

*Notes:* The values allowed for *unum* and *name* may depend on your computer system. If STATUS = 'SCRATCH', then the specifier FILE = *name* should be omitted.

---

## ENDFILE Statement

ENDFILE (UNIT = unum)

**Interpretation:** The ENDFILE statement writes a special end-of-file record onto the file specified by *unum*. No data may be written following the end-of-file record.

---

**CLOSE Statement**

$$CLOSE \ (UNIT \ = \ unum)$$

**Interpretation:** The CLOSE statement disconnects the file specified by *unum* from the program. Unit number *unum* can be reconnected to another file.

---

The internal binary forms of data are stored directly in an unformatted file. In a formatted file, each data item is represented as a string of characters. Consequently, formatted files are easier to move from one computer to another. On the other hand, the particular computer system used to create an unformatted file determines the form of each record of that file. Although they are less portable, unformatted files can generally be processed more quickly.

On some systems, file properties may be specified using special control statements as well as the OPEN statement. You should find out what special statements, if any, are required on your system.

**Exercise 8.16:** Write the OPEN statement needed to connect an existing, sequential, unformatted file named 'TEXT' to unit number 3.

### 8.6.3  Reading an Existing File

We mentioned that one motivation for files was to enable the output generated by one program to be used as data by another program. This process is illustrated next.

**Example 8.18:** The program in Fig. 8.2 reads the file 'ODDNUM' created by the execution of the program in Fig. 8.1 and echo prints each file record. Note that the OPEN statement is the same as before except that the value of STATUS is now 'OLD'. The ENDFILE statement is no longer needed since the file is being read, not written. The unit number for file 'ODDNUM' is still specified as 3 although another unit number could be used instead.

```
C READ THE FILE ODDNUM
C
      INTEGER NEXNUM
C
C OPEN FILE ODDNUM FOR INPUT
      OPEN (UNIT = 3, FILE = 'ODDNUM', FORM = 'FORMATTED',
     Z        ACCESS = 'SEQUENTIAL', STATUS = 'OLD')
C
C READ AND PRINT EACH RECORD
      PRINT*, 'LIST OF FILE DATA'
      WHILE (.TRUE.) DO                              20 CONTINUE
         READ (UNIT = 3, FMT = 25, END = 35) NEXNUM
   25    FORMAT (I3)
         PRINT*, NEXNUM
      ENDWHILE                                       GO TO 20
```

(Continued)

```
C
C END-OF-FILE REACHED
   35 CONTINUE
      PRINT*, 'FILE READ COMPLETED'
C
      STOP
      END
```

**Fig. 8.2**   Echo printing the file 'ODDNUM'.

Within the WHILE loop, each record file 'ODDNUM' is read (using format 19) into the variable NEXNUM. A list-directed output statement prints each value of NEXNUM. The loop repetition condition is always true; however, we want the loop to terminate when all file records have been read, or the end of the file is reached. The optional *control specifier* END = 35 causes a transfer to label 35 when the end-of-file record is read. Since the repeat condition is always true, this is the only method provided for loop exit.

The general READ and WRITE statements are described in the next display.

---

**General READ/WRITE**

> READ (UNIT = *unum, control list) input list*
> or   WRITE ( UNIT = *unum, control list) output list*

**Interpretation:** The READ or WRITE operation is performed on the file associated with unit *unum*. The variables receiving data (for READ) are provided in *input list*; the variables and values to be written (for WRITE) are provided in *output list*. The optional *control list* is described next.

FMT = *flab*:   Specifies the label of the FORMAT statement to be used with a formatted file (omitted if the file is unformatted). FMT = * specifies list-directed input/output.

END = *endlab*: Specifies the label to transfer to when the end-of-file record is read. This specifier may not appear in a WRITE statement or with a direct access file.

REC = *recnum*: Specifies the record number (a positive integer) of a direct access file (omitted if the file is sequential).

*Note:* There are additional control specifiers; however, they are beyond the scope of this discussion.

---

### 8.6.4   Creating a File from Card or Terminal Input

The program in Fig. 8.1 generated data to be stored on a sequential file. It is also possible to read or enter data from the primary input device (such as the card reader or terminal) and save this data on a file for later use. This process is illustrated in the next example.

**Example 8.19:**  The inventory for a bookstore is read from data cards or data lines (at a terminal) by the program in Fig. 8.3a. Each input record consists of the stock number, author name, title, cost, and quantity on hand, as shown below.

         1 'TY COBB' 'NO PLACE LIKE HOME' 5.95 6

The list-directed **READ** statement

         READ*, STOCK, AUTHOR, TITLE, PRICE, QUANT

is used to read each data record, and a formatted **PRINT** statement (format 35) echo prints the data record. The statement

         WRITE (UNIT = 1) STOCK, AUTHOR, TITLE, PRICE, QUANT

writes this record to the file (INVEN) associated with unit number 1; there is no format specified since the file is unformatted.

The WHILE loop in Fig. 8.3a terminates when the sentinel card (beginning with a stock number of 0) is read. The total number of books is counted and displayed after loop exit. A sample of the output for this program is shown in Fig. 8.3b.

```
C CREATE THE BOOK STORE INVENTORY
C
      INTEGER STOCK, QUANT, TOTAL
      REAL PRICE
      CHARACTER * 20 AUTHOR, TITLE
      CHARACTER * 10 DATE
C
C OPEN FILE INVEN FOR OUTPUT
      OPEN (UNIT = 1, FILE = 'INVEN', FORM = 'UNFORMATTED',
    Z       ACCESS = 'SEQUENTIAL', STATUS = 'NEW')
C
C PRINT HEADING AND INITIALIZE TOTAL
      READ*, DATE
      PRINT*, 'BOOK INVENTORY FOR ', DATE
      PRINT 31, '0STOCK NO.', 'AUTHOR', 'TITLE', 'PRICE', 'QUANTITY'
   31 FORMAT (A, 6X, A, 16X, A, 10X, A, 2X, A)
C
C READ AND PRINT EACH DATA RECORD
C COPY IT TO FILE INVEN AND ACCUMULATE TOTAL INVENTORY AMOUNT
      TOTAL = 0
      READ*, STOCK, AUTHOR, TITLE, PRICE, QUANT
      WHILE (STOCK .NE. 0) DO              30 IF (STOCK .EQ. 0)
         PRINT 35, STOCK, AUTHOR, TITLE,    Z GO TO 39
    Z        PRICE, QUANT
   35    FORMAT ('0', 2X, I4, 2X, A, 2X, A, 1X, F6.2, 2X, I5)
         WRITE (UNIT = 1) STOCK, AUTHOR, TITLE, PRICE, QUANT
         TOTAL = TOTAL + QUANT
         READ*, STOCK, AUTHOR, TITLE, PRICE, QUANT
      ENDWHILE                             GO TO 30
C                                          39 CONTINUE
```

*(Continued)*

```
          C PRINT RESULTS, TERMINATE AND CLOSE FILE
                PRINT*, ' '
                PRINT*, 'TOTAL NUMBER OF BOOKS = ', TOTAL
                PRINT*, 'FILE INVEN CREATED'
                ENDFILE (UNIT = 1)
                CLOSE (UNIT = 1)
          C
                STOP
                END
```

**Fig. 8.3a**   Creating the inventory file 'INVEN'.

```
BOOK INVENTORY FOR 12/25/80
STOCK NO.          AUTHOR              TITLE            PRICE   QUANTITY
      1   TY COBB            NO PLACE LIKE HOME     5.95       6
      2   PETE ROSE          GREATEST HITS          6.34       3
      3   JIM RICE           BOMBED NEW YORK        4.99       1
      4   HOYT WILHEM        KNUCKLING UNDER        3.44       8
      5   BILLY MARTIN       USA TRAVEL GUIDE       2.50     500

TOTAL NUMBER OF BOOKS = 518
FILE INVEN CREATED
```

**Fig. 8.3b**   Sample run of program in Fig. 8.3a.

**Example 8.20:**   The program in Fig. 8.4 could be used to update file INVEN by adding additional data records to this file. The first WHILE loop advances to the current end of the file. It does this by reading each record until the end-of-file record is read. After loop exit, the statement

BACKSPACE (UNIT = 1)

repositions the file to the start of the last record read (the current end-of-file record). The additional data records are then written onto the file INVEN, erasing the original end-of-file record. A new end-of-file record is written following the last data record added to the file.

```
C EXTEND EXISTING INVENTORY FILE
C
      INTEGER STOCK, QUANT, TOTAL
      REAL PRICE
      CHARACTER * 20 AUTHOR, TITLE
C
C OPEN FILE INVEN FOR EXTENSION
      OPEN (UNIT = 1, FILE = 'INVEN', FORM = 'UNFORMATTED',
     Z      ACCESS = 'SEQUENTIAL', STATUS = 'OLD')
C
C FIND CURRENT END OF FILE
      WHILE (.TRUE.) DO                        20 CONTINUE
          READ (UNIT = 1, END = 25) STOCK, AUTHOR,
     Z      TITLE, PRICE, QUANT
          PRINT*, STOCK, AUTHOR, TITLE, PRICE, QUANT
      ENDWHILE                                 GO TO 20
C
```

*(Continued)*

```
C ADD NEW DATA TO THE END OF FILE
    25 CONTINUE
       BACKSPACE (UNIT = 1)
       READ*, STOCK, AUTHOR, TITLE, PRICE, QUANT
       WHILE (STOCK .NE. 0) DO                     31 IF (STOCK .EQ. 0)
          WRITE (UNIT = 1) STOCK, AUTHOR,          Z  GO TO 39
    Z        TITLE, PRICE, QUANT
          READ*, STOCK, AUTHOR, TITLE, PRICE, QUANT
       ENDWHILE                                        GO TO 31
C                                                   39 CONTINUE
C TERMINATE AND CLOSE FILE
       PRINT*, 'EXTENSION OF INVEN COMPLETE'
       ENDFILE (UNIT = 1)
       CLOSE (UNIT = 1)
C
       STOP
       END
```

**Fig. 8.4**   Adding more records to file 'INVEN'.

Another command

$$REWIND \ (UNIT = unum)$$

may be used to reposition a sequential file to its first record. This enables the data on a sequential file to be processed more than once during the execution of a program.

**Exercise 8.17:**   Write a program that reads a name and a list of three exam scores for each student in a class, and copies this information onto a sequential file called GRADES. Test your program on the following data:

| | | | |
|---|---|---|---|
| IVORY | 47 | 82 | 93 |
| CLARK | 86 | 42 | 77 |
| MENACE | 99 | 88 | 92 |
| BUMSTEAD | 88 | 74 | 81 |

**Exercise 8.18:**   Write a program that reads and echo prints the file GRADES.

**Exercise 8.19:**   Write a program that adds a fourth exam score to each component of the file GRADES listed in Exercise 8.17. Assume that each student's name and fourth exam score are provided as input data.

### 8.6.5   Merging Two Sequential Files

A common problem when working with files is to update one file (master file) by merging in information from a second file (update file). This process is illustrated in the following problem.

**Problem 8A:**   The Junk Mail Company has recently received a new mailing list (file UPDATE) that it wishes to merge with its master file (file OLDMST). Each of these files is in alphabetical order by name. The company wishes to produce a new master file (NEWMST) that is also in alphabetical order. Each client name and address on either mailing list is represented by four character strings as shown below:

                    'CLAUS, SANTA'
                    '1 STAR LANE'
                    'NORTH POLE'
                    'ALASKA, 99999'

There is a sentinel name and address at the end of each of the files UPDATE and OLDMST. The sentinel is the same for both files; one copy should be written at the end of the NEWMST file. The sentinel entry consists of four character strings containing all Z's. We shall assume that there are no names that appear on both files OLDMST and UPDATE.

**Discussion:** In addition to the two input files (OLDMST and UPDATE), we will need an output file (NEWMST) that will contain the merged data from OLDMST and UPDATE. NEWMST will then serve as the new master file of mailing labels. The files information is summarized in the data table shown below.

**Data Table for Merge Problem (8A)**

| *Input files* | *Output files* |
|---|---|
| OLDMST: The original mailing list in alphabetical order by name | NEWMST: The final mailing list, formed by merging OLDMST and UPDATE |
| UPDATE: The additions to be made to OLDMST, also in alphabetical order by name | |

The level one flow diagram is shown in Fig. 8.5a. The program must read one name and address entry at a time from each input file. These two entries are compared and the one that comes first alphabetically is copied to the output file (NEWMST). Another entry is then read from the file containing the entry just copied and the comparison process is repeated.

When the end of one of the input files (OLDMST or UPDATE) is reached, the program should copy the remaining information from the other input file to NEWMST and then write the sentinel record.

To simplify the implementation of the algorithm, we will use two string arrays, OLDATA and UPDATA, to hold the current client data from OLDMST and UPDATE, respectively. The layout of these arrays is shown below.

|  | 1 | 2 | 3 | 4 |
|---|---|---|---|---|
| OLDATA (from OLDMST) | $name_1$ | $street_1$ | $city_1$ | $state_1$ & $zip_1$ |
| UPDATA (from UPDATE) | $name_2$ | $street_2$ | $city_2$ | $state_2$ & $zip_2$ |

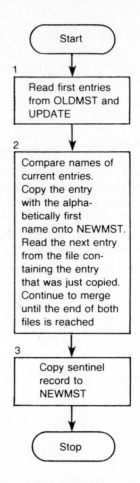

**Fig. 8.5a**   Level one flow diagram for merge problem (8A).

The additional data table entries are shown below. The refinement of Step 2 of Fig. 8.5a is shown in Fig. 8.5b.

### Additional Data Table Entries for Problem 8A

*Input variables*

OLDATA: String array to hold
current client data from OLDMST
(character*20, size 4)

UPDATA: String array to hold
current client data from UPDATE
(character*20, size 4)

It is important to verify that all the remaining data on one file will be copied into NEWMST when the end-of-file record on the other file has been read.

Just before reaching the end-of-file record on UPDATE (or OLDMST), the sentinel record will be read into UPDATA (or OLDATA). Since the sentinel name
(all Z's) alphabetically follows any other client name, the remaining client data
on the unfinished file will be copied to NEWMST as desired. When loop repetition terminates, both OLDATA and UPDATA will contain the sentinel record.

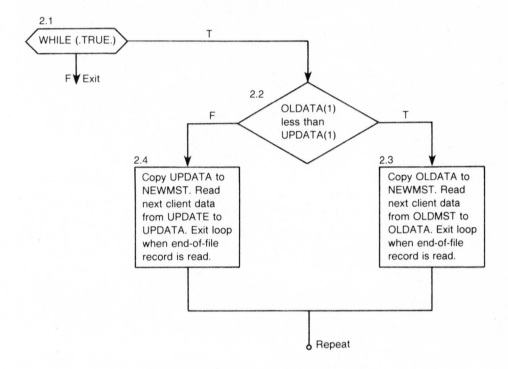

**Fig. 8.5b**  Refinement of Step 2 of Fig. 8.5a.

In the program shown in Fig. 8.6, a transfer to label 94 or 95 occurs if either UPDATE or OLDMST is empty (a file consisting of an end-of-file record
only). In this case, there is no need to perform the merge operation since the other file contains the required alphabetical list.

**Exercise 8.20:**  Modify the program for Problem 8A to handle the situation in which
the UPDATE file may contain some of the same names as the OLDMST file. In this case
only one address should appear on the NEWMST file; the address in file UPDATE
should be used as it is more recent. Also, print a count of the number of file entries in
each of the three files.

**Exercise 8.21:**  Let FILEA and FILEB be two files containing the name and identification number of the students in two different programming classes. Assume that these files
are arranged in ascending order by student number and that no student is in both classes.
Write a program to read the information on FILEA and FILEB, and merge them onto a
third file (FILEC) retaining the ascending order.

```
C MERGE OLDMST AND UPDATE TO NEWMST
C
      CHARACTER * 20 OLDATA(4), UPDATA(4)
C
C OPEN THREE SEQUENTIAL FILES
      OPEN (UNIT = 1, FILE = 'OLDMST', FORM = 'UNFORMATTED',
     Z       ACCESS = 'SEQUENTIAL', STATUS = 'OLD')
      OPEN (UNIT = 2, FILE = 'UPDATE', FORM = 'UNFORMATTED',
     Z       ACCESS = 'SEQUENTIAL', STATUS = 'OLD')
      OPEN (UNIT = 3, FILE = 'NEWMST', FORM = 'UNFORMATTED',
     Z       ACCESS = 'SEQUENTIAL', STATUS = 'NEW')
C
C READ FIRST DATA ENTRIES FROM EACH INPUT FILE
      READ (UNIT = 1, END = 94) OLDATA
      READ (UNIT = 2, END = 95) UPDATA
C
C MERGE ALPHABETICALLY SMALLER DATA TO NEWMST AND
C GET NEXT RECORD FROM INPUT FILE WITH SMALLER
C DATA.  CONTINUE UNTIL END OF EITHER FILE IS READ
      WHILE (.TRUE.) DO                                    20 CONTINUE
C         COPY RECORD WITH SMALLER NAME
          IF (OLDATA(1) .LT. UPDATA(1)) THEN
             WRITE (UNIT = 3) OLDATA
             READ (UNIT = 1, END = 40) OLDATA
          ELSE
             WRITE (UNIT = 3) UPDATA
             READ (UNIT = 2, END = 40) UPDATA
          ENDIF
      ENDWHILE                                             GO TO 20
C
C COPY SENTINEL RECORD TO NEWMST AND CLOSE FILES
   40 CONTINUE
      WRITE (UNIT = 3) UPDATA
      PRINT*, 'MERGE COMPLETED'
      ENDFILE (UNIT = 3)
      CLOSE (UNIT = 1)
      CLOSE (UNIT = 2)
      CLOSE (UNIT = 3)
      STOP
C
C PRINT ERROR MESSAGE IF INPUT FILE IS EMPTY
   94 CONTINUE
      PRINT*, 'FILE OLDMST IS EMPTY'
      STOP
   95 CONTINUE
      PRINT*, 'FILE UPDATE IS EMPTY'
C
      STOP
      END


*** PROGRAM EXECUTION OUTPUT ***

MERGE COMPLETED
```

**Fig. 8.6**  Merging two files (Problem 8A).

## 8.7   DIRECT ACCESS FILES

### 8.7.1   Record Length and Record Number

In a direct access file, the records may be accessed in a random, rather than a fixed, order. Consequently, the operating system must be able to easily locate any record in a file. To facilitate this, each record is assigned a unique *record number* (analogous to an array index) and all records must be the same size or length.

The record length must be specified by a file specifier in the OPEN statement (RECL = *len*). The procedure used for determining the integer value required for *len* is dependent on the computer system.

When using random access files, each READ or WRITE statement must contain a control specifier indicating which record is to be read or written (REC = *recnum*). The record number, *recnum*, is an integer expression with a positive value. An example of the processing of direct access files is given next.

### 8.7.2   Updating a Direct Access File

A direct access file has the property that any record may be read or written without disturbing the rest of the file. For this reason, direct access files are used when only selected records of a file are likely to be modified during a file update. Sequential files are used when most, or all, of the records are likely to be changed during a file update and the records are modified in serial order.

**Problem 8B:**   A program is needed to update a bookstore inventory file at the end of each day. The input data for the program consist of the stock number (STCKNO) and amount sold (ORDER) for each book that is purchased. These data are not arranged in any special order.

**Discussion:**   Considering the random nature of the input data, a direct access file (DIRINV) should be used. We will assume that all inventory records contain the same information about a book as the records shown in Example 8.19, Section 8.6.4 (stock number, title, author, price, quantity). This information must be read from each inventory record that is to be modified; the inventory amount (QUANT) should then be updated and the new inventory record written to the file. The data table is shown next and the flow diagrams are given in Fig. 8.7a.

**Data Table for Problem 8B**

*Program parameter*

MAXSTK = 1000, maximum record number for file DIRINV

*In-out files*

DIRINV: Direct ac-
cess inventory file

*Input variables*                                   *Output variables*

STCKNO: The stock
number of book pur-
chased (integer)

ORDER: The quantity
purchased (integer)

$\left.\begin{array}{l} \text{STOCK} \\ \text{AUTHOR} \\ \text{TITLE} \\ \text{PRICE} \\ \text{QUANT} \end{array}\right\}$ Used for storage of each inventory file record

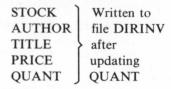 $\left.\begin{array}{l} \text{STOCK} \\ \text{AUTHOR} \\ \text{TITLE} \\ \text{PRICE} \\ \text{QUANT} \end{array}\right\}$ Written to file **DIRINV** after updating QUANT

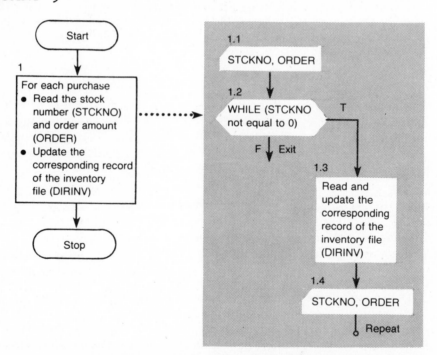

**Fig. 8.7a**    Flow diagrams for file update problem (8B).

In Step 1.3, the stock number of the book purchased (STCKNO) will be used to select the record of **DIRINV** accessed (REC = STCKNO). Before reading the file, we must check that STCKNO is in range; after reading the file, we must validate the stock number (STOCK) of the record accessed by comparing it to STCKNO. Finally, we must verify that the new inventory amount is positive before replacing the old record in **DIRINV** with the new one. These validations are shown in the refinement of Step 1.3 drawn in Fig. 8.7b and the FORTRAN program (See Fig. 8.8).

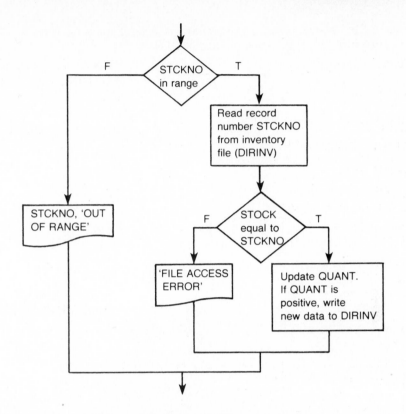

**Fig. 8.7b**   Refinement of Step 1.3 in Fig. 8.7a.

---

**Program Style**

*Validating a file read operation*

     Each record of the file **DIRINV** read in Fig. 8.8 contains a stock number that should always be the same as its record number (e.g., record number 1 has stock number 1). Although it may seem that the stock number is redundant, we have included it in order to validate that the file was read correctly. There are other techniques for detecting file input/output errors; however, they are beyond the scope of this discussion.

---

     It is important to realize that the sequential file **INVEN** created in Example 8.19 cannot be updated by the program in Fig. 8.8. It first would be necessary to copy this data to a direct access file (**DIRINV**) before attempting the update; this is left as an exercise.

     The record length specified in Fig. 8.8 corresponds to the number of characters (54) described by format 21.

**Exercise 8.22:**   Write a program that creates the file **DIRINV** that is updated by the program in Fig. 8.8.

**Exercise 8.23:** Write a program that creates a new direct access file whose records are the odd numbered records only of an existing direct access file.

```
C UPDATE AN INVENTORY FILE
C
C PROGRAM PARAMETERS
      INTEGER MAXSTK
      PARAMETER (MAXSTK = 1000)
C
C VARIABLES
      INTEGER STOCK, STCKNO, QUANT, ORDER
      REAL PRICE
      CHARACTER * 20 AUTHOR, TITLE
C
C OPEN FILE 'DIRINV' FOR UPDATE
      OPEN (UNIT = 2, FILE = 'DIRINV', FORM = 'FORMATTED',
     Z       ACCESS = 'DIRECT', RECL = 54 , STATUS = 'OLD')
C
C UPDATE FILE 'DIRINV' TO REFLECT DAILY SALES
      READ*, STCKNO, ORDER
      WHILE (STCKNO .NE. 0) DO                     22 IF (STCKNO .EQ. 0)
C        UPDATE INVENTORY RECORD STCKNO              Z GO TO 29
C        IF STCKNO IS IN RANGE AND IS VALID
         IF ((STCKNO .GT. 0) .AND.
     Z       (STCKNO .LE. MAXSTK)) THEN
            READ (UNIT = 2, FMT = 21, REC = STCKNO)
     Z         STOCK, AUTHOR, TITLE, PRICE, QUANT
   21       FORMAT (I4, A20, A20, F6.2, I4)
            IF (STCKNO .EQ. STOCK) THEN
               QUANT = QUANT - ORDER
               IF (QUANT .GE. 0) THEN
                  WRITE (UNIT = 2, FMT = 21, REC = STCKNO)
     Z              STOCK, AUTHOR, TITLE, PRICE, QUANT
               ELSE
                  PRINT*, 'INVENTORY AMOUNT ', QUANT, ' IS NEGATIVE'
                  PRINT*, 'IGNORE SALE OF ITEM ', STCKNO,
     Z               ' AMOUNT ', ORDER
               ENDIF
            ELSE
               PRINT*, 'FILE ERROR ', STOCK,
     Z            ' DOES NOT MATCH RECORD NUMBER ', STCKNO
            ENDIF
         ELSE
            PRINT*, STCKNO, 'OUT OF RANGE - FILE NOT ACCESSED'
         ENDIF
         READ*, STCKNO, ORDER
      ENDWHILE                                      GO TO 22
C                                                29 CONTINUE
      STOP
      END
```

**Fig. 8.8** Updating direct access file.

## 8.8  COMMON PROGRAMMING ERRORS

### 8.8.1  Format Errors

There are a number of very common errors that can be made in working with formatted input and output. Some of the errors are described in the list below. Errors 3, 5, and 6 are not unique to formatted input and output but can just as easily be made in working with list-directed input and output.

1. Type mismatches in the correspondence between variable names and edit descriptors will result in execution-time errors. A diagnostic message may be printed, and many FORTRAN compilers will immediately terminate execution of your program. Failure to provide a sufficient number of descriptors to accommodate an input or output list may result in type mismatches.
2. If an integer is not right-justified in its field, the blanks on the right may be interpreted as trailing zeros during input. This may change the value of the input data. Similarly, embedded blanks in an integer field may be read as zeros. These comments also apply to real numbers that are punched without a decimal point.
3. If apostrophes in character strings are not carefully paired, the format specification will not be interpreted correctly, and a compile-time syntax error will occur.
4. Attempting to print a number in a field that is too small will result in an execution-time error on some systems. Some compilers will print only part of the number along with one or more asterisks. No explicit diagnostic will be printed.
5. Not providing sufficient data to satisfy the input variable list will result in an execution-time error. The diagnostic message will indicate that there was insufficient input data or the end of the input file was reached.
6. Failure to provide a line control character (1, +, 0, or blank) in an output format for the line printer will produce unpredictable program output. This failure may be noted by a warning diagnostic but usually will go undetected during compilation.

### 8.8.2  File Errors

When working with files, you should use the OPEN statement to connect each file before it is processed. Check with your instructor to determine the local restrictions on file names and unit numbers. If the file is formatted, you must be careful to use the same format to read it as was used to create it.

Make sure that an END = *lab* control specifier is used in each READ statement associated with a sequential file. This is generally the only way to determine when all the file records have been processed; otherwise, an *attempt to read beyond end of file* error may occur. When writing a sequential file, make sure that you always append an end-of-file record before disconnecting the file. When performing a READ/WRITE operation with a direct access file, the record number of the file being accessed must always be specified via a REC = *recnum* control specifier. You should also verify that the integer expression for *recnum* is within range for the file being processed. Otherwise, a file-access error will occur.

## 8.9  SUMMARY

A detailed description of the meaning of the more frequently used FOR-TRAN format descriptors was given. The descriptors X, I, F, E, A, and/were described. The use of formatted READ statements was also discussed. This is most helpful when reading character data or data that have been previously punched or formatted. In most other cases, list-directed input is preferred since there is little to be gained by using formats.

However, this is not the case with list-directed output. No list-directed output facility allows the degree of horizontal and vertical space control that can be achieved with formats. Formatted output permits direct control over the use of every print position of every line printed on a page. When such detailed control is necessary for the generation of precisely spaced output, formats must be used, and the programmer must provide all details of spacing, data types, and data widths.

We also provided an introduction to file processing in FORTRAN. We introduced both sequential files and direct access files. The essential difference between them is that the records of sequential files are always processed in a fixed, serial order; whereas the records of a direct access file may be processed in random order.

We described the OPEN, CLOSE and ENDFILE statements and the file parameters that must be specified in the OPEN statement. We also discussed the general READ/WRITE in FORTRAN and the control specifiers that are needed for file input/output operations. The new statements for this chapter are described in Table 8.3.

| *Example* | *Effect* |
|---|---|
| *FORMAT Statement*<br>`10 FORMAT (I3, 3X, F4.1)` | Describes two data fields (I3 and F4.1) and one space field (3X). |
| *OPEN Statement*<br>`OPEN (UNIT = 1, FILE = 'ERZA',`<br>`Z STATUS = 'OLD'`<br>`Z FORM = 'UNFORMATTED',`<br>`Z ACCESS = 'SEQUENTIAL')` | Connects an existing sequential file named ERZA to the program. The file is associated with unit number 1. |

*ENDFILE Statement*
```
ENDFILE (UNIT = 1)
```
Appends an end-of-file record to the file associated with unit number 1.

*CLOSE Statement*
```
CLOSE (UNIT = 1)
```
Disconnects the file associated with unit number 1.

*General READ*
```
READ (UNIT = 1, FMT = 10,
Z END = 11) N, A
```
Reads two data values from the file with unit number 1 using format 10. If the end-of-file record is read, a transfer to label 11 occurs.

*General WRITE*
```
WRITE (UNIT = 1,
Z REC = I/2) N, A
```
Writes the value of N and A as a record on the file with unit number 1. The record number is determined by the value of the integer expression I/2.

**Table 8.3 FORTRAN Statements Introduced in Chapter 8.**

## PROGRAMMING PROBLEMS

**8C** Write a program to read in a deck of cards containing the addresses of all the students in the class, and print each address on an envelope. You may assume that a "skip to the top of the next page" operation implies that the next envelope is put in position for the first line of an address.

Each address will consist of three lines. The information to be printed on the first line is in columns 1–20 of each input card. The second line of information is in columns 21–40, and the third line is contained in columns 41–80. For example:

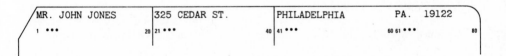

| MR. JOHN JONES | 325 CEDAR ST. | PHILADELPHIA PA. 19122 |

**8D** Especially during the colder months of the year, weather forecasters frequently will inform us not only of the degrees F temperature (TEMP) reading at a given hour, but also of the wind-chill factor (WCF) at that time. This factor is used to indicate the relative degree of coldness that we are likely to experience if we are outside, and its calculation is based not only on the thermometer reading (TEMP), but also upon the velocity (V) of the wind at the time. Write a program to compute the wind-chill factor for temperatures ranging from $-50°$ to $+50°$ in increments of $5°$, and for wind velocities of from 5 to 60 miles per hour, also in increments of $5°$. Your output should appear similar to the following table.

```
WIND-CHILL FACTOR TABLE (DEGREES F)
TEMPERATURE                    WIND VELOCITY (MILES PER HOUR)
READING (DEG F)        5  10  15  20  25  30  35  . . .
        -50
        -45
        -40
         .                              .
         .                              .
         .                              .
          0
          5
         .
         .
         .
         20              . . .              -16
         .
         .
         50
```

The formula for computing the wind-chill factor is

$$WCF = 91.4 - (.486 + .305\sqrt{V} - .020V) \times (91.4 - TEMP)$$

Your answers should be rounded to the nearest whole degree. The WCF of $-16$ has been given as a test value. It is the WCF when TEMP is 20 degrees and V is 25. (You might wish to use the PRINT and FORMAT statements from Exercise 8.14.)

**8E**    (Base 2 addition, using the multiple-alternative decision structure) Write a program to read two 15-digit binary numbers (strings of 0's and 1's) into the arrays I and J (both of size 15), using the 15I1 format. Then compute the decimal (base 10) representation of these numbers, and print both the binary and decimal representations.

Next, compute the column-by-column sum of these two numbers, moving *right to left*. Use the variable CARRY to indicate whether or not a previous addition contained a CARRY. A zero value for CARRY should indicate that the previous addition had no carry; a one value should be used when a carry occurs. (Initially, CARRY is 0.) Store the column-by-column sum in the array SUMIJ of size 16. The value of each element SUMIJ $(L+1)$ and the next value of CARRY is determined by adding the values of elements I(L), J(L), and CARRY as shown next.

For all L from 15 to 1:

If only one of I(L), J(L), or CARRY is 1, then

SUMIJ(L+1) is 1, and CARRY must be set to 0.

If all three of I(L), J(L), and CARRY are 1, then

SUMIJ(L+1) is 1 and CARRY must be set to 1.

If any two of I(L), J(L), or CARRY are 1, then

SUMIJ(L+1) is 0 and CARRY must be set to 1.

Otherwise, SUMIJ(L+1) is 0 and CARRY must be set to 0.

This will define the values of SUMIJ(16) through SUMIJ(2), from I(15), J(15) through I(1) and J(1), respectively. SUMIJ(1) can be defined directly from CAR-RY, once the other "additions" are done. Compute the decimal representation of the SUM, and print the binary and decimal representations.

Test your program on the following strings:

```
I = 00000 00010 11011
J = 00000 01010 11110
I = 00000 00010 00110
J = 00000 00001 11101
I = 10001 00010 00100
J = 11100 11101 11001
```

Your output, for example, for the second set of data, might appear as follows:

```
      I = 00000 00010 00110     70
      J = 00000 00001 11101     61
SUMIJ = 000000 00100 00011     131
```

Use functions and subroutines as needed, and draw a program system chart.

**8F**  *Determining the collating sequence on your computer.* On a single card, punch the following character string (starting in column 1):

```
+JB4PUD(OFY3RH2MCAT)=7Z N.W8KLG56$9QIV*/X-E,O'1S
 1        10        20        30        40
```

There should be a total of 48 characters in your card. Write a program that will read these characters into an array CHARS, one character per array element, and sort them. Print the string before the sort and after. The result will tell you the exact collating sequence for these characters on the computer you are using. *Hint:* Use the subroutine SORT shown in Chapter 7, Fig. 7.11, or an equivalent subroutine of your own to perform the sort. Remember to change the declarations in this subroutine: you will be using it to sort character strings and not real numbers. Use the edit descriptor A1. *WARNING to instructors:* For desired results, these hints may require modification on some computers.

**8G**  Complete the questionnaire shown in Fig. 8.9, by filling in the blanks as instructed. Each blank shown has a number below it. These numbers indicate the columns of the card in which your responses will be keypunched for processing by the computer. Write a program which will read in the responses to the questionnaire for all students in your class and tabulate the results as follows:

Compute and print the total number of responses, and a breakdown according to class and according to age: less than 18; 18–22; over 22.
Compute and print the number of Yes and No answers to each of questions 4–10.

Label all output appropriately, and use formats for all input and output.

---

POLITICS AS USUAL—A PREFERENCE POLL

1. Name:

$\overline{1}\ \overline{2}\ \ \overline{3}\ \overline{4}\ \overline{5}\ \overline{6}\ \overline{7}\ \overline{8}\ \overline{9}\ \overline{10}\ \overline{11}\ \overline{12}$,   $\overline{14}\ \overline{15}\ \overline{16}\ \overline{17}\ \overline{18}\ \overline{19}\ \overline{20}\ \overline{21}$   $\overline{23}$
      Last                              First                      M.I.

2. Academic year:

$\overline{25}\ \ \overline{26}$

(Fr, So, Jr, Sr, Use 0 for other)

3. Age:

$\overline{28}\ \ \overline{29}$

For items 4. through 10 , answer yes (Y) or no (N).

4. Have you ever voted in a presidential election?

$\overline{\phantom{xx}}$
31

5. Do you think that most politicians are basically honest?

$\overline{\phantom{xx}}$
32

6. Do you think that most politicians are responsive to the needs of their constituents?

$\overline{\phantom{xx}}$
33

7. Do you think that the Federal government has taken steps sufficient to prevent another Watergate?

$\overline{\phantom{xx}}$
34

8. Have you ever taken a Political Science course?

$\overline{\phantom{xx}}$
35

9. Are you very interested in national politics?

$\overline{\phantom{xx}}$
36

10. Have you ever paid any Federal income taxes?

$\overline{\phantom{xx}}$
37

---

**Fig. 8.9**  Questionnaire for Problem 8G.

*Hint:* Read the answers to questions 4 through 10 into an array (ANSWER) of size 7, using 7A1 format. To compare the responses to the letter Y, use a statement such as

```
IF (ANSWER(I) .EQ. 'Y') THEN
```

**8H**  Write a program to print a table (with headings) for values of $i$, $i^2$, $i^3$, and $\sqrt{i}$ and $\sqrt[3]{i}$, where $i$ is an integer that ranges from 1 to 100 in steps of 1. Use formats.

**8I**  Write a program which, given the size of an angle in degrees, computes the size in radians, and then computes the sine, cosine, and tangent of the angle. The program should print a neatly arranged, appropriately labelled 5-column table for degrees, radians, sine, cosine, and tangent of angles from $-90°$ to $+90°$ in steps of $1°$. Note that the SIN, COS, and TAN functions all require real arguments in radians, and that tan $90°$ and tan $-90°$ are not mathematically defined. You should note undefined computations in a meaningful way in your table. Keep your answers accurate to 5 decimal places, and use the following formula for degrees-to-radians conversion:

Number of radians = 0.01745 * Number of degrees.

**8J** We can consider a single sheet of printer paper as a piece of graph paper containing a grid of 50 × 100—50 rows and 100 columns (with space left over).

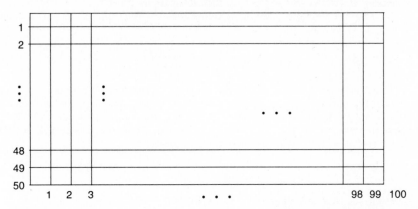

We can use this grid to plot a function $f$ on $x,y$-axes, in much the same way as we would plot $f$ on a piece of graph paper. To do this, we set up two 100-element real arrays, Y and XLIST, and a CHARACTER*1 array LINE of size 100. LINE will be used to define 50 lines of 100 print positions each. Initially, LINE is to contain all blanks. Y will be used to store 100 values of $f$ (one for each of 100 values of $x$ along the $x$-axis). XLIST will be used to define the index of each $x$ in the 100-item list used to compute $f(x)$. Thus, initially, $\text{XLIST}(i) = i$ for all values of $i$ between 1 and 100. These indices will be used to indicate which of the 100 horizontal grids corresponds to each value of $x$ for which $f(x)$ was computed.

Now proceed as shown in the following diagram.

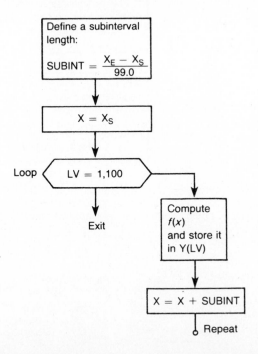

This will compute a value of $f$ for each of the 100 values of X used. Each $x$ index and the corresponding value $f(x)$ will be stored in corresponding elements of XLIST and Y, respectively.

Next we sort, in descending order, the array Y, exchanging the contents of the elements in XLIST in parallel with the exchanges in Y.

Then we determine 50 $y$-axis values (one for each line of output) as follows: Compute YINT as

$$\text{YINT} = \frac{Y(1) - Y(100)}{49.0}$$

where $Y(1)$ now contains the largest value of $f$, and $Y(100)$ contains the smallest value. We must print each of the 50 lines of the grid. The first line corresponds to the value $Y(1)$, the second $Y(1) - \text{YINT}$, the third to $Y(1) - 2 * \text{YINT}, \ldots,$ and the 50th to $Y(1) - 49 * \text{YINT}$, or $Y(100)$. (For example, if $Y(1) = 1000.0$ and $Y(100) = 100.$, then the lines would correspond to

```
1000.00, 981.63, 693.26, . . . , 136.74, 118.37, 100.00.
```

Note that 900./49. = 18.37.)

We do this as follows.

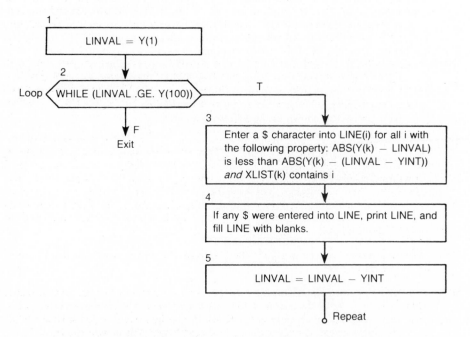

The implementation of Step 5 is simplified since the data in array Y has been sorted. We illustrate by example, using the Y-axis values computed earlier. Suppose the first few elements in Y and XLIST are defined as follows.

| XLIST(1) | XLIST(2) | XLIST(3) | XLIST(4) | XLIST(5) |
|----------|----------|----------|----------|----------|
| 37 | 53 | 36 | 54 | 35 |

| Y(1) | Y(2) | Y(3) | Y(4) | Y(5) |
|------|------|------|------|------|
| 996.3 | 992.2 | 987.3 | 980.1 | 970.5 |

The first time through loop 2, LINVAL will be equal to 1000. Since both 996.3 and 992.2 are closer in value to 1000. than to 981.63, a $ will be entered into LINE(37) and LINE(53). The next time through loop 2, LINVAL will be 981.63. Since 987.3 and 980.1 are closer to 981.63 than to 963.26, a $ will be placed in LINE(36) and LINE(54). Therefore, the test for the condition

```
ABS(Y(k) - LINVAL) < ABS(Y(k)-(LINVAL - YINT))
```

must be made only upon consecutive elements in Y, starting where the previous test failed, and continuing until the test fails again.

   Write a program to implement the above algorithms. Use subroutines whenever practicable. Test your program on the function

$$f(x) = (x - 1)^4/(x - 6),$$

where $x$ ranges from $-1$ to $+6$ ($[-1,6]$). Provide a program system chart to describe the components of your system.

**8K**    Write a program that merges two direct access files onto a third direct access file.

**8L**    In the inventory file for Problem 8B, the entries on this file were in sequence according to stock number, ordered from 1, 2, and so on. Write a program to read the stock entries for a dozen or so inventory items and build a sequential file containing these items. (You are not to make any assumptions concerning the ordering of the stock numbers of these items).

**8M**    Write a program to read the sequential file created in Example 8.19, sort the file in ascending order according to stock number and write the results on a new file. You may assume that the entire sequential file will fit in memory at once (Use an array large enough to accommodate the sequential file entries that you made in Problem 8L).

**8N**    Create a sequential file SALMEN containing the salaries of 10 men, and a second sequential file SALWOM containing the salaries of 10 women. For each employee on these files, there is an employee number (four digits), an employee name (a string) and an employee salary. Each file is arranged in ascending order by employee number. Write a program that will read each of these files and merge the contents onto a third file, SALARY, retaining the ascending order of employee numbers. For each employee written to the file SALARY, write an "M" (for male) or an "F" (for female) following the employee number.

**8O**    Write a program to read and print the file SALARY and compute the average salary for all employees.

**8P**    Assume you have a file of records each containing a person's last name, first name, birth date, and sex. Create a new file containing only first names and sex. Also, print out the complete name of every person whose last name begins with the letter A, C, F, or P through Z and was born in a month that begins with the letter J.

**8Q**    Write a program that prints every record of a direct access file whose record number ends with a zero.

# LOGICAL AND CHARACTER STRING DATA

9

## 9.1   INTRODUCTION

We have already introduced the character and logical data types in Chapter 4. We have also used character and logical data in many of the programs we have written. Character string constants have been used in PRINT statements (to clarify output data) and for format specification. Simple logical expressions have been used as conditions in WHILE loops and decision statements, and logical variables have been used as program flags.

In this chapter, we will describe the syntax of more complicated logical expressions and learn how to evaluate logical expressions containing logical operators such as .AND., .OR., and .NOT.. We will also learn how to manipulate character strings in FORTRAN and study some typical character string manipulation problems.

## 9.2   LOGICAL EXPRESSIONS

### 9.2.1   Syntax Rules for Logical Expressions

Since the beginning of the text, we have been using logical expressions to specify the condition in decision structures and WHILE loops. We have also used logical variables as program flags to signal to one program segment the results of the prior execution of another. Examples of statements that involve logical expressions are listed below.

```
FOUND = .TRUE.
MORE = .FALSE.
IF (FOUND) THEN
WHILE (MORE) DO
IF (COUNTR .GT. NRITMS) GO TO 20
IF (PINS(FIRST) + PINS(FIRST+1) .EQ. 10) THEN
IF (GRADE .LT. 0.0 .OR. GRADE .GT. 100.0) THEN
IF (NAME .LT. 'J') THEN
```

As shown in these examples, logical expressions consist of logical constants (.TRUE. or .FALSE.), logical variables used by themselves or arithmetic or character operands connected by the relational operators (.LT., .LE., .NE., .EQ., .GT., .GE.). As introduced in Chapter 4, compound logical expressions may also be formed from simple logical expressions using the logical operators .AND., .OR., and .NOT.. In this chapter, we will introduce the logical operators .EQV. and .NEQV., which may be used to compare logical expressions, and we will learn more about the use and evaluation of logical expressions and logical operators. We begin by presenting the rules of formation for logical expressions.

---

**Rules of Formation for Logical Expressions**

    1. A logical expression may be a logical constant (.TRUE., .FALSE.) or a logical variable.

    2. A logical expression may be an expression of the form:

$$e_1 \; relop \; e_2$$

where $e_1$ and $e_2$ are both arithmetic expressions or both character expressions, and *relop* is a relational operator (.GT., .GE., .LT., .LE., .EQ., .NE.).

    3. A logical expression may be formed by using the logical operators in combination with other logical expressions. The binary logical operators .AND., .OR. may be used to write logical expressions of the form:

$$lex_1 \, . \text{AND} . \; lex_2$$
$$lex_1 \, . \text{OR} . \; lex_2$$

where $lex_1$, $lex_2$ are logical expressions.
The unary logical operator .NOT. may be used to write logical expressions of the form

$$. \text{NOT} . \; lex$$

    4. Logical expressions may be compared using the logical operator .EQV. (logical equivalence) or .NEQV. (logical nonequivalence). These operators can be used to form new logical expressions of the form:

$$lex_1 \, . \text{EQV} . \; lex_2$$
$$lex_1 \, . \text{NEQV} . \; lex_2$$

where $lex_1$, $lex_2$ are logical expressions.

---

    The new operators in Rule 4 (.EQV. and .NEQV.) are used to test logical expressions and values for equivalence or nonequivalence, respectively. They are analogous to the relational operators .EQ. and .NE. that are used to test for equality or inequality, respectively, of arithmetic or character data.

**Example 9.1:**    The decision structure below prints a message based on the values of the logical variables **FLAG** and **SWITCH**.

```
IF (FLAG .NEQV. SWITCH) THEN
    PRINT*, 'FLAG AND SWITCH ARE DIFFERENT'
ELSE IF (FLAG .EQV. .TRUE.) THEN
    PRINT*, 'FLAG AND SWITCH ARE BOTH TRUE'
ELSE
    PRINT*, 'FLAG AND SWITCH ARE BOTH FALSE'
ENDIF
```

The condition (FLAG .EQV. .TRUE.) could also be written as (FLAG).

**Example 9.2:** Assume that X, Y, and Z are type real, NAME and KEY are type character, and FLAG is type logical. The following are all legal logical expressions:

1. (X .GT. 2.0) .AND. (Y .GT. 2.0)
2. (X + Y / Z) .LE. 3.5
3. .NOT. ((X .GT. Y) .OR. (X .GT. Z))
4. (.NOT. FLAG) .OR. ((Y + Z) .LE. (X − Z))
5. .NOT. FLAG
6. NAME .NE. KEY
7. (0.0 .LT. X) .AND. (X .LT. 1.0)
8. Any combination of (1) through (7) using the logical operators .AND. and .OR.—for example:

$$\underbrace{((X + Y / Z) .LE. 3.5)}_{(2)} .AND. \underbrace{(.NOT.((X .GT. Y) .OR. (X.GT.Z)))}_{(3)}$$

$$\underbrace{(.NOT. FLAG)}_{(5)} .OR. \underbrace{(NAME .NE. KEY)}_{(6)}$$

9. Any combination of (1) through (8) using the logical operators .EQV. and .NEQV.—for example:

$$\underbrace{(.NOT. FLAG)}_{(5)} .EQV. \underbrace{(NAME .NE. KEY)}_{(6)}$$

$$\underbrace{((X .GT. 2.0) .AND. (Y .GT. 2.0))}_{(1)} .NEQV. \underbrace{((X + Y / Z) .LE. 3.5)}_{(2)}$$

Some common errors in writing logical expressions are illustrated below.

10. FLAG .EQ. .TRUE.
Logical expressions (including logical constants or variables) cannot be used as operands of a relational operator (use either FLAG by itself or .EQV. instead of .EQ.)

11. X .GE. NAME
Real and character data may not be mixed as operands of a relational operator.

12. 0.0 .LT. X .LT. 1.0
The logical expression 0.0 .LT. X cannot be used as an operand of the second relational operator .LT.. (See 7 for the correct form of the condition "X lies between 0.0 and 1.0".)

13. X .AND. Y .GT. 2.0
Real variable X cannot be used as an operand of the logical operator .AND.. (See 1 for the correct form of the condition "X and Y are both greater than 2.0".)

14. .NOT. (X .GT. Y .OR. Z)

Real variable Z cannot be an operand of the logical operator .OR.. (See 3 for the correct form.)

Some points illustrated in this example are:

- A relational operator may be used to compare arithmetic data or character data.
- A relational operator cannot be used to compare logical data.
- A logical operator can manipulate logical data only.

**Exercise 9.1:** Identify and correct the errors in the following illegal expressions. Assume X, Y, Z are type real, I is type integer, and FLAG1, FLAG2 are type logical.

a) I .LT. 1 .AND. 2 .AND. 3
b) X .EQ. Y .OR. Z
c) X .OR. Y .LT. Z
d) FLAG1 .OR. (FLAG2 .EQ. .TRUE.)
e) (FLAG1 .EQ. FLAG2) .OR. (X .NE. Y)

## 9.2.2 Evaluating Logical Expressions

We know how to evaluate simple logical expressions involving only a relational operator. In order to evaluate compound logical expressions, we must review the properties of the logical operators and learn more about the precedence of operators.

Table 9.1 contains a summary of the properties of the logical operators with two operands. Each row of the table represents a different pair of values for the logical expressions $lex_1$, $lex_2$. (T and F are abbreviations for .TRUE. and .FALSE., respectively.) There are separate columns of values for each of the logical operators .AND., .OR., .EQV., and .NEQV. applied to $lex_1$, $lex_2$.

| $lex_1$ | $lex_2$ | $lex_1$ .AND. $lex_2$ | $lex_1$ .OR. $lex_2$ | $lex_1$ .EQV. $lex_2$ | $lex_1$ .NEQV. $lex_2$ |
|---------|---------|----------------------|---------------------|----------------------|-----------------------|
| T | T | T | T | T | F |
| T | F | F | T | F | T |
| F | T | F | T | F | T |
| F | F | F | F | T | F |

**Table 9.1 Properties of the logical operators**

The logical operator .NOT. forms the complement of its single logical operand as shown below.

| $lex_1$ | .NOT. $lex_1$ |
|---------|---------------|
| T | F |
| F | T |

The precedence of arithmetic operators was discussed in Section 4.3.2. The display below lists the precedence of all operators discussed so far. A higher precedence operator will be evaluated before a lower precedence operator in the same subexpression.

---

**Precedence of Operators**

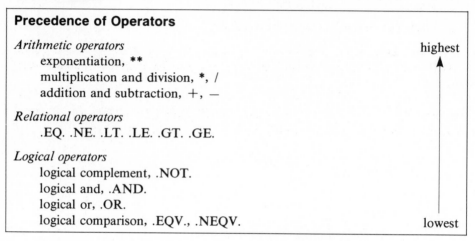

*Arithmetic operators*                                                   highest
    exponentiation, **
    multiplication and division, *, /
    addition and subtraction, +, −

*Relational operators*
    .EQ. .NE. .LT. .LE. .GT. .GE.

*Logical operators*
    logical complement, .NOT.
    logical and, .AND.
    logical or, .OR.
    logical comparison, .EQV., .NEQV.                                lowest

---

As is the case with arithmetic expressions, when in doubt it is best to use parentheses to clearly specify the desired grouping of operands and operators.

**Example 9.3:** The evaluation of a sample logical expression is given in Fig. 9.1. The value of each subexpression is written as T or F on the line coming out of the operator circle. We will assume that X and Y are real variables containing 2.0 and 3.0, respectively:

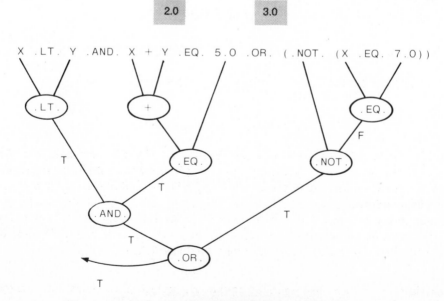

**Fig. 9.1** Evaluation of a logical expression.

**Exercise 9.2:**   Evaluate the expression in Fig. 9.1 for X = 3.0, Y = 2.0; for X = 7.0, Y = 12.0.

**Exercise 9.3:**   Evaluate all of the legal expressions in Example 9.2 for X = 1.0, Y = 2.5, Z = −1.0, FLAG = .TRUE., NAME = 'JOHN' and KEY = 'KING'.

### 9.2.3   Application of Logical Expressions

Logical expressions are most frequently used to specify conditions in WHILE loops or decision statements. They can also be assigned to logical variables.

**Example 9.4:**   Use of the logical variables L1, L2, and L3 in Fig. 9.2 simplifies the listing of conditions in the multiple-alternative decision structure.

```
REAL X, Y
LOGICAL L1, L2, L3
READ*, X, Y
PRINT*, 'X = ', X, '      Y = ', Y
L1 = X .GT. Y
L2 = X .GE. 0.0
L3 = Y .GE. 0.0
IF (L1 .AND. L3) THEN
    PRINT*, 'X IS BIGGER THAN Y. BOTH ARE POSITIVE.'
ELSE IF (L1 .AND. L2) THEN
    PRINT*, 'X IS BIGGER THAN Y. X IS POSITIVE, Y IS NEGATIVE.'
ELSE IF (L1) THEN
    PRINT*, 'X IS BIGGER THAN Y. BOTH ARE NEGATIVE.'
ELSE
    PRINT*, 'Y IS GREATER THAN OR EQUAL TO X.'
ENDIF
STOP
END


*** PROGRAM EXECUTION OUTPUT ***

X =   10.50000000000         Y =    21.00000000000
Y IS GREATER THAN OR EQUAL TO X.
```

**Fig. 9.2**   Using logical variables in conditions.

L1, L2, and L3 are assigned the value .TRUE. or .FALSE., depending on the values of X, Y prior to executing the multiple-alternative decision structure. Based on these values, one of the four messages in the multiple-alternative structure will be printed.

Another common use of logical variables is as a program flag to communicate to one program segment the results of the execution of another. This application has been illustrated earlier (Sections 4.5.2 and 7.5.2).

**Example 9.5:**   As another example of the use of logical variables as program flags, we will rewrite the program for searching an array (see Fig. 5.10) as a subroutine. The program flag, FOUND, is used to communicate the results of the search to the calling program.

   We will use a WHILE loop to control the search process (rather than a DO loop as shown in Fig. 5.10). This loop is repeated as long as the array has not been exhausted and the target item has not been located.

### Data Table for Search Subroutine (SEARCH)

*Input arguments*

A: Represents the array being searched (integer)

SIZE: Represents the number of elements of A that are to be searched (integer)

KEY: Represents the item to be located (integer)

*Output arguments*

FOUND: Represents a program flag— set to .TRUE. if the item is found; set to .FALSE. if the item is not found (logical)

WHERE: Represents the subscript of the first element of A that contains KEY if it is found (integer)

*Program variables*

NEXT: Subscript of the next element in A to be compared to KEY

   The flow diagram and subroutine for SEARCH are shown in Figs. 9.3 and 9.4, respectively. As shown in Step 2.3 (Fig. 9.3), WHERE is not defined when the KEY is missing.

   A common error in writing the WHILE loop shown in Fig. 9.4 is using the logical expression

```
NEXT .LE. SIZE
```

instead of

```
NEXT .LT. SIZE
```

in the repeat condition. This substitution would cause the loop body to be executed when NEXT is equal to SIZE (changing NEXT to SIZE+1); the repeat condition would then be evaluated for NEXT equal to SIZE+1. Since dummy array A is declared to have only SIZE elements, A(NEXT) would be out of bounds and a subscript range error would result.

**Example 9.6:**  The body of the search subroutine (excluding the declarations) may be rewritten using a DO loop as shown in Fig. 9.5.

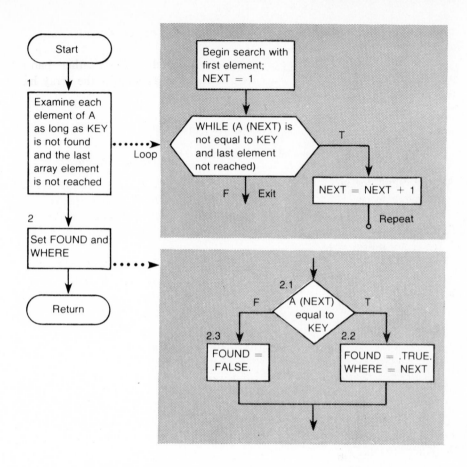

**Fig. 9.3**  Flow diagram for SEARCH subroutine (Example 9.5).

```
       SUBROUTINE SEARCH (A, SIZE, KEY, FOUND, WHERE)
C
C SEARCH AN INTEGER ARRAY FOR A SPECIFIED DATA ITEM
C
C ARGUMENT DEFINITIONS --
C   INPUT ARGUMENTS
C     A - ARRAY TO BE SEARCHED
C     SIZE - NUMBER OF ELEMENTS OF ARRAY TO BE EXAMINED
C     KEY - ITEM TO BE FOUND
C   OUTPUT ARGUMENTS
C     FOUND - DEFINED TO BE TRUE IF ITEM FOUND - OTHERWISE FALSE
C     WHERE - SUBSCRIPT OF ARRAY ELEMENT CONTAINING KEY (IF FOUND)
C
       INTEGER SIZE
```

*(Continued)*

```
        INTEGER A(SIZE), KEY
        LOGICAL FOUND
        INTEGER WHERE
C
C LOCAL VARIABLES
        INTEGER NEXT
C
C SEARCH WHILE KEY IS NOT FOUND AND LAST ELEMENT IS NOT REACHED
        NEXT = 1
        WHILE ((A(NEXT) .NE. KEY) .AND. (NEXT .LT. SIZE)) DO
            NEXT = NEXT + 1
        ENDWHILE
C
C SET FOUND AND WHERE
        IF (A(NEXT) .EQ. KEY) THEN
            FOUND = .TRUE.
            WHERE = NEXT
        ELSE
            FOUND = .FALSE.
        ENDIF
C
        RETURN
        END
```

**Fig. 9.4**   Subroutine SEARCH (Example 9.5).

```
C
C SEARCH ALL ELEMENTS TO FIND THE KEY
C SET WHERE AND RETURN IF KEY FOUND
        DO 40 NEXT = 1, SIZE
            IF (A(NEXT) .EQ. KEY) THEN
                FOUND = .TRUE.
                WHERE = NEXT
                RETURN
            ENDIF
   40 CONTINUE
C
C KEY NOT FOUND
        FOUND = .FALSE.
        RETURN
```

**Fig. 9.5**   Body of Subroutine SEARCH with DO loop.

**Program Style**

*Multiple returns from a subroutine*

The DO loop implementation of subroutine SEARCH contains two RE-TURN statements. Many computer scientists feel there should only be one return point from a subroutine. Consequently, they would prefer the WHILE loop implementation shown in Fig. 9.4. Of course, this option is only available if your compiler supports the WHILE loop (not part of the 1977 FORTRAN Standard).

**Exercise 9.4:**   Add a fourth input argument, START, to subroutine SEARCH, to permit a calling program to indicate where the search is to begin. (In the current version of SEARCH, the search always begins with the first element of the array.) What changes must be made to the search subroutine shown in Fig. 9.4 to accommodate the change? Note that START must always be less than or equal to SIZE and larger than 0.

**Exercise 9.5:**   Write the level one flow diagram and refinements for the subroutine segment shown in Fig. 9.5.

## 9.3   CHARACTER STRING MANIPULATION

### 9.3.1   Introduction

So far, we have seen limited use of character data. Character variables have appeared in the list portion of data initialization, READ and PRINT statements, and they have been used for storing character strings that were later displayed to identify program output. Strings have also been used in PRINT statements to annotate program output and for format specification.

Many computer applications are concerned with the manipulation of textual data rather than numerical data. Computer-based *word processing systems* enable a user to compose letters, term papers, newspaper articles and even books at a computer terminal instead of using a typewriter. The advantage of using such a system is that words and sentences can be modified, whole paragraphs can be moved, etc. and then a fresh copy can be printed without mistakes or erasures.

Additional applications include the use of computerized typesetters in the publishing industry; text editors are used to update telephone directories and annual reports on a regular basis; computers are used in the analysis of great works of literature.

In the sections that follow, we will introduce some fundamental operations that can be performed on character data. We will describe how to reference a character substring and how to concatenate (or join) two strings. We will learn how to search for a substring in a larger string, and how to delete a substring or replace it with another.

### 9.3.2   Character String Declaration and Length

We have already used the CHARACTER statement to declare character strings and string arrays. For example, the declaration

```
CHARACTER * 10 A(20), FLOWER
```

allocates storage space for 21 character strings consisting of ten characters each. (Twenty of these strings are elements of the array A.)

The notion of the length of a character string is important to the discussion of character-type data. We will introduce this concept by defining what is meant by the length of a character string constant or variable. The definition of the length of other character entities will be given as they are introduced in later sections.

---

**Length of Character String Constants and Variables**

1. The length of a character string constant is equal to the number of characters in the constant (excluding the apostrophes used to delimit the constant) except that a pair of adjacent apostrophes within the constant is counted as a single character.
2. The length of a character variable is defined to be the length given to the variable when it is declared.

*Note:* The length of any character entity is a positive number and may not change. Zero length character strings are not permitted in FORTRAN.

---

As indicated above, the length of a character variable remains unchanged although character data of different lengths may be assigned to the variable. Any character string that is shorter than the declared length of a variable will be padded on the right with blanks when assigned to that variable; any string that is too long cannot be stored in its entirety (see Section 9.5.2.).

FORTRAN provides a library function LEN that can be used to determine the length of its character string argument. We will describe this function and provide examples of its use in Section 9.7. These examples will help provide a better understanding of the length properties of character entities.

## 9.4  SUBSTRINGS

We frequently need to reference substrings of a longer character string. For example, we might want to examine the day, '25', in the string 'JUNE 25, 1981' or remove the substring 'Machinery' from the string 'Association for Computing Machinery'. In this section, we will see how to use special features of FORTRAN to segment a character string into substrings or to extract part of a longer string.

To specify a substring of a character variable or character array element, we write the substring name in the form shown below.

---

**Substring References**

$$cname \ (exp_1: exp_2)$$

**Interpretation:** *Cname* is a character variable or character array element, and $exp_1$, $exp_2$ are substring expressions. The values of $exp_1$ and $exp_2$ should be type integer. $exp_1$ and $exp_2$ are used to specify which substring of *cname* should be referenced. The value of $exp_1$ indicates the position in *cname* of the first character of the substring; the value of $exp_2$ indicates the position in *cname* of the last character of the substring.

*Note:* The reference *cname* $(exp_1: exp_2)$ is called the *substring name*. The integer values of $exp_1$ and $exp_2$ must satisfy the following constraints:

$$1 \ \leq \ exp_1 \ \leq \ exp_2 \ \leq \ \text{length of } cname$$

If $exp_1$ is omitted, it is considered to be 1; if $exp_2$ is omitted, it is considered to be the same as the length of *cname*. The substring length is defined as $exp_2 - exp_1 + 1$.

---

**Example 9.7:**  The names of three substrings of the character variable PRES are shown below:

```
CHARACTER * 18 PRES
DATA PRES /'ADAMS, JOHN QUINCY'/
```

```
        PRES(1 : 5)        PRES(13 : )
                PRES(8 : 11)
```

**Example 9.8:**  The program segment below

```
CHARACTER * 11 SOSSEC
CHARACTER * 3 SSN1
CHARACTER * 2 SSN2
CHARACTER * 4 SSN3
READ*, SOSSEC
SSN1 = SOSSEC(1 : 3)
SSN2 = SOSSEC(5 : 6)
SSN3 = SOSSEC(8 : 11)
```

reads a character string representing a social security number. If the string '042–30–0786' is read, this program segment breaks the Social Security number into substrings as shown below:

| SOSSEC | SSN1 | SSN2 | SSN3 |
|---|---|---|---|
| 042-30-0786 | 042 | 30 | 0786 |

The assignment statements in this program are *character assignment statements;* they each assign a character string to a character variable. We will discuss character assignments in more detail in Section 9.5.2.

**Example 9.9:**  The program in Fig. 9.6 prints each word of the sentence SENTNC on a separate line. It assumes that a single blank occurs between words.

The statement

```
CHARACTER * (LENGTH) SENTNC
```

declares a character variable whose length is determined by the parameter LENGTH. Any expression involving integer constants and parameters may be used to specify the length of a character variable. The expression must be enclosed in parentheses.

The program variable FIRST always points to the start of the current word and is initialized to one. During each execution of the loop, the condition

```
(SENTNC(NEXT : NEXT) .EQ. BLANK)
```

tests to see whether the next character is a blank. If it is, the statements

```
PRINT*, SENTNC(FIRST : NEXT)
FIRST = NEXT + 1
```

cause all characters in the current word (from FIRST through the blank) to be printed, and FIRST to be reset to point to the first character following the blank.

The statement

$$PRINT*, \quad SENTNC(FIRST \ : \ LENGTH)$$

following the loop is used to print the last word. This is only necessary when the last character (SENTNC(LENGTH : LENGTH)) is not a blank.

```
C PRINT EACH WORD IN A SENTENCE
C
C PROGRAM PARAMETERS
      INTEGER LENGTH
      CHARACTER * 1 BLANK
      PARAMETER (LENGTH = 80, BLANK = ' ')
C
C VARIABLES
      CHARACTER * (LENGTH) SENTNC
      INTEGER FIRST, NEXT
C
C ENTER DATA
      READ '(A)', SENTNC
      PRINT*, SENTNC
C
C PRINT EACH WORD (THE CHARACTERS BETWEEN THE BLANKS)
      FIRST = 1
      DO 10 NEXT = 1, LENGTH
         IF (SENTNC(NEXT : NEXT) .EQ. BLANK) THEN
            PRINT*, SENTNC(FIRST : NEXT)
            FIRST = NEXT + 1
         ENDIF
   10 CONTINUE
C
C PRINT THE LAST WORD IF IT IS NOT YET PRINTED
      IF (SENTNC(LENGTH : LENGTH) .NE. BLANK) THEN
         PRINT*, SENTNC(FIRST : LENGTH)
      ENDIF
C
      STOP
      END


*** PROGRAM EXECUTION OUTPUT ***

THE QUICK BROWN FOX JUMPED
THE
QUICK
BROWN
FOX
JUMPED
```

**Fig. 9.6**  Program to print words in a sentence.

The formatted READ statement

```
READ '(A)', SENTNC
```

is used to enter the data in SENTNC as one long string, whose length is the same as the declared length (80) of SENTNC.

**Exercise 9.6:**   Given the character variables SSN1 and PRES (defined in Examples 9.7 and 9.8), list the characters that would be printed by the statements

```
1)   PRINT*, SSN1(1 : 3)
2)   PRINT*, PRES(6 : )
3)   PRINT*, PRES( : 6)
4)   PRINT*, PRES( : )
```

**Exercise 9.7:**   Indicate how you could modify the program in Fig. 9.6 to convert a sentence to a primitive form of "Pig Latin" in which the first letter of each word is moved to the end of the word, followed by the letters AY. The string "THE QUICK BROWN FOX JUMPED" would become "HETAY UICKQAY ROWNBAY OXFAY UMPEDJAY". *Hint:* It is only necessary to change the two PRINT statements.

**Exercise 9.8:**   Write a program to read in a sentence (containing no punctuation and with one blank between words) and print the first letters of the words all together on one line. You may assume a maximum of ten words in the sentence.

**Exercise 9.9:**   Modify Example 9.9 so that the restriction of a single blank between words is removed. Your program will have to skip over a group of consecutive blanks.

## 9.5   CHARACTER EXPRESSIONS

Until now, character expressions have consisted of individual character variables, substrings, or character string constants. In FORTRAN, we can join strings using the *concatenation operator* // (two consecutive slashes). The concatenation operator comes between the arithmetic operators and the relational operators in the precedence table shown in Section 9.2.2.

---

**The Concatenation Operator**

$$S_1 // S_2$$

**Interpretation.** The character string $S_1$ is concatenated with the character string $S_2$. This means the string $S_2$ is joined to the right end of the string $S_1$. The length of the resulting string is equal to the sum of the lengths of $S_1$ and $S_2$.

---

**Example 9.10:**

a) The expression

```
'ABC' // 'DE'
```

concatenates the strings 'ABC' and 'DE' together to form one string of length 5, 'ABCDE'.

b) Given a string MESSAG with eleven or more characters, the expression

```
MESSAG (1 : 5) // '*****' // MESSAG (11 : )
```

creates a new string that differs from MESSAG only in character positions 6 through 10 (replaced with asterisks).

    c) Given the string

$$\text{'ADAMS, JOHN QUINCY'}$$

stored in the character variable PRES (length 18), the expression

```
PRES(8 : 12) // PRES(13 : 13) // '. ' // PRES( : 5)
```

forms the string

$$\text{'JOHN Q. ADAMS'}$$

of length thirteen $(5 + 1 + 2 + 5)$.

**Exercise 9.10:**  Given the string PRES

$$\text{'ADAMS, JOHN QUINCY'}$$

write the strings formed by the following expressions:

1)  PRES(8 : 11) // ' ' // PRES( : 5)
2)  PRES(1 : 5) // PRES(7 : 8) // '.' // PRES(13 : 13) // '.'

### 9.5.2   Use of Character Expressions

    Character expressions may be used in FORTRAN in character assignment statements, as operands of relational operators in logical expressions, in PRINT statements, and as arguments in subprogram calls. In this section we will describe the rules for the first two of these uses of character expressions; character string arguments will be discussed in Section 9.7.2.

    The character assignment statement is described in the next display.

---

**Character Assignment Statement**

$$cname = expression$$

**Interpretation:** *Cname* may be a character variable or array element, or a substring name. A character *expression* consists of a sequence of character string constants, character variables, character array elements, substrings or character-valued functions connected by the character operator, //.

*Note 1:* If the length of *cname* exceeds the length of *expression, expression* will be padded on the right with blanks before being stored.

*Note 2:* If the length of *cname* is less than the length of expression, the extra characters at the right of expression will be discarded.

*Note 3:* If *cname* is a substring name, only the specified substring is defined by the assignment; all other characters in the string are unchanged e.g., NAME(2 : 4) = *expression* changes only characters two through four of NAME.

*Note 4:* None of the character positions being defined in *cname* may be part of *expression* e.g., NAME(1 : 2) = NAME(1 : 1) // 'A' is illegal since NAME(1 : 1) is being defined. (NAME(2 : 2) = NAME(1 : 1) would be legal).

---

**Example 9.11:** Consider the sample program segment below.

```
CHARACTER * 8 NAME, HERS, HIS
CHARACTER * 4 FIRST, FIRSTA, FIRSTB, INITLS
CHARACTER * 9 LSTFST
NAME = 'JOHN DOE'
FIRST = 'JIM'
FIRSTA = NAME(1 : 2)
FIRSTB = NAME
HIS = FIRST // NAME(6 : )
LSTFST = NAME(6 : ) // ', ' // NAME( : 4)
INITLS = NAME(1 : 1) // '.' // NAME(6 : 6) // '.'
HERS = NAME
HERS(3 : 3) = 'A'
```

The execution of the assignment statements in this segment will result in the following string assignments.

| NAME | FIRST | FIRSTA | FIRSTB | LSTFST | INITLS |
|------|-------|--------|--------|--------|--------|
| JOHN □ DOE | JIM □ | JO □□ | JOHN | DOE, □ JOHN | J.D. |

| HIS | HERS |
|-----|------|
| JIM □ DOE □ | JOAN □ DOE |

**Example 9.12:** Consider the following sample program segment.

```
CHARACTER * 17 BIGGER
CHARACTER * 8 SMALLR
CHARACTER * 12 SAME
BIGGER = 'EXTRASENSORY'
SMALLR = BIGGER
SAME = BIGGER
```

The result of executing the assignment statements is shown next.

| BIGGER | SAME | SMALLR |
|--------|------|--------|
| EXTRASENSORY □□□□□ | EXTRASENSORY | EXTRASEN |

The assignment statement:

```
BIGGER = SMALLR(1 : 5) // SAME
```

would change BIGGER as shown below.

BIGGER

EXTRAEXTRASENSORY

This result could also be obtained using the assignment statement

```
BIGGER(6 : ) = SAME
```

but

```
BIGGER(6 : ) = BIGGER(1 : 12)
```

would be illegal. (Why?)

**Exercise 9.11:**  Let HIPPO, QUOTE, QUOTE1, QUOTE2, and BIGGER be declared and initialized as:

```
CHARACTER * 12 HIPPO, BIGGER
CHARACTER * 30 QUOTE1
CHARACTER * 24 QUOTE2, QUOTE
DATA HIPPO, BIGGER / 'HIPPOPOTAMUS', 'SMALL' /
DATA QUOTE1 / 'STRUCTURED PROGRAMS ARE BETTER' /
```

Carry out each of the following assignment statements in sequence. Indicate if any are illegal.

a) `QUOTE( : 24) = QUOTE1(21 : 24) // QUOTE1( : 20)`
b) `QUOTE2 = QUOTE(21 : 24) // QUOTE( : 20)`
c) `QUOTE(1 : 24) = QUOTE2`
d) `HIPPO(4 : ) = 'S'`
e) `BIGGER = 'LARGE'`
f) `BIGGER(1 : 6) = BIGGER(7 : 12)`
g) `BIGGER(6 : 7) = 'ST'`

## 9.6    STRING COMPARISONS

### 9.6.1    The Collating Sequence

The comparison of character strings was discussed in Section 4.4.2. We learned that the order relationship between two strings is based on the FORTRAN collating sequence as described below.

---

**Collating Sequence**

The blank character precedes (is less than) all of the digits (0,1,2, . . . , 9) and all of the letters (A,B,C, . . . , Z).

The letters are ordered lexigraphically (in dictionary sequence); i.e., A precedes B, B precedes C, . . . , Y precedes Z.

The digits follow their normal numeric sequence; i.e., the character 0 precedes 1, 1 precedes 2, . . . , 8 precedes 9.

Digits and letters may not be intermixed. Either all digits must precede all letters, or vice versa.

---

It is important to note what is not specified by the collating sequence. In particular:

- The collating sequence for special characters (not A–Z, 0–9, or blank) is not specified, nor are the relationships between these characters and A–Z, 0–9, and blank specified.
- The relationship between the letters A–Z and the characters 0–9 are not specified.

As mentioned in Section 4.4.2, the collating sequence ensures that order comparisons between strings of letters will follow their dictionary sequence; however, little else is guaranteed.

**Example 9.13:** The relationship between the pairs of character strings shown below are not specified.

| | |
|---|---|
| `'ALLEN, R.'` and `'ALLEN, RICHARD'` | (collating sequence of the period is not specified) |
| `'X123'` and `'XYZ'` | (collating sequence of 0–9 is not specified in relation to that of A–Z) |
| `'./*'` and `'.*//*='` | (collating sequence for special characters is not specified) |
| `'***'` and `'*'` | (collating sequence for a blank, in relation to the special characters, is not specified) |

All relationships not specified are defined separately for each compiler. It is important that you learn the collating sequence defined for the compiler that you are using. The special functions described next may be of some help in doing this.

### 9.6.2 The Functions CHAR and ICHAR

FORTRAN has a library function, ICHAR, which can be used to determine the relative position of a character in the collating sequence. For example, the statement

```
PRINT*, ICHAR(' '), ICHAR('A'), ICHAR('0')
```

prints the relative positions in the collating sequence of the three characters shown.

The function CHAR is the *inverse* of the function ICHAR; CHAR can be used to determine the character in a specified position of the collating sequence. The statement

```
PRINT*, CHAR(0)
```

will print the first character of the collating sequence (position 0).

The functions CHAR and ICHAR are described in the displays that follow.

---

**Function ICHAR**

ICHAR(*character*)

**Interpretation:** The integer function ICHAR determines the relative position of *character* in the FORTRAN collating sequence.

---

---

### Function CHAR

CHAR(*pos*)

**Interpretation:** The character function CHAR determines the character at relative position *pos* in the collating sequence. The value of *pos* must be between 0 and n−1 where n is the number of characters in the collating sequence.

---

**Example 9.14:** The program in Fig. 9.7 prints the position in the collating sequence of each character in the FORTRAN character set. The character set is stored in the character variable SET. The PRINT statement prints a character, SET (POS : POS), followed by its position in the collating sequence, ICHAR (SET(POS : POS)). Try this program on your computer.

```
C PRINT THE RELATIVE POSITIONS IN THE COLLATING SEQUENCE
C OF ALL CHARACTERS IN THE FORTRAN CHARACTER SET
C
C PROGRAM PARAMETERS
      INTEGER SETSIZ
      PARAMETER (SETSIZ = 49)
C
C VARIABLES
      INTEGER POS
      CHARACTER * (SETSIZ) SET
C
C FORTRAN CHARACTER SET
      DATA SET
     Z /' ABCDEFGHIJKLMNOPQRSTUVWXYZ0123456789+-*/()=$,.'':'/
C
C PRINT EACH CHARACTER AND ITS RELATIVE POSITION
      PRINT*, 'COLLATING SEQUENCE POSITIONS FOR FORTRAN CHARACTERS'
      PRINT*, ' '
      PRINT*, 'CHARACTER      POSITION'
      DO 10 POS = 1, SETSIZ
         PRINT '(5X, A, 11X, I3)',
     Z        SET(POS : POS), ICHAR(SET(POS : POS))
   10 CONTINUE
C
      STOP
      END
```

**Fig. 9.7**   Position of FORTRAN characters in the collating sequence.

**Example 9.15:** The program in Fig. 9.8a counts and prints the number of occurrences of each character in an 80 character string (TEXT). The array CHRCNT (subscripts 0 through 255) is used to keep track of the number of occurrences of each character. This array is initialized to all zeros. Since ICHAR(NEXCHR) is the position of character NEXCHR in the collating sequence, the statement

```
CHRCNT(ICHAR(NEXCHR)) = CHRCNT(ICHAR(NEXCHR)) + 1
```

increments the array element corresponding to character **NEXCHR**. The statement

```
PRINT '(4X, A, 17X, I2)', CHAR(POS), CHRCNT(POS)
```

in **DO** loop 30 prints each character followed by its number of occurrences.

```
C FIND THE NUMBER OF OCCURRENCES OF EACH CHARACTER IN TEXT
C
      INTEGER COLLEN, TXTLEN
      PARAMETER (COLLEN = 255, TXTLEN = 45)
C
      INTEGER CHRCNT(0 : COLLEN), POS, NEXT
      CHARACTER * (TXTLEN) TEXT
      CHARACTER * 1 NEXCHR
C
C INITIALIZE ARRAY OF COUNTERS AND ENTER DATA STRING
      DATA CHRCNT /COLLEN * 0, 0/
      READ '(A)', TEXT
      PRINT*, 'THE INPUT STRING IS ...'
      PRINT '(1X, A/)', TEXT
C
C INCREASE COUNT OF EACH CHARACTER IN TEXT
      DO 20 NEXT = 1, TXTLEN
         NEXCHR = TEXT(NEXT : NEXT)
         CHRCNT(ICHAR(NEXCHR)) = CHRCNT(ICHAR(NEXCHR)) + 1
   20 CONTINUE
C
C PRINT RESULTS
      PRINT*, 'CHARACTER          OCCURRENCE'
      DO 30 POS = 0, COLLEN
         IF (CHRCNT(POS) .NE. 0)
     Z      PRINT '(4X, A, 17X, I2)', CHAR(POS), CHRCNT(POS)
   30 CONTINUE
C
      STOP
      END
```

**Fig. 9.8a**   Counting character occurrences in a string.

     The program shown in Fig. 9.8a works regardless of the collating sequence used on a particular computer. The decision step inside loop 30 ensures that counts are printed only for those characters that actually appear in the text string. The output for the text string

```
THE QUICK BROWN FOX JUMPED OVER THE LAZY DOG.
```

is shown in Fig. 9.8b.

```
THE INPUT STRING IS ...
THE QUICK BROWN FOX JUMPED OVER THE LAZY DOG.

CHARACTER                OCCURRENCE
                             8
    .                        1
    A                        1
    B                        1
    C                        1
    D                        2
    E                        4
    F                        1
    G                        1
    H                        2
    I                        1
    J                        1
    K                        1
    L                        1
    M                        1
    N                        1
    O                        4
    P                        1
    Q                        1
    R                        2
    T                        2
    U                        2
    V                        1
    W                        1
    X                        1
    Y                        1
    Z                        1
```

**Fig. 9.8b**   Sample output for program in Fig. 9.8a.

## 9.7   STRING LENGTH AND SEARCH FUNCTIONS

### 9.7.1   The Function LEN

The FORTRAN library function LEN determines the length (number of characters) in its character string argument as described in the next display.

**Example 9.16:**

a)    LEN ('ABCDE') returns a value of 5.

b)    LEN ('MY' // ' NAME') returns a value of 7.

c)    If WORD is declared as a character variable of length 10, LEN(WORD) always returns a value of 10. Thus the sequence of statements

```
WORD = 'ABCDE'
I = LEN(WORD)
PRINT*, I
```

would result in the printing of the value 10, the length of the string (after padding with 5 blanks) that is stored in WORD.

---

**String Length Function LEN**

LEN(*string*)

**Interpretation:** *String* may be any character expression (including character string constants and variables). The value returned is an integer denoting the length of *string*.

*Note 1:* If *string* is a character string constant, its length is determined by counting the characters inside the enclosing apostrophes (e.g., LEN ('ABC') is three; LEN('A''S') is also 3 (Why?).)

*Note 2:* If *string* is a character variable (or array element), its length is defined by the variable (or array) declaration statement.

*Note 3:* If *string* is a substring of the form $S(exp_1 : exp_2)$, its length is equal to $exp_2 - exp_1 + 1$.

*Note 4:* If *string* is a character expression involving the concatenation operator, $//$, its length is equal to the sum of the individual string lengths of the operands of $//$.

---

### 9.7.2   The Function GETLEN

As indicated above, the length of a character variable is independent of the data stored in it. There are times when we would like to know exactly how many characters are stored in a string, excluding any blank padding.

**Example 9.17:**   The user-defined function GETLEN in Fig. 9.9 determines the "actual" length of an input string, excluding any blank padding. It does this by starting with the last character and skipping over blanks until the first nonblank character is reached. If the last character in string S is not a blank, the value of GETLEN(S) is the same as LEN(S). If all characters are blank, GETLEN returns a value of zero. (The statement

```
CHARACTER * (*) STRING
```

in Fig. 9.9 is explained in the next section.)

```
      INTEGER FUNCTION GETLEN (STRING)
C
C DETERMINE LENGTH OF STRING EXCLUDING ANY BLANK PADDING
C
C ARGUMENT DEFINITIONS --
C    INPUT ARGUMENTS
C       STRING - STRING WHOSE LENGTH IS TO BE DETERMINED
C
         CHARACTER * (*) STRING
C
C FUNCTION PARAMETERS
         CHARACTER * 1 BLANK
         PARAMETER (BLANK = ' ')
```

*(Continued)*

```
C
C LOCAL VARIABLES
      INTEGER NEXT
C
C START WITH THE LAST CHARACTER AND FIND THE FIRST NON-BLANK
      DO 10 NEXT = LEN(STRING), 1, -1
         IF (STRING(NEXT : NEXT) .NE. BLANK) THEN
            GETLEN = NEXT
            RETURN
         ENDIF
   10 CONTINUE
C
C ALL CHARACTERS ARE BLANKS
      GETLEN = 0
C
      RETURN
      END
```

**Fig. 9.9**  Function GETLEN (Example 9.17).

The GETLEN function is very useful when concatenating strings. If SHORTS is a character variable, there may be a number of extraneous blanks before the letter M in the character string formed by the expression

```
        SHORTS // ' MORE STUFF'
```

There is only one blank before the letter M in the character string formed by

```
    SHORTS(1 : GETLEN(SHORTS)) // ' MORE STUFF'
```

**Exercise 9.12:**  Rewrite function GETLEN so that it only has one RETURN statement. *Hint*: See Fig. 9.4.

### 9.7.3  Length of Character Arguments

Character arguments in a subprogram call can be character expressions (including character string constants or character variables). Character arguments used in a subprogram call must, of course, correspond to type character dummy arguments.

The type declaration of a character dummy argument specifies its length. The declared length of a character dummy argument may not exceed the length of its associated actual argument. If the length, L, of a dummy argument is less than the actual argument length, then only the leftmost L characters of the actual argument will be associated with the dummy argument.

In most cases, we wish the length of a character dummy argument to be the same as its corresponding actual argument. However, the dummy argument lengths can not be predetermined since the subprogram arguments (and their lengths) change from one call to the next. Consequently, FORTRAN allows the programmer to use the symbols (*) to declare the length of a character dummy argument. The statement

```
CHARACTER * (*) STRING
```

in Fig. 9.9 indicates that the dummy argument **STRING** will assume the length of its corresponding actual argument. The actual argument length may vary from one call to the next and may be determined by the function **LEN**, as illustrated in the DO loop header in Fig. 9.9.

### 9.7.4   Searching for a Substring

In this section, we will introduce a library function that is very helpful in examining and manipulating a string stored in memory. The string search function, **INDEX**, can be used to search a string (the *subject string*) for a desired substring (the *target string*). For example, if SENTNC is the subject string shown below

SENTNC

WHAT NEXT

we could use this function to determine whether or not a target string 'AT' is a substring in SENTNC; the INDEX function should tell us that 'AT' appears at position 3 (counting from the left). If we tried to locate a target string 'IT', the INDEX function would return a value of zero since 'IT' is not a substring of SENTNC. The function INDEX is described next.

---

**String Search Function INDEX**

INDEX(*subject, target*)

**Interpretation:** The function **INDEX** returns an integer value indicating the starting position in the character string *subject* of a substring matching the string *target.*
*Note 1:* If there is more than one occurrence of *target*, the starting position of the first occurrence is returned.
*Note 2:* If *target* does not occur in *subject*, the value 0 is returned.

---

**Example 9.18:** The program in Fig. 9.10 replaces all occurrences of the string AIN'T in SENTNC with the string IS NOT. The statement

```
COPY = SENTNC(1 : POSIT-1) // 'IS NOT' // SENTNC(POSIT+5 : )
```

replaces the substring (AIN'T) in positions POSIT through POSIT+4 of SENTNC with the string IS NOT. A more general version of a string replacement program is described in Section 9.8.3.

```
C REPLACE AIN'T WITH IS NOT
C
      CHARACTER * 80 SENTNC
      CHARACTER * 96 COPY
      INTEGER POSIT
C
C ENTER DATA STRING
      READ '(A)', SENTNC
      PRINT*, 'OLD SENTENCE: ', SENTNC
C
C FIND EACH OCCURRENCE OF AIN'T IN SENTNC
      POSIT = INDEX(SENTNC, 'AIN''T')
      WHILE (POSIT .NE. 0) DO              10 IF (POSIT .EQ. 0)
          COPY = SENTNC(1 : POSIT-1) //       Z GO TO 19
     Z        'IS NOT' // SENTNC(POSIT+5 : )
          SENTNC = COPY
          POSIT = INDEX(SENTNC, 'AIN''T')
      ENDWHILE                                GO TO 10
C                                          19 CONTINUE
C PRINT RESULTS
      PRINT*, 'NEW SENTENCE: ', SENTNC
C
      STOP
      END


*** PROGRAM EXECUTION OUTPUT ***

OLD SENTENCE: OCCURENCE AIN'T SPELLED OCCURRENCE, IS IT?
NEW SENTENCE: OCCURENCE IS NOT SPELLED OCCURRENCE, IS IT?
```
**Fig. 9.10**   Program and output for Example 9.18.

## 9.8   TEXT PROCESSING APPLICATIONS

### 9.8.1   Generating Cryptograms

In the previous sections, we introduced the FORTRAN string manipulation features and provided several examples of their use. We will now illustrate the application of these features in the solution of three sample problems. The first problem is a program for generating cryptograms; the second problem involves a subroutine for processing a DO loop header; the third problem is a text editor program.

**Problem 9A:**   A cryptogram is a coded message formed by substituting a code character for each letter of an original message. The substitution is performed uniformly throughout the original message, i.e., all A's might be replaced by Z, all B's by Y, etc. We will assume that all punctuation (including blanks between words) remains unchanged.

**Discussion:**   The program must examine each character in a message, MESSAG, and insert the appropriate substitution for that character in the cryptogram, CRYPTO. This can be done by using the position of the original character in the alphabet string ALFBET as an index to the string of code symbols, CODE (e.g., the code symbol for the letter A should always be the first symbol in CODE; the code symbol for letter B should be the second symbol in CODE, etc.). The data table is shown below; the flow diagrams are drawn in Fig. 9.11 and the program is shown in Fig. 9.12.

**Data Table for Problem 9A**

*Program parameters*

ALFBET = 'ABC . . . Z', the alphabet string (character)

| *Input variables* | *Program variables* | *Output variables* |
|---|---|---|
| CODE: Replacement code followed by punctuation marks (character) | POS: Position of original character in string ALFBET, used as an index to CODE (integer) | CRYPTO: The cryptogram (character) |
| MESSAG: Original message (character) | NEXT: Loop control variable, indicates next character in MESSAG to encode (integer) | |

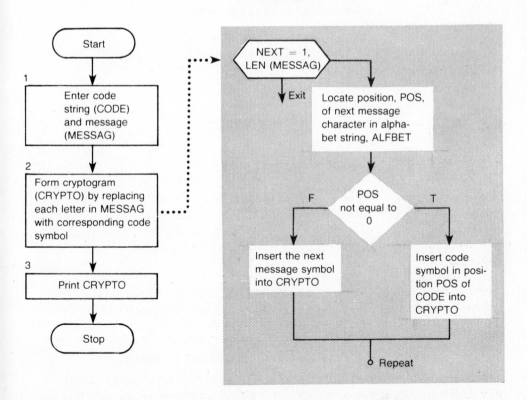

**Fig. 9.11**  Flow diagrams for cryptogram generator (Problem 9A).

```
C CRYPTOGRAM GENERATOR
C
C PROGRAM PARAMETERS
      CHARACTER * 26 ALFBET
      PARAMETER (ALFBET = 'ABCDEFGHIJKLMNOPQRSTUVWXYZ')
C
C VARIABLES
      INTEGER POS, NEXT
      CHARACTER * 26 CODE
      CHARACTER * 80 MESSAG, CRYPTO
C
C ENTER CODE AND MESSAGE
      READ '(A)', CODE
      PRINT*, 'ALPHABET: ', ALFBET
      PRINT*, 'CODE    : ', CODE
      PRINT*, ' '
      READ '(A)', MESSAG
C
C SUBSTITUTE THE CODE SYMBOL FOR EACH LETTER
      DO 10 NEXT = 1, LEN(MESSAG)
C        LOCATE CURRENT MESSAGE CHARACTER IN ALFBET
         POS = INDEX(ALFBET, MESSAG(NEXT : NEXT))
         IF (POS .NE. 0) THEN
C           INSERT CORRESPONDING CODE SYMBOL IN CRYPTO
            CRYPTO(NEXT : NEXT) = CODE(POS : POS)
         ELSE
C           INSERT MESSAGE SYMBOL IN CRYPTO
            CRYPTO(NEXT : NEXT) = MESSAG(NEXT : NEXT)
         ENDIF
   10 CONTINUE
C
C PRINT MESSAGE AND CRYPTOGRAM
      PRINT*, 'MESSAGE   : ', MESSAG
      PRINT*, 'CRYPTOGRAM: ', CRYPTO
C
      STOP
      END


*** PROGRAM EXECUTION OUTPUT ***

ALPHABET: ABCDEFGHIJKLMNOPQRSTUVWXYZ
CODE    : ZYXWVUTSRQPONMLKJIHGFEDCBA

MESSAGE   : ENCODE THIS MESSAGE.
CRYPTOGRAM: VMXLWV GSRH NVHHZTV.
```

Fig. 9.12 Program and sample output for Problem 9A.

In Fig. 9.12, the statement

```
POS = INDEX(ALFBET, MESSAG(NEXT : NEXT))
```

locates the current message symbol in the string ALFBET and the statement

```
CRYPTO(NEXT : NEXT) = CODE(POS : POS)
```

inserts the corresponding code symbol in the cryptogram. The statement

```
CRYPTO(NEXT : NEXT) = MESSAG(NEXT : NEXT)
```

inserts any message symbol that is not a letter directly in the cryptogram.

### 9.8.2  Scanning a DO loop Header

The next problem involves scanning a character string and extracting substrings.

**Problem 9B:**  We can consider the DO loop structure header as a character string of the form

$$\text{DO}\ \ lab\ lcv = initial,\ limit,\ step$$

For example,

```
DO 35 I = FIRST, LAST, 5
```

One of the tasks of a compiler in translating this statement might be to separate the substrings representing the loop parameters *initial, limit,* and *step* from the rest of the string and to save these substrings in separate character variables for later reference. We will write a program to perform this substring separation.

**Discussion:**  The task of our program is to identify and copy each of the DO loop parameters—*initial, limit,* and *step*—into the character variables INIT, LIMIT, and STEP, respectively.

The most difficult subtask for our program involves determining the starting and ending positions of the loop parameter strings. This, in turn, requires the identification of the positions in the header string of the equal sign (POSEQL) and the first and second commas (POS1CM and POS2CM) beyond the equal sign. If the second comma (and third parameter) in the header statement is missing, the character '1' is stored in STEP. If either the equal sign or the first comma is missing, an error message is printed and program execution is terminated. For simplicity, we will assume that the DO loop parameters contain no array element or function references.

Once the positions of the equal sign and the commas have been located (using the INDEX function), the substrings *delimited* by them (including all blanks) must be copied into INIT, LIMIT, and STEP. In each case, the copy can be performed using a simple character assignment statement.

The data table for the main program follows; the flow diagram is shown in Fig. 9.13.

## Data Table for Scanning a DO Loop Header (Problem 9B)

*Program parameters*

EQUAL : the character '='
COMMA : the character ','
ONE : the character '1'

| *Input variables* | *Program variables* | *Output variables* |
|---|---|---|
| HEADER: DO loop header (character*80) | POSEQL: Position of '=' in HEADER (integer) | INIT: Initial value loop parameter (character*80) |
| | POS1CM: Position of first ',' in HEADER beyond '=' (integer) | LIMIT: Limit value loop parameter (character*80) |
| | POS2CM: Position of second ',' in HEADER beyond '=' (integer) | STEP: Step value loop parameter (character*80) |

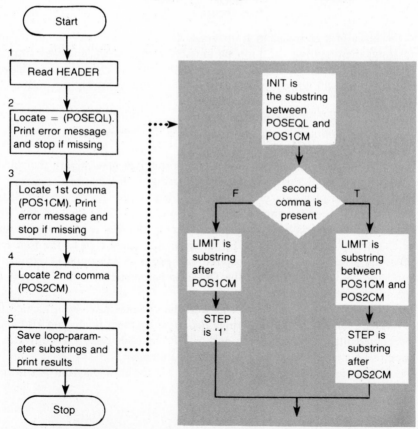

**Fig. 9.13**    Flow diagrams for the DO loop scanner (Problem 9B).

The function index in the statement

```
POS1CM = INDEX(HEADER(POSEQL+1 : ), COMMA) + POSEQL
```

searches for the first comma in the substring of **HEADER** following the equal sign. The position of this comma relative to the start of the string **HEADER** is obtained by adding **POSEQL** to the result of the substring search. To locate the second comma, the substring **HEADER(POS1CM + 1 : )** must be searched. The program is shown in Fig. 9.14.

```
C SEPARATE AND SAVE SUBSTRINGS OF DO LOOP HEADER
C
C PROGRAM PARAMETERS
      CHARACTER * 1 COMMA, EQUAL, ONE
      PARAMETER (COMMA = ',', EQUAL = '=', ONE = '1')
C
C VARIABLES
      CHARACTER * 80 HEADER
      CHARACTER * 80 INIT, LIMIT, STEP
      INTEGER POSEQL, POS1CM, POS2CM
C
C ENTER DATA
      READ '(A)', HEADER
C
C SEARCH FOR EQUAL SIGN - PRINT ERROR MESSAGE IF EQUAL SIGN MISSING
      POSEQL = INDEX(HEADER, EQUAL)
      IF (POSEQL .EQ. 0) THEN
         PRINT*, ' = SIGN MISSING'
         STOP
      ENDIF
C
C SEARCH FOR FIRST COMMA FOLLOWING = SIGN
C PRINT ERROR MESSAGE IF COMMA MISSING
      POS1CM = INDEX(HEADER(POSEQL+1 : ), COMMA) + POSEQL
      IF (POS1CM .EQ. POSEQL) THEN
         PRINT*, ' COMMA IS MISSING'
         STOP
      ENDIF
C
C SEARCH FOR SECOND COMMA FOLLOWING  FIRST COMMA
      POS2CM = INDEX(HEADER(POS1CM+1 : ), COMMA) + POS1CM
C
C SAVE LOOP PARAMETER SUBSTRINGS
      INIT = HEADER(POSEQL+1 : POS1CM-1)
      IF (POS2CM .NE. POS1CM) THEN
         LIMIT = HEADER(POS1CM+1 : POS2CM-1)
         STEP = HEADER(POS2CM+1 : )
      ELSE
         LIMIT = HEADER(POS1CM+1 : )
         STEP = ONE
      ENDIF
C
C PRINT RESULTS
      PRINT*, 'DO LOOP HEADER: ', HEADER
```

*(Continued)*

```
          PRINT*, 'INITIAL VALUE EXPRESSION: ', INIT
          PRINT*, 'LIMIT VALUE EXPRESSION: ', LIMIT
          PRINT*, 'STEP VALUE EXPRESSION: ', STEP
C
          STOP
          END
```

*** PROGRAM EXECUTION OUTPUT ***

```
DO LOOP HEADER:       DO 20 INDEX = FIRST, LAST-1, 10
INITIAL VALUE EXPRESSION:  FIRST
LIMIT VALUE EXPRESSION:  LAST-1
STEP VALUE EXPRESSION:  10
```

**Fig. 9.14**   Program and sample output for Problem 9B.

**Exercise 9.13:**   Of what relevance is the assumption made in the discussion of Problem 9B that array element or function references should not appear in the DO loop control parameters?

### 9.8.3   Text Editing Problem

**Problem 9C:**   There are many applications for which it is useful to have a computerized text editing program. For example, if you are preparing a laboratory report (or a textbook), it would be convenient to edit or modify sections of the report (improve sentence and paragraph structure, change words, correct spelling mistakes, etc.) at a computer terminal and then have a fresh, clean copy of the text typed at the terminal without erasures or mistakes.

**Discussion:**   A Text Editor System is a relatively sophisticated system of subprograms that can be used to instruct the computer to perform virtually any kind of text alteration. At the heart of such a system is a subprogram that replaces one substring in the text with another substring. As an example, consider the following sentence prepared by an overzealous member of the Addison-Wesley advertising group.

```
          'THE BOOK BY FRIEDMEN AND KOFFMAN
           IN FRACTURED PROGRAMING IS GRREAT?'
```

To correct this sentence we would want to specify the following edit operations:

1) Replace 'MEN' with 'MAN'
2) Replace 'IN' with 'ON'
3) Replace 'FRAC' with 'STRUC'
4) Replace 'AM' with 'AMM'
5) Replace 'RR' with 'R'
6) Replace '?' with '!'

The result is now at least grammatically correct.

```
          'THE BOOK BY FRIEDMAN AND KOFFMAN
           ON STRUCTURED PROGRAMMING IS GREAT!'
```

We will write the replacement program module as the subroutine REPLAC. The argument portion of the data table is shown below.

**Data Table for Subroutine REPLAC**

| *Input arguments* | *In-out arguments* |
|---|---|
| MAXLEN: Maximum length of TEXT (integer) | TEXT: Character string being edited (character) |
| OLD: Character string to be replaced (character) | CURLEN: Current length of TEXT excluding blank padding (integer) |
| OLDLEN: Length of OLD (integer) | |
| NEW : Character string to be inserted (character) | |
| NEWLEN: Length of NEW (integer) | |

MAXLEN is a Text Editor System parameter that is defined to be equal to the maximum length of the text string. CURLEN would be defined when the string to be edited is first placed in TEXT (probably in the main program) and would be redefined each time a change was made to TEXT.

The first task to be performed by REPLAC is to locate the first occurrence in TEXT of the substring to be replaced, OLD ( : OLDLEN). This can be accomplished using the function INDEX.

The additional data table entry required for REPLAC is shown next. The flow diagrams are drawn in Fig. 9.15a.

**Additional Data Table Entries for REPLAC**

*Program variables*

POSOLD: The position of the first character of OLD in TEXT if OLD is found (integer)

Before we can write the subroutine, a further refinement of Step 2.2 is needed. If NEWLEN is larger than OLDEN, it is possible that the length of the revised version of TEXT, REVLEN, would exceed MAXLEN. In this case, an error message should be printed and the replacement operation ignored; otherwise, a copy of TEXT can be made by concatenating the substring preceding OLD (the head of TEXT), NEW, and the substring following OLD (the tail of TEXT). The new data table entries follow. The refinement of Step 2.2 is shown in Fig. 9.15b.

*Additional program variables*

REVLEN: Length of edited text
(integer)

COPY: Temporary copy of the
edited text (character * 1000)

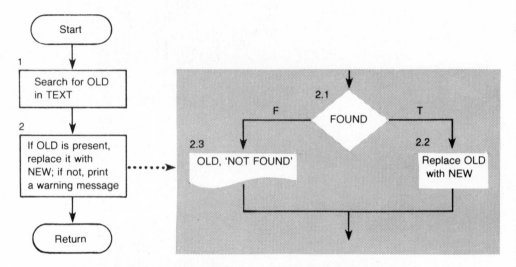

**Fig. 9.15a**   Flow diagrams for string replacement problem (9C).

In forming COPY (Step 2.2.4), the program must check for two special cases: TEXT starts with OLD or TEXT ends with OLD. In the former case, NEW becomes the head of COPY; in the latter case, NEW becomes the tail. The substring NEW ( : NEWLEN) represents the actual character data in NEW, exluding blank padding.

Subroutine REPLAC is given in Fig. 9.16.

**Exercise 9.14:**   Consider the subroutine in Fig. 9.16. Why is the character variable COPY needed in this program?

**Exercise 9.15:**   Explain the purpose of each condition of the multiple-alternative decision used to implement Step 2.2.4 of Fig. 9.15b. Why can't we just use the ELSE task in all cases and eliminate this decision structure? Modify this structure to handle as a separate alternative the case: the length of the new string (NEWLEN) is equal to the length of the old string (OLDLEN).

**Exercise 9.16:**   For each of the editing operations listed below, write a call statement to REPLAC. Write a main program segment that contains all declarations and data entry statements.

   a)   Replace 'FRAC' with 'STRUC'

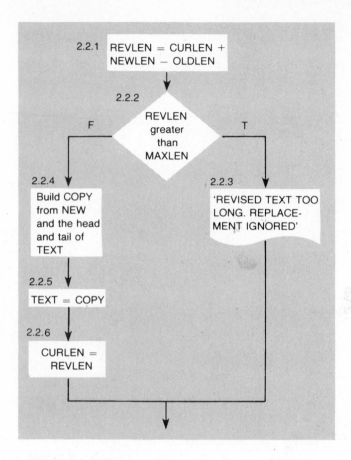

**Fig. 9.15b** Refinement of Step 2.2 in subroutine REPLAC.

b)   Replace the 'I' in 'IN' with an 'O'
c)   Replace 'BOOK' by 'TEXT'
d)   Insert an extra 'M' into 'PROGRAMING'
e)   Delete an 'R' from 'GRREAT'

**Exercise 9.17:**   From Exercise 9.16, parts (d) and (e), it is clear that REPLAC can be used to perform both insertions into and deletions from TEXT simply by providing enough contextual information in the arguments representing the new and the old strings. Nevertheless, we might wish to write subroutines DELETE and INSERT to handle all deletions and insertions.

a)   Using the REPLAC subroutine as a guide, write a subroutine DELETE to delete the first occurrence of a string OLD of length OLDLEN from TEXT.
b)   We can write a subroutine INSERT to insert a character string NEW of length NEWLEN into TEXT. In addition to NEW and NEWLEN, this subroutine will need a third input argument that marks the exact position in TEXT in which the insertion is to be performed. For example, the statements

```
        SUBROUTINE REPLAC (TEXT, MAXLEN, CURLEN,
   Z    OLD, OLDLEN, NEW, NEWLEN)
C
C REPLACE SUBSTRING OLD WITH STRING NEW IN TEXT
C
C ARGUMENT DEFINITIONS --
C   INPUT ARGUMENTS
C     MAXLEN - MAXIMUM LENGTH OF STRING TEXT
C     OLD - STRING TO BE REPLACED
C     OLDLEN - LENGTH OF OLD
C     NEW - REPLACEMENT STRING
C     NEWLEN - LENGTH OF NEW
C   IN-OUT ARGUMENTS
C     TEXT - STRING BEING EDITED
C     CURLEN - CURRENT LENGTH OF TEXT
C
        CHARACTER * (*) TEXT, OLD, NEW
        INTEGER MAXLEN, CURLEN, OLDLEN, NEWLEN
C
C LOCAL VARIABLES
        INTEGER POSOLD, REVLEN
        CHARACTER * 1000 COPY
C
C SEE IF OLD IS IN TEXT. IF SO, REPLACE IT. IF NOT, IGNORE REQUEST
        POSOLD = INDEX(TEXT, OLD( : OLDLEN))
        IF (POSOLD .NE. 0) THEN
C           CHECK REVISED LENGTH BEFORE REPLACEMENT
            REVLEN = CURLEN + NEWLEN - OLDLEN
            IF (REVLEN .GT. MAXLEN) THEN
                PRINT*, ' REVISED TEXT TOO LONG. REPLACEMENT IGNORED'
            ELSE
C               BUILD COPY BY REPLACING OLD WITH NEW IN TEXT
                IF (POSOLD .EQ. 1) THEN
C                   REPLACE HEAD OF TEXT WITH NEW
                    COPY = NEW( : NEWLEN) // TEXT(POSOLD+OLDLEN : )
                ELSE IF (POSOLD + OLDLEN .EQ. CURLEN) THEN
C                   REPLACE TAIL OF TEXT WITH NEW
                    COPY = TEXT( : POSOLD-1) // NEW( : NEWLEN)
                ELSE
C                   REPLACE OLD IN MIDDLE OF TEXT WITH NEW
                    COPY = TEXT( : POSOLD-1) // NEW( : NEWLEN) //
   Z                        TEXT(POSOLD+OLDLEN : )
                ENDIF
                TEXT = COPY
                CURLEN = REVLEN
            ENDIF
        ELSE
            PRINT*, OLD, ' NOT FOUND, REPLACEMENT IGNORED'
        ENDIF
C
        RETURN
        END
```

**Fig. 9.16** Subroutine REPLAC (Problem 9C).

```
CALL INSERT ('ELLIOT ', 7, 26)
CALL INSERT ('FRANK ', 6, 13)
```

would insert the strings 'ELLIOT ' and 'FRANK ' in front of 'KOFFMAN' and 'FRIEDMAN', respectively, in the original version of TEXT. Again using REPLAC as a guide, write the subroutine INSERT.

## 9.9  COMMON PROGRAMMING ERRORS

Now that we know how to manipulate different types of data, we must be careful not to misuse these data types in expressions. Only logical expressions can be operands of logical operators (.AND., .OR., .NOT.). Character strings can be operands of the character operator (concatenation, //) and relational operators (.GT., .LE., etc.). Remember that character variables and character constants can be manipulated only with other character data.

Misspelling the name of a logical or character variable (or neglecting to declare it) may result in compiler detection of syntax errors. This is because the type declaration intended for that variable will not be recognized if the variable name is spelled incorrectly. Consequently, the compiler will follow the implied type convention and assume that the variable is type integer or real. Since arithmetic variables cannot be operands of logical or character operators, diagnostic messages may be generated.

In specifying substrings, it is illegal to reference a character whose position is outside of the string. For example, if the substring name were

```
PRES(1 : IX)
```

and the value of IX exceeded the declared length of PRES, a diagnostic message would be printed. This is analogous to an array range error and may be caused by a loop that does not terminate properly or simply an incorrect substring expression. If in doubt, print out the values of suspect substring expressions, or compare these values with the declared bounds.

An additional source of error in character assignment statements has been mentioned earlier. That involves referencing the same character position on both sides of the assignment operator ('=' sign). The statement

```
PRES(1 : 5) = PRES(4 : 9)
```

is illegal, because positions 4 and 5 are both defined and referenced in this statement.

## 9.10  SUMMARY

In this chapter, we have provided a description of logical and character data manipulation. We explained how compound logical expressions may be formed and evaluated, and we have expanded on the use of program flags to communicate information between a subprogram and a calling module.

We also reviewed earlier work with character strings and introduced several new functions (LEN, INDEX, CHAR, and ICHAR) and a new operator for concatenation (//). We also discussed how to name and search for substrings. These new features are summarized in Table 9.2.

Many examples of these new features for manipulating character strings were presented. We have applied these features to generate cryptograms, to solve a problem that might arise in compiler design (processing a DO loop header), and in the design of a text editor replacement subroutine.

These kinds of problems are called nonnumerical problems and they are among the most challenging in computer science. The techniques presented in this chapter should give you a better idea of how to use the computer to solve nonnumerical problems.

| *Statement* | *Effect* |
|---|---|
| Logical operations | |
| FLAG .EQV. SWITCH | Compares logical variables |
| FLAG .NEQV. SWITCH | FLAG and SWITCH for equality |
| Substring | |
| NEW (1 : NEWLEN) | Denotes the substring of NEW consisting of the first NEWLEN characters. |
| Concatenation | |
| NEW // OLD | Concatenates (joins) strings NEW and OLD |
| Functions with character arguments or values | |
| LEN(NEW) | Returns the declared length of NEW |
| CHAR(0) | Returns the character in the first position (position zero) of the collating sequence for your compiler. |
| ICHAR('A') | Returns the position of A in the collating sequence |
| INDEX(SUBJCT, KEY) | Searches the string SUBJCT for the first occurrence of substring KEY. Returns the position of the first character of KEY in SUBJCT if found; otherwise, returns zero. |

**Table 9.2 FORTRAN features introduced in Chapter 9.**

# PROGRAMMING PROBLEMS

**9D**    Write a program to read in a set or words (given below) represented as character strings of 10 characters or less, and determine whether or not each word falls between the words in FIRST and LAST (the words DINGBAT and WOMBAT, respectively). Print the words in FIRST and LAST and print each word read in along with the indentifiers 'BETWEEN' or 'NOT BETWEEN,' whichever applies. Use the following data:

> HELP
> ME
> STIFLE
> THE
> DINGBAT
> AND
> THE
> WOMBATS
> BEFORE
> IT
> IS
> TOO
> LATE

**9E**    Write a program to read a character string of length 20 into BUFFER and a character string into ITEM. Then search BUFFER for the string contained in ITEM. Print

```
STRING FOUND
```

if the string in ITEM is a substring of BUFFER. Print

```
STRING NOT FOUND
```

if the string in ITEM is not a substring of BUFFER. Test your program on the following strings:

```
BUFFER: IS A MAN IN THE MOON
ITEMS: MAN , NUT , N T , MEN, I

BUFFER: IS DUST ON THE MOON
ITEMS: THEM, UST , DUST , IN , ON , O
```

*Hint:* Use the function GETLEN (Fig. 9.9) to determine the length of each string.

**9F**    Write a subroutine which will search a subject string for a specified target substring starting with a designated character position in the subject string (three input arguments). The subroutine will determine whether or not the substring is present and the position of its first occurrence if found.

**9G**    Assume a set of data cards is to be processed. Each card contains a single character string that consists of a sequence of words, each separated by one or more blank spaces. Write a program that will read these cards and count the number of words with one letter, two letters, etc., up to ten letters.

**9H**    Write a subroutine that will scan a string TEXT of length N and replace all multiple occurrences of a blank with a single occurrence of a blank. You may assume that TEXT and N are input arguments. You should also have an output argument, COUNT, which will be used to return the number of occurrences of multiple blanks found in each call of the subroutine.

**9I**    Write a program to read in a collection of cards containing character strings of length less than or equal to 80 characters. For each card read, your program should do the following:

a) Find and print the actual length of the string, excluding trailing blanks;
b) Count the number of occurrences of four-letter words in each string;
c) Replace each four-letter word with a string of four asterisks,****;
d) Print the new string.

**9J**    *A simple payroll problem.* Write a data table, flow diagram and program that will process the employee record cards described in Table 9.3, and perform the following tasks.

a) For each employee compute the gross pay:
   Gross pay = Hours worked * Hourly pay +
                    Overtime hours worked * Hourly pay * 1.5
b) For each employee compute the net pay as follows:
   Net pay = Gross pay − Deductions

Deductions are computed as follows:

Federal Tax = (gross pay − 13 * no. of dependents) * .14
FICA = gross pay * .065

$$\text{City Tax} = \begin{cases} \$0.00 \text{ if employee is works in the suburbs} \\ 4\% \text{ of gross pay if employee works in city} \end{cases}$$

$$\text{Union Dues} = \begin{cases} 0.00 \text{ if employee is not a union member} \\ 6.75\% \text{ of gross pay otherwise} \end{cases}$$

| Columns | Data description |
|---------|------------------|
| 1–6 | Employee number (an integer) |
| 7–19 | Employee last name |
| 20–27 | Employee first name |
| 28–32 | Number of hours worked (to the nearest 1/2 hour) for this employee |
| 33–37 | Hourly pay rate for this employee |
| 38 | Contains a C if employee works in the City Office and an S for the Suburban Office |
| 39 | Contains an M if the employee is a union member |
| 40–41 | Number of dependents |
| 42–46 | Number of overtime hours worked (if any) (also to the nearest 1/2 hour) |

**Table 9.3 Employee record card for Problem 9J**

For each employee, print a line of output containing:

1. Employee number
2. First and last name
3. Number of hours worked
4. Hourly pay rate
5. Overtime hours
6. Gross pay
7. Federal tax
8. FICA

9. City wage tax (if any)
10. Union dues (if any)
11. Net pay

Also compute and print:
1. Number or employees processed
2. Total gross pay
3. Total federal tax withheld
4. Total hours worked
5. Total overtime hours worked

Use formats and provide appropriate column headings for employee output and labels for totals.

**9K**  Shown below is the layout of a card that the registrar uses as input for a program to print the end-of-the-semester final grade report for each student.

| *Card columns* | *Data description* |
|---|---|
| 1–6 | Student number |
| 7–18 | Last name |
| 20–26 | First name |
| 27 | Middle initial |
| 28–29 | Academic year—FR, SO, JR, SR |
| 30–32 | First course—Department ID (3 letters) |
| 33–35 | First course—Number (3 digits) |
| 36 | First course—Grade A, B, C, D, or F |
| 37 | First course—Number of credits: 0–7 |
| 40–42 43–45 46 47 | Second course: data as described above |
| 50–52 53–55 56 57 | Third course data |
| 60–62 63–65 66 67 | Fourth course data |
| 70–72 73–75 76 77 | Fifth course data |

Write a data table, flow diagram and program to print the following grade report sheet for each student.

```
Line    1                  MAD RIVER COLLEGE
Line    2                 YELLOW GULCH, OHIO
Line    3
Line    4           GRADE REPORT, SPRING SEMESTER 1981
Line    5
Line    6    (student number)    (year)        (student name)
             - - - - - - - -    - year -      - - - - - - -
```

```
Line      7
Line      8                          GRADE  SUMMARY
Line      9            COURSE
Line     10       DEPT    NMBR         CREDITS              GRADE
Line     11    1. − − −  − − −            −                  −
Line     12    2. − − −  − − −            −                  −
Line     13    3. − − −  − − −            −                  −
Line     14    4. − − −  − − −            −                  −
Line     15    5. − − −  − − −            −                  −
Line     16
Line     17     SEMESTER  GRADE  POINT  AVERAGE  =  − − − −
```

Compute the grade-point average as follows:

i)    Use 4 points for an A, 3 for a B, 2 for a C, 1 for a D, and 0 for an F
ii)   Compute the product of points times credits for each course
iii)  Add together the products computed in (ii)
iv)   Add together the total number of course credits
v)    Divide (iii) by (iv) and print the result rounded off to two decimal places.
      *Hint:* Rounding is easy when formats are used for printing.

Use formats for all input and output. Your program should work for students taking anywhere from 1 to 5 courses. You will have to determine the number of courses taken by a student from the input data.

**9L**    Write a program to read in a string of up to 10 characters representing a number in the form of a Roman numeral. Print the Roman numeral form and then convert to Arabic form (a standard **FORTRAN** integer). The character values for Roman numerals are

| | |
|---|---|
| M | 1000 |
| D | 500 |
| C | 100 |
| L | 50 |
| X | 10 |
| V | 5 |
| I | 1 |

Test your program on the following input.

| | |
|---|---|
| LXXXVII | 87 |
| CCXIX | 219 |
| MCCCLIV | 1354 |
| MMDCLXXIII | 2673 |
| MDCDLXXXI | ? |

Use formats for all input and output.

**9M**    *Continuation of 9L.* Write a program to read in an integer and print the integer and its Roman numeral representation.

**9N**    Write an arithmetic expression translator which compiles fully-parenthesized arithmetic expressions involving the operators **\***, /, +, and −. For example, given the input string

$$( ( A + ( B * C ) ) - ( D / E ) )$$

the compiler would print out:

$$Z = (B*C)$$
$$Y = (A+Z)$$
$$X = (D/E)$$
$$W = (Y-X)$$

Assume only the letters A through F can be used as variable names. *Hint:* Find the first right parenthesis. Remove it and the four characters preceding it and replace them with the next unused letter (G–Z) at the end of the alphabet. Print out the assignment statement used. For example, the following is a summary of the sequence of steps required to process expression (i).

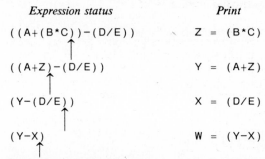

|      *Expression status*      |      *Print*      |
| :---------------------------: | :---------------: |
| $((A+(B*C))-(D/E))$           | $Z = (B*C)$       |
| $((A+Z)-(D/E))$               | $Y = (A+Z)$       |
| $(Y-(D/E))$                   | $X = (D/E)$       |
| $(Y-X)$                       | $W = (Y-X)$       |

**9O**  Write a subprogram, BLNKSP which removes all of the blanks from a character string and "compacts" all nonblank characters in the string. Assume the last character of the input string is a dollar sign. You should only have to scan the input string once from left to right.

**9P**  Write a program system (with appropriate documentation) which reads a FORTRAN program or subprogram and classifies each statement according to the following statement types:

1. Subroutine or function header
2. Type declaration (INTEGER, REAL, LOGICAL, or CHARACTER)
3. Data-initialization statement
4. Comment statement
5. Assignment statement
6. Decision-structure header IF (—) THEN
7. Loop structure header (DO loop or WHILE)
8. Structure terminator (CONTINUE, ENDIF, ENDWHILE)
9. IF statement
10. Transfer statement (GO TO, RETURN, STOP)
11. END statement
12. Decision-structure alternative header (ELSE, ELSEIF(—) THEN)
13. Input/output statement (READ or PRINT)
14. Subroutine call
15. None-of-the-above (possible error)

Assume that each statement fits on a single card. Print each statement and its type in a legible form. *Hint:* You may find the BLNKSP subroutine (Problem 9O) helpful here.

**9Q**  *Continuation of 9P.* Add a collection of subroutines to your program system for Problem 9P. These subroutines should each break up the statements of types 1–15 into their basic parts, as shown below. Omit types 4, 8, 10, 11, and, of course, 15.

1.  SUBROUTINE name (argument list)
    $\underbrace{\phantom{name}}_{A}$ $\underbrace{\phantom{(argument\ list)}}_{B}$

    type FUNCTION name (argument list)
    $\underbrace{\phantom{name}}_{A}$ $\underbrace{\phantom{(argument\ list)}}_{B}$

2.  type variable list
    $\underbrace{\phantom{variable\ list}}_{A}$

3.  DATA name list / constant list /
    $\underbrace{\phantom{name\ list}}_{A}$ $\underbrace{\phantom{constant\ list}}_{B}$

5.  variable or array element name = expression
    $\underbrace{\phantom{variable\ or\ array\ element\ name}}_{A}$ $\underbrace{\phantom{expression}}_{B}$

6.  IF (condition) THEN
    $\underbrace{\phantom{(condition)}}_{A}$

7.  DO sn loop-control variable = initial, end, step
    $\underbrace{\phantom{loop-control\ variable}}_{A}$ $\underbrace{\phantom{initial}}_{B}$, $\underbrace{\phantom{end}}_{C}$, $\underbrace{\phantom{step}}_{D}$

    WHILE (condition) DO
    $\underbrace{\phantom{(condition)}}_{A}$

9.  IF (condition) dependent statement
    $\underbrace{\phantom{(condition)}}_{A}$ $\underbrace{\phantom{dependent\ statement}}_{B}$

12. ELSEIF (condition) THEN
    $\underbrace{\phantom{(condition)}}_{A}$

13. READ*, list or PRINT*, list
    $\underbrace{\phantom{list}}_{A}$ $\underbrace{\phantom{list}}_{A}$

14. CALL name (argument list)
    $\underbrace{\phantom{name}}_{A}$ $\underbrace{\phantom{(argument\ list)}}_{B}$

*Warning:* This problem is best done by small groups of people rather than by one person. For students working alone, 2 or 3 of the subroutines required should be more than sufficient.

**9R**    Use the subroutine REPLAC (Problem 9C) and subroutines DELETE and IN-SERT (Exercise 9.17), and write a simple Text Editor System to perform the following tasks:

a)  Delete the first occurrence of a character string from TEXT;
b)  Replace the first occurrence of a character string with another string;
c)  Insert a character string at a specified position of TEXT;

The program system chart for the Text Editor is as follows:

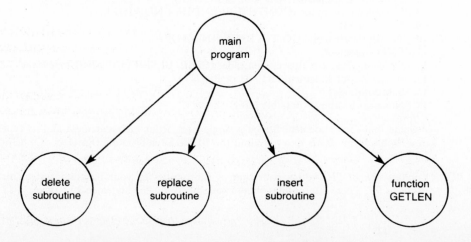

Test your system with the following input data:

```
'THE ORGANIZATION OF A PROGRAM IS A VERY MISERABLE EX-
PERIENCE. EVERY PROGRAMMER HAS HIS OWN INEFFICIENT WAY
OF GOING ABOUT THE DEVELOPMENT PROCESS. PROGRAMMING IS
STILL A WASTE OF TIME, BUT EVEN TEACHERS WASTE TIME.'

'REPLACE' 'MISERABLE' 'PERSONAL'
'DELETE' 'THIS IS NONSENSE'
'INSERT' 'DESIGN AND' 5
'REPLACE' 'A WASTE OF TIME' 'AN ART.'
'DELETE' 'INEFFICIENT'
'DELETE' 'BUT EVEN TEACHERS WASTE TIME.'
'PRINT'
'QUIT'
```

For this problem, TEXT should be a character string of maximum length 300. The main program should begin by reading the character string into TEXT.

Next, the main program should enter a loop in which text edit directives, RE-PLACE, INSERT, DELETE, PRINT and QUIT are procesed. The main program should read a directive, then read the data associated with that directive, and call the appropriate subroutine to perform the indicated edit. PRINT should be used to print the current string TEXT. If QUIT is read, the program should terminate.

# MULTIDIMENSIONAL ARRAYS

10

## 10.1 INTRODUCTION

We have been introduced to several types of data: real, integer, logical, and character. In addition, we have used one data structure, the array, for the identification and referencing of a collection of data items of the same type. The array enables us to save a list of related data items in memory. All of these data items are referred to by the same name, and the array subscript is used to distinguish among the individual array elements.

In this chapter, the use of the array will be extended to facilitate the convenient organization of related data items into tables and lists of more than one dimension. For example, we will see how a two-dimensional array with three rows and three columns can be used to represent a tic-tac-toe board. This array has nine elements, each of which can be referenced by specifying the row subscript (1, 2, or 3) and column subscript (1, 2, or 3), as shown in Fig. 10.1. Similarly, we shall see that arrays of three or more dimensions can be used to represent collections of data items that can be conveniently described in terms of a multidimensional picture.

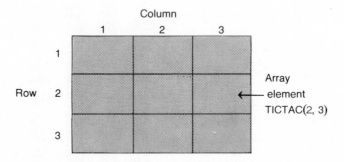

**Fig. 10.1** Representation of a tic-tac-toe board as a two-dimensional array, TICTAC.

## 10.2 DECLARATION OF MULTIDIMENSIONAL ARRAYS

The general form of an array declaration can be expanded to handle arrays of two or more dimensions, as shown in the display at the top of the next page.

**Example 10.1:**

```
CHARACTER * 1 TICTAC(3, 3)
REAL RECORD(7, 5, 6)
```

The array TICTAC is a two-dimensional array consisting of nine elements (3 × 3). Each subscript may take on the value 1, 2, or 3. The array RECORD consists of three dimensions. The first subscript may take on values from 1 to 7; the second from 1 to 5; and the third from 1 to 6. There are a total of 7 × 5 × 6 or 210 elements in the array RECORD.

---

**Array Declaration (for Multidimensional Arrays)**

$$type\ name(range_1,\ range_2,\ \ldots,\ range_n)$$

**Interpretation:** *Type* is any of the four data types INTEGER, REAL, CHAR-ACTER * k, or LOGICAL. In the restricted form above, $range_i$ represents the range of dimension i. $Range_i$ can be specified by an integer constant, parameter, or expression, or by a dummy argument if *name* is a subprogram dummy argument.

*Note 1: $Range_i$* specifies the range of permissible subscript values for dimension i. The total number of array elements (array size) is determined by the product

$$range_1 \times range_2 \times \ldots \times range_n$$

*Note 2:* For a two-dimensional array, $range_1$ refers to the number of rows and $range_2$ to the number of columns in the array.

*Note 3:* The general form of an n-dimensional array declaration is

$$type\ name(lower_1:upper_1,\ lower_2:upper_2,\ \ldots,\ lower_n:upper_n)$$

where $lower_i$ and $upper_i$ are the smallest and largest subscript values for dimension i.

---

## 10.3  MANIPULATION OF MULTIDIMENSIONAL ARRAYS

### 10.3.1  Manipulation of Individual Array Elements

Since the computer can manipulate only individual memory cells, we must be able to identify the individual elements of a multidimensional array. This is accomplished by using a subscripted reference to the array, as shown next.

---

**Subscripted Array Reference (Multidimensional Arrays)**

$$name\ (S_1,\ S_2,\ ,\ \ldots,\ S_n).$$

**Interpretation:** Each of the $S_i$ is a subscript expression. The forms permitted in FORTRAN 77 are discussed in Section 5.3. Any integer valued expression may be used.

---

In the case of two-dimensional arrays, the first subscript of an array reference is considered the *row subscript* and the second subscript the *column subscript*. Consequently, the subscripted array reference

```
TICTAC(2, 3)
```

selects the element in row 2, column 3 of the array TICTAC shown in Fig. 10.1. (This row/column convention is derived from the area of mathematics called *matrix algebra*. A *matrix M* is a two-dimensional arrangement of numbers. Each element in $M$ is referred to by the symbol $M_{ij}$, where $i$ is the number of its row and $j$ is the number of its column.)

**Example 10.2:**   Consider the array TICTAC drawn below.

This array contains three blank elements (TICTAC(1, 2), TICTAC(2, 1), TIC-TAC(2, 3)); three elements with value 'X' (TICTAC(1, 1), TICTAC(3, 1), TIC-TAC(3, 2)); and three elements with value 'O' (TICTAC(1, 3), TICTAC(2, 2), TICTAC(3, 3)).

**Example 10.3:**   A university offers 50 courses at each of five campuses. We can conveniently store the enrollments of these courses in an array declared as

```
INTEGER ENROLL(50, 5)
```

This array consists of 250 elements; ENROLL(I, J) represents the number of students in course I at campus J.

If we wish to have this enrollment information broken down further according to student rank (freshman, sophomore, junior, senior), we would need a three-dimensional array with 1000 elements:

```
INTEGER ENRANK(50, 5, 4)
```

The subscripted array reference ENRANK(I, J, K) would represent the number of students of rank K taking course I at campus J (see Fig. 10.2). We will assume that K must have a value between 1 and 4 and that rank 1 is associated with freshmen, rank 2 with sophomores, rank 3 with juniors, and rank 4 with seniors.

In Fig. 10.2, the circled element, ENROLL(1, 3), has a value of 33. The numbers shown in the array ENRANK represent the number of students of each rank in course 1 on campus 3. The following program segment computes the total number of students in course 1 at campus 3 regardless of rank. You should verify that CSUM will have the value 33 at the completion of the execution of this loop.

```
        CSUM = 0
        DO 10 K = 1, 4
            CSUM = CSUM + ENRANK(1, 3, K)
     10 CONTINUE
```

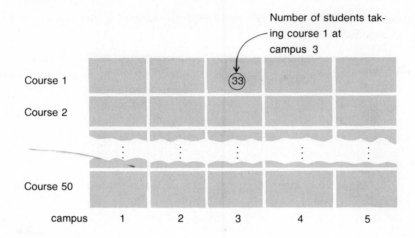

(a) Two-dimensional Array ENROLL

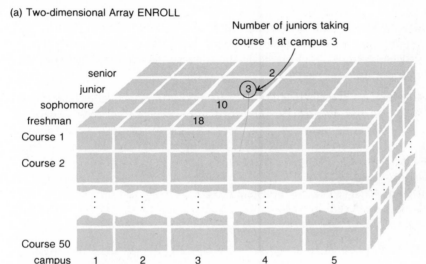

(b) Three-dimensional array ENRANK

**Fig. 10.2**   (a) Two-dimensional array ENROLL. (b) Three-dimensional array ENRANK.

**Exercise 10.1:**   Given the array ENRANK shown in Fig. 10.2, write program segments to perform the following operations:

   a)   Find the number of juniors in all classes at campus 3. Students will be counted once for each course in which they are enrolled.

   b)   Find the number of sophomores on all campuses who are enrolled in course 25.

   c)   Compute and print the number of students at campus 1 enrolled in each course and the total number of students at campus 1 in all courses. Students will be counted once for each course in which they are enrolled.

   d)   Compute and print the number of upper-class students in all courses at each campus, as well as the total number of upper-class students enrolled. (Upper-class students are juniors and seniors.) Again, students will be counted once for each course in which they are enrolled.

### 10.3.2  Relationship between Loop Control Variables and Array Subscripts

Sequential referencing of array elements is frequently required when working with multidimensional arrays. This process often requires the use of nested loops, since more than one subscript must be incremented in order to process all or a portion of the array elements. It is very easy to become confused in this situation and interchange subscripts, or nest the loops improperly. If you are in doubt as to whether or not your loops and subscripts are properly synchronized, you should include extra PRINT statements to display the subscript and array element values.

Exercise 10.1, especially parts (c) and (d), provides some experience in writing nested loops to process multiple-dimension arrays. The following problem, which processes the array TICTAC (described earlier), provides further illustration.

**Problem 10A:**  Write a subroutine which will be used after each move is made in a computerized tic-tac-toe game to see if the game is over. When the game is over, the subroutine should indicate the winning player or the fact that the game ended in a draw.

**Discussion:**  To see whether a player has won, the subroutine must check each row, column, and diagonal on the board to determine if all three squares are occupied by the same player. A draw occurs when all squares on the board are occupied but neither player has won. The flow diagrams for this problem are shown in Fig. 10.3. The data table follows.

**Data Table for Tic-Tac-Toe Problem (10A)**

*Input arguments*

TICTAC: Represents an array which shows the current state of the tic-tac-toe board after each move (character * 1, size $3 \times 3$)

*Output arguments*

OVER: Represents a flag used to indicate whether the game is over (OVER will be defined as true if the game is over; otherwise it will be false) (logical)

WINNER: Represents an indicator used to define the winner of the game ('X', 'O', or 'D' for draw when the game is over (character * 1)

*Program parameters*

DRAW = 'D' (character * 1)

BLANK = ' ' (character * 1)

(Continued on page 443)

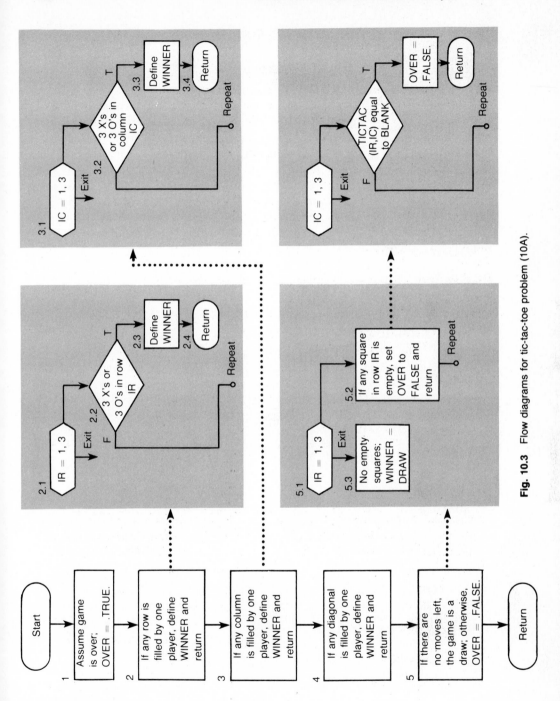

**Fig. 10.3**  Flow diagrams for tic-tac-toe problem (10A).

```
       SUBROUTINE CHKOVR (TICTAC, OVER, WINNER)
C
C CHECK IF TIC-TAC-TOE GAME IS OVER AND DETERMINE WINNER (IF ANY)
C
C ARGUMENT DEFINITIONS --
C   INPUT ARGUMENTS
C    TICTAC - REPRESENTS THE CURRENT STATE OF THE GAME BOARD
C   OUTPUT ARGUMENTS
C    OVER - INDICATES WHETHER OF NOT GAME IS OVER
C    WINNER - INDICATES THE WINNER (O OR X) OR A DRAW (D)
C
       CHARACTER * 1 TICTAC(3,3), WINNER
       LOGICAL OVER
C
C SUBROUTINE PARAMETERS
       CHARACTER * 1 BLANK, DRAW
       PARAMETER (BLANK = ' ', DRAW = 'D')
C
C FUNCTIONS USED
       LOGICAL SAME
C
C LOCAL VARIABLES
       LOGICAL DSAME
       INTEGER IR, IC
C ASSUME GAME IS OVER AT START
       OVER = .TRUE.
C
C CHECK FOR A WINNER
C CHECK ROWS FOR A WINNER
       DO 10 IR = 1, 3
          IF (SAME(TICTAC(IR,1), TICTAC(IR,2), TICTAC(IR,3))) THEN
             WINNER = TICTAC(IR,1)
             RETURN
          ENDIF
   10 CONTINUE
C NO WINNER BY ROWS, CHECK COLUMNS FOR WINNER
       DO 20 IC = 1, 3
          IF (SAME(TICTAC(1,IC), TICTAC(2,IC), TICTAC(3,IC))) THEN
             WINNER = TICTAC(1,IC)
             RETURN
          ENDIF
   20 CONTINUE
C NO WINNERS BY ROWS OR COLUMNS. CHECK DIAGONALS
       DSAME = SAME(TICTAC(1,1), TICTAC(2,2), TICTAC(3,3))
     Z .OR. SAME(TICTAC(1,3), TICTAC(2,2), TICTAC(3,1))
       IF (DSAME) THEN
          WINNER = TICTAC(2,2)
          RETURN
       ENDIF
C
C NO WINNER AT ALL. SEE IF GAME IS A DRAW
C CHECK EACH ROW FOR AN EMPTY SPACE
       DO 40 IR = 1, 3
          DO 45 IC = 1, 3
```

*(Continued)*

```
            IF (TICTAC(IR,IC) .EQ. BLANK) THEN
                OVER = .FALSE.
                RETURN
            ENDIF
    45      CONTINUE
    40 CONTINUE
C
C NO BLANK FOUND, GAME IS A DRAW
        WINNER = DRAW
C
        RETURN
        END
```

**Fig. 10.4**   Subroutine for Problem 10A.

*Program variables*

IR: Row subscript for array TICTAC;
used as lcv (integer)

IC: Column subscript for array TIC-
TAC; used as lcv (integer)

The refinements of Steps 2.2, 3.2 and 4 (Fig. 10.3) all involve the same operation—a comparison of the contents of three elements of the array TICTAC to see whether they are identical. (In Step 4 there are two diagonals to be checked.) To perform this operation, we will use a logical function SAME which will return a value of true if the three array elements are the same (and not blank), and will return false otherwise. The input arguments for SAME will be the three elements of TICTAC that are to be compared. With this lowest-level detail now handled, we can write the subroutine for Problem 10A (see Fig. 10.4). Additional data table entries are shown below.

**Additional Data Table Entries (Problem 10A):**

*Program variables*

DSAME: Defined to be true when a
diagonal is filled with 3 X's, or
3 O's. (logical)

*Subprograms referenced*

SAME    (logical function): Tests a row, column, or diagonal; returns a value of
true if all three elements are the same ('X' or 'O'); otherwise, returns a
value of false.

| *Arguments* | *Definition* |
|---|---|
| 1, 2, 3 | The arguments are the elements of a row, column, or diagonal of TICTAC. The order in which these elements are specified is immaterial. |

**Exercise 10.2:**   Write the function SAME (include all appropriate comments). *Note:* Make sure SAME properly handles the situation in which all three items being compared

are blank; the value returned should be true only if all three items are 'X' or all three are 'O'.

### 10.3.3    Reading, Printing, and Initializing Multidimensional Arrays

The use of an array name without subscripts in an input list causes the entire array to be filled with data items. Similarly, if the array name appears in an output list without subscripts, the entire array contents will be printed. A subscripted reference to an array in a READ or PRINT statement will affect only one element of the array.

If a two-dimensional array appears in an output list without subscripts, this array will be printed on a column-by-column basis (first column, second column, etc.). This means that the statement

$$\text{PRINT*, TICTAC}$$

would print the array TICTAC as shown next.

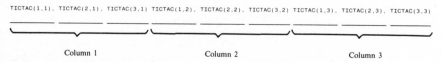

The compiler will position as many of these values as it can fit on each line in the order shown. The board in Example 10.2 would be printed as:

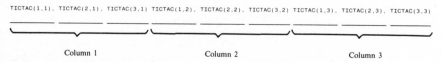

where the blanks represent empty spaces in the board.

Usually, however, we would prefer the contents of an array to be printed on a row-by-row basis. This, too, can be done fairly easily. The program segment shown below displays the array contents in the form of a tic-tac-toe board.

```
        PRINT*, '--------'
        DO 10 ROW = 1, 3
            PRINT 30, (TICTAC(ROW,COL), COL = 1, 3)
   30       FORMAT (1X, 3(1X, A1))
            PRINT*, '-------'
   10 CONTINUE
```

The PRINT statements in loop 10 will be executed once for each value of the row subscript, ROW. The implied DO loop causes the three elements of each row to be printed across the output line. The program output is shown in Fig. 10.5.

```
        -------
        X  O
        -------
          O
        -------
        X X O
        -------
```

**Fig. 10.5**  Printing the array TICTAC.

A similar problem exists in entering data. Normally, we prefer to enter one row of values per data card (or line). However, the compiler stores data items on a column-by-column basis if an unsubscripted array reference appears in an input list. Therefore, explicit and/or implied loops will be required to enter data on a row-by-row basis.

**Example 10.4:**   The subroutine below reads data items into one row of an integer array at a time and echoes these values.

```
         SUBROUTINE ENTER (MATRIX, NROWS, NCOLS)
C
C ENTER INTEGER DATA INTO AN ARRAY OF TWO DIMENSIONS
C ON A ROW-BY-ROW BASIS
C
C ARGUMENT DEFINITIONS --
C   INPUT ARGUMENTS
C      NROWS - NUMBER OF ROWS
C      NCOLS - NUMBER OF COLUMNS
C   OUTPUT ARGUMENTS
C      MATRIX - ARRAY TO RECEIVE DATA ITEMS
C
         INTEGER NROWS, NCOLS
         INTEGER MATRIX(NROWS, NCOLS)
C
C LOCAL VARIABLES
         INTEGER I, J
C
C READ CARDS ON A ROW BY ROW BASIS
         DO 10 I = 1, NROWS
            READ*, (MATRIX(I, J), J = 1, NCOLS)
            PRINT*, (MATRIX(I, J), J = 1, NCOLS)
      10 CONTINUE
C
         RETURN
         END
```

**Example 10.5:**   In Example 10.3, the array ENROLL (50 rows, 5 columns) is used for storing the course enrollment figures for each of fifty courses at five branch campuses of a large university. If a separate data card is keypunched for each course, with the enrollment figures broken down by campus, the statement

```
         CALL ENTER(ENROLL, 50, 5)
```

would properly process the data cards shown next.

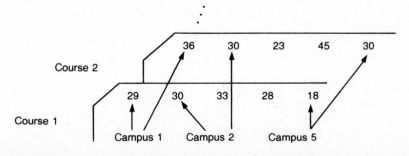

If the enrollment figures are keypunched by campus instead of by course, then the statement

```
READ*, ENROLL
```

would suffice. The data cards should be prepared as follows:

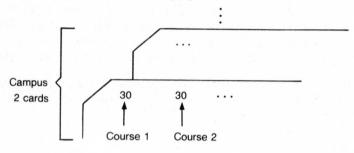

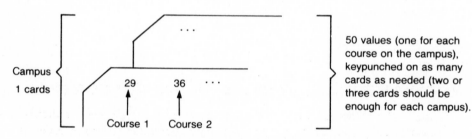

50 values (one for each course on the campus), keypunched on as many cards as needed (two or three cards should be enough for each campus).

The points illustrated in these examples are summarized in the next display.

> *In general, if a multidimensional array appears unsubscripted in a variable list of a READ or PRINT statement, the compiler will assume that the first subscript changes first and most often, then the second, then the third, etc. To alter this sequence, the programmer must utilize a suitable combination of implied or explicit DO loops.*

What we mean by the statement "the first subscript changes first and most often" is that the compiler processes the array as if the first subscript were the innermost loop control variable in a nest of loops, and the last subscript were the outermost loop control variable. Consequently, while all other subscripts are still set at their initial value, the first subscript cycles through all of its values. The second subscript is then incremented, and the first subscript cycles through all of its values again. This continues until subscript 2 reaches its final value. Then subscript 3 is incremented to its next value and the process repeats, with subscripts 1 and 2 cycling through their respective ranges.

**Example 10.6:**    The processing just described also applies for a data initialization statement. The statements:

```
  DATA TICTAC /'X', ' ', 'X', ' ', 'O', 'X', 'O', ' ', 'O'/
  DATA ((TICTAC(I, J), J = 1, 3), I = 1, 3)
Z          /'X', ' ', 'O', ' ', 'O', ' ', 'X', 'X', 'O'/
```

will both initialize the array TICTAC to the configuration shown in Fig. 10.5. The first statement initializes the array on a column-by-column basis. (The row subscript changes first and most often.) The second statement initializes the array on a row-by-row basis, using a pair of nested implied DO loops. The column subscript, J, is considered the inner loop control variable (J changes first and most often); the row subscript, I, is considered the outer loop control variable.

**Exercise 10.3:**  Given a square array TABLE with M rows and M columns, describe the effect of the following program segments. (Does every element get printed at least once? Do any elements get printed more than once?)

a)  `PRINT*, (TABLE(I, I), I = 1, M)`

b)
```
    DO 10 I = 1, M
        PRINT*, (TABLE(I, J), J = 1, I)
 10 CONTINUE
```

c)
```
    DO 20 I = 1, M
        PRINT*, (TABLE(I, J), J = I, M)
 20 CONTINUE
```

## 10.4  COMPILER ROLE FOR MULTIDIMENSIONAL ARRAYS*

Certainly the compiler has no capability to modify the physical configuration of memory or reorganize memory cells into a rectangular pattern in order to represent a two-dimensional array. Memory consists of a linear sequence of individual memory cells. The compiler allocates a block of adjacent cells for storing a multidimensional array, and must associate each element of the array with a particular cell, as shown in Fig. 10.6. The allocation of the block of memory is done on a column-by-column basis, so that the first column of the array occupies the first subblock of cells, the second column occupies the next subblock, etc.

The actual location in memory of each array element is determined by computing its *offset* from the *base address*. (The base address is the address of the memory cell assigned to the first element of the array.) The formula for this computation is

$$\text{array element address} = \text{base address of array} + \text{offset.}$$

The offset for TICTAC (I, J) is determined from the formula

$$\text{offset} = (I - 1) + (J - 1) * 3,$$

where 3 represents the number of rows in TICTAC. In general for two-dimensional arrays, the formula for computing the offset for element (I, J) is

$$\text{offset} = (I - \text{lower}_1) + (J - \text{lower}_2) * (\text{upper}_1 - \text{lower}_1 + 1)$$

---

*This section may be omitted.

where upper$_i$ represents the largest subscript value for dimension i and lower$_i$ represents the smallest subscript value. A similar formula can be derived for arrays of dimensions greater than two.

From either of these formulas, we can compute the offsets as shown in the following table.

| Array element | Offset | Address |
|---|---|---|
| TICTAC (1, 1) | 0 + 0*3 = 0 | base address |
| TICTAC (2, 2) | 1 + 1*3 = 4 | 4th cell after base |
| TICTAC (3, 3) | 2 + 2*3 = 8 | 8th cell after base |

A comparison of the offsets in this table with Fig. 10.6 verifies the correctness of these formulas.

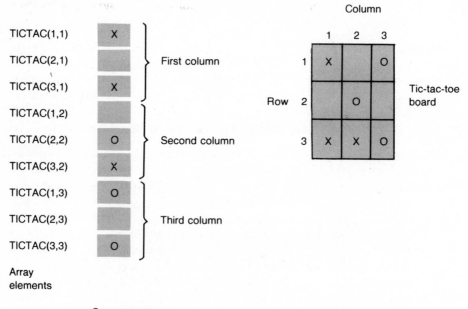

**Fig. 10.6**   Storage of a two-dimensional array in a contiguous memory block.

## 10.5   MULTIDIMENSIONAL ARRAYS AS SUBPROGRAM ARGUMENTS

Any array that is used as an argument in a subprogram must be declared in the subprogram as well as in the calling program. In the initial declaration, the range of each subscript must be specified using an integer constant, parameter, or constant expression. However, within the subprogram it is possible to use an integer dummy argument to specify the range of a subscript.

An argument that defines the range of a multidimensional array subscript must have the same value as was specified in the declaration of the actual array. This rule must be followed even when only a portion of an array is to be processed by a subprogram.

The reason for this rule is that subprograms are compiled separately and the translations will involve instructions for computing the address of each memory cell corresponding to an array reference. As shown in the preceding section, the memory cell address for a particular element in a two-dimensional array is a function of the range of the row subscript for that array. Consequently, the correct value of this range must be available to all subprograms that reference the array. Since the last dimension does not enter into this computation, it is not necessary to pass its range as a subprogram argument. FORTRAN 77 permits the use of an * to represent the upper bound (or range) of the last dimension instead of a constant or dummy argument. Thus for one-dimensional arrays, only the '*' is required in specifying the range.

**Example 10.7:**   The subroutine EXCHNG creates a modified version of any array such that row one of the original will be column one of the new array, column one of the original will be row one of the new array, etc. If the array represents a matrix, the modified version is called the *transpose* of the original matrix. A matrix and its transpose are shown in Fig. 10.7. Note that the old array (matrix) has two rows and four columns, while the new array (transpose) has four rows and two columns.

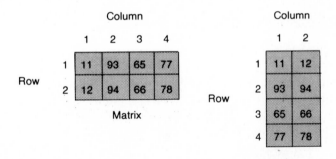

Fig. 10.7   A matrix and its transpose.

The subroutine EXCHNG is provided in Fig. 10.8.

Within loop 20 of EXCHNG, I serves as the row subscript for NEW and the column subscript for OLD; J serves as the column subscript for NEW and the row subscript for OLD. Column I of OLD is copied into row I of NEW, as J is sequenced from 1 to M in inner loop 20. This subroutine will transpose any two-dimensional integer array.

```
          SUBROUTINE EXCHNG (OLD, NEW, M, N)
C
C FORM THE TRANSPOSE OF AN INTEGER MATRIX
C ARGUMENT DEFINITIONS --
C   INPUT ARGUMENTS
C      OLD - ORIGINAL ARRAY (M X N)
C      M, N - RANGE OF ARRAY SUBSCRIPTS
C   OUTPUT ARGUMENTS
C      NEW - TRANSPOSE OF OLD (N X M)
C
          INTEGER M ,N
          INTEGER OLD(M, N), NEW(N, M)
C
C LOCAL VARIABLES
          INTEGER I, J
C
C FORM EACH ROW OF NEW
          DO 10 I = 1, N
C      COPY A COLUMN OF OLD INTO A ROW OF NEW
             DO 20 J = 1, M
                NEW(I, J) = OLD(J, I)
   20        CONTINUE
   10 CONTINUE
C
          RETURN
          END
```

**Fig. 10.8**  Subroutine EXCHNG for transposing a matrix.

# 10.6  APPLICATIONS OF MULTIDIMENSIONAL ARRAYS

## 10.6.1  Room Scheduling

To further illustrate the use of multidimensional arrays we will present two solved problems in which this data structure plays a central role.

**Problem 10B:**  The little red high school building in Sunflower, Indiana, has three floors, each with five classrooms of various sizes. Each semester the high school runs 15 classes that must be scheduled for the rooms in the building. We will write a program that, given the capacity of each room in the building and the size of each class, will attempt to find a satisfactory room assignment that will accommodate all 15 classes in the building. For those classes that cannot be satisfactorily placed, the program will print a "ROOM NOT AVAILABLE" message. The program will also indicate the number of leftover seats in each room as well as the total number of leftover seats for the entire building.

**Discussion:**  As part of the data table definition, we must decide how the table of room capacities is to be represented in the memory of the computer. Since the building may be pictured as a two-dimensional structure with three floors (vertical dimension) and five rooms (horizontal dimension), a two-dimensional array should be a convenient structure for representing the capacities of each room in the building. We will read the room capacities into a 3x5 array, RMCAP, as shown in Fig. 10.9.

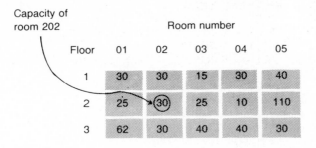

**Fig. 10.9** Room capacities for Sunflower High.

By using a two-dimensional array, we will be able to determine the number of the room assigned to each class directly from the indices of the array element that represents that room. For example, if a class is placed in a room with capacity given by RMCAP (2, 4), we know that the number of this room is 204. In general, RMCAP (I, J) represents the capacity of the room whose number is the value of the expression

$$I \ * \ 100 \ + \ J$$

The data table follows; the level one algorithm for Problem 10B is shown in Fig. 10.10.

### Data Table for Problem 10B

*Parameters*

NUMCLS = 15, number of classes scheduled
NUMROW = 3, number of floors
NUMCOL = 5, number of rooms per floor

| *Input variables* | *Output variables* |
|---|---|
| RMCAP: An array used to store the capacities for each room (integer, size 3 × 5) | RMCAP: Revised room capacity table |
| | LFTOVR: Number of seats left over after all classes are assigned (integer) |

The program system chart for this problem is shown in Fig. 10.11, and the main program is shown in Fig. 10.12. We will handle Steps 2 and 4 of the algorithm through the use of a subroutine, PRTCAP, which will print a two-dimensional integer table in a readable form (see Exercise 10.5). Step 3 will be

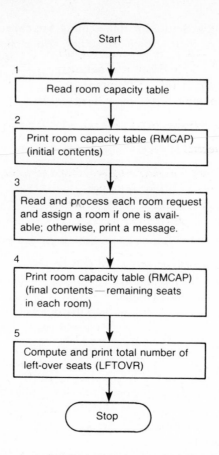

**Fig. 10.10** Level one flor diagram for room scheduling problem (10B).

performed by the subroutine PROCES, which will read and process room requests and print the room assigned, if any. When PROCES assigns a room to a class, it will determine the number of left-over seats in the room, and store this result back into RMCAP, in place of the original size of the room. Step 5 will be taken care of in the subroutine TOTAL. These subroutines are described next.

## Additional Data Table Entries for Problem 10B

*Subroutines referenced*

PRTCAP:   Prints the contents of the room capacity table

  *Arguments (all integers)*

    1. Array to be printed (input)
    2. Number of rows in the array (input)
    3. Number of columns in the array (input)

PROCES:   Reads and processes each room request consisting of a class ID and a class size. Determines the room number to be assigned (if one is available) and prints the room number.

*Arguments (all integers)*

1. Number of classes to be assigned (input)
2. Room capacity table (input and output)
3. Number of rows in the array (input)
4. Number of columns in the array (input)

TOTAL:    Computes the sum of leftover seats for the entire building

*Arguments (all integers)*

1. Array containing the remaining seats for each room (input)
2. Number of rows in the array (input)
3. Number of columns in the array (input)
4. Total number of leftover seats (output)

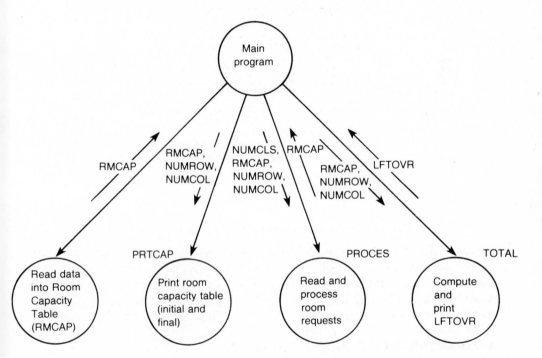

**Fig. 10.11**   Program system chart for room scheduling problem (10B).

For each room request, subroutine PROCES will read a pair of data items representing the course identification number and class size. Subroutine PROCES should find a room that is large enough to hold each class if one is available. (The ideal situation would be to find a room whose capacity exactly matches the class size.) For each class, PROCES will print, in tabular form, the class ID (CLASID) and size (CLSIZE), and the number (ROOMNO) and capacity (as stored in RMCAP) of the room assigned to the class. The local variables for PROCES are shown in the following data table; the level one algorithm is shown in Fig. 10.13.

```
C SCHEDULE CLASSES FOR THE LITTLE RED HIGH SCHOOL BUILDING
C
C PROGRAM PARAMETERS
      PARAMETER (NUMROW = 3, NUMCOL = 5, NUMCLS = 15)
C
C VARIABLES
      INTEGER RMCAP(NUMROW, NUMCOL)
      INTEGER LFTOVR
C
C READ IN ROOM CAPACITY TABLE
      READ*, RMCAP
C
C PRINT INITIAL CONTENTS OF ROOM CAPACITY TABLE
      PRINT '(A)', '1ROOM CAPACITY TABLE, INITIAL CONTENTS'
      CALL PRTCAP(RMCAP, NUMROW, NUMCOL)
C
C READ AND PROCESS ROOM REQUESTS
      PRINT '(A)', '0ROOM ASSIGNMENT TABLE'
      PRINT '(A)', ' CLASS ID     SIZE      ROOM      CAPACITY'
      CALL PROCES(NUMCLS, RMCAP, NUMROW, NUMCOL)
C
C PRINT FINAL CONTENTS OF ROOM CAPACITY TABLE (LEFT OVER SEATS)
      PRINT '(///A)', ' REMAINING SEATS IN EACH ROOM'
      CALL PRTCAP(RMCAP, NUMROW, NUMCOL)
C
C COMPUTE AND PRINT TOTAL NUMBER OF LEFT OVER SEATS
      CALL TOTAL(RMCAP, NUMROW, NUMCOL, LFTOVR)
      PRINT '(A, I3)', '0THE NUMBER OF REMAINING SEATS IS ',
     Z LFTOVR
C
      STOP
      END
```

**Fig. 10.12**   Program for Problem 10B.

## Data Table for Subroutine PROCES

*Input arguments*

NUMCLS: Number of classes to be processed

RMCAP: Initial room capacity table

NUMROW: Number of rows in RMCAP

NUMCOL: Number of columns in RMCAP

*Output arguments*

RMCAP: Final room capacity table

| *Input variables* | *Program variables* | *Output variables* |
|---|---|---|
| CLASID: Identification code for each class (character * 4) | CLASS: Loop control variable for loop to process each class (integer) | ROOMNO: Number of the room assigned to each class (also required as output are the capacity of the room assigned and CLASID and CLSIZE) (integer) |
| CLSIZE: Size of each class (integer) | | |

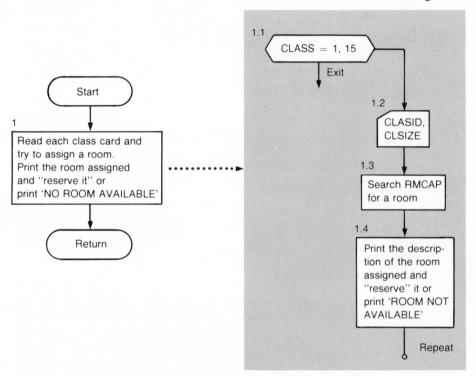

**Fig. 10.13**   Flow diagrams for Subroutine PROCES.

A third level subroutine, ASSIGN, will be called by PROCES to perform Step 1.3. This subroutine will search the room capacity table to find a room of size, CLSIZE, or greater. It will return the indices of an assigned room, if one is found, and indicate success or failure by setting a program flag, FOUND, to TRUE or FALSE. The additional data table entries for PROCES are shown next, along with a description of subroutine ASSIGN. The data flow between PROCES and ASSIGN is summarized in the program system chart drawn in Fig. 10.14.

**Additional Data Table Entries for Subroutine PROCES**

*Program variables*

FOUND: Program flag defined
by call to ASSIGN.
Set to .TRUE. if a room is
available for a class;
otherwise, set to .FALSE.
(logical)

FLOOR, ROOM: Indices specifying room to be assigned
to a class; returned by
ASSIGN if a room is found
(integer)

*Subroutine referenced*

ASSIGN: Assigns a suitable room if one is available

*Arguments*

1 :   Room capacity table (input, integer array)
2,3 : Number of rows, columns in argument 1 (input, integer)
4 :   Size of class being assigned (input, integer)
5,6 : Row, column index of assigned room (output, integer)
7 :   Program flag indicating success or failure of room
        assignment (output, logical)

The refinement of Step 1.4 of PROCES is shown in Fig. 10.15.

There are probably many ways to resolve the problem indicated in Step
1.4.3 of Fig. 10.15. Once a room (with capacity RMCAP(i,j)) is assigned, we
must ensure that it cannot be reassigned to another class. We will provide this
protection simply by negating the number of remaining seats (to be saved in
RMCAP(i,j)) for the room when the assignment is made. Exactly why this works
will become clearer when subroutine ASSIGN is written. We can now write
PROCES (see Fig. 10.16).

The only step left is the specification of subroutine ASSIGN. The algorithm
that we will use to find a room for a given class size (CLSIZE) may be summarized as follows:

Search the array RMCAP to find the smallest room that is
greater than or equal to CLSIZE and is still not assigned.

This is called the *best fit* algorithm because the unassigned room with the least
excess capacity is chosen for each class. The ideal situation is to find a room that
fits exactly. This algorithm assigns as many classes to suitable rooms as is physically possible without later juggling room assignments. The implementation of
this search requires two nested loops with loop control variables ROW and

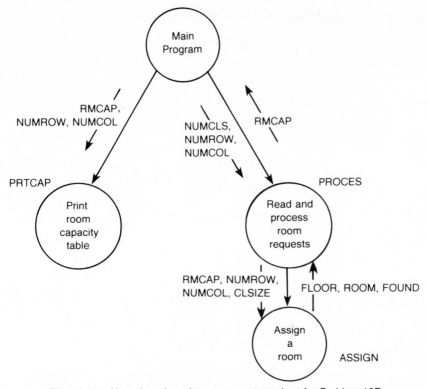

**Fig. 10.14**  Altered portion of program system chart for Problem 10B.

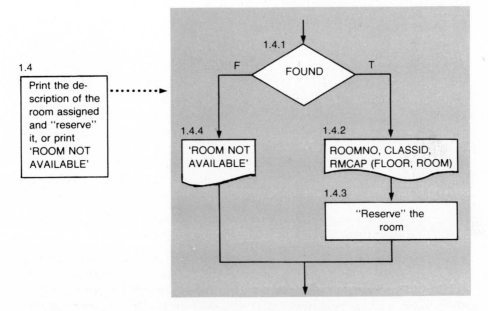

**Fig. 10.15**  Refinement of Step 1.4 of PROCESS.

```
        SUBROUTINE PROCES (NUMCLS, RMCAP, NUMROW, NUMCOL)
C
C READ AND PROCESS ROOM REQUESTS
C
C ARGUMENT DEFINITIONS --
C   INPUT ARGUMENTS
C     NUMCLS - NUMBER OF CLASSES TO BE PROCESSED
C     NUMROW - NUMBER OF ROWS IN THE ARRAY
C     NUMCOL - NUMBER OF COLUMNS IN THE ARRAY
C   IN-OUT ARGUMENTS
C     RMCAP - INITIAL AND FINAL ROOM CAPACITY ARRAY
C
        INTEGER NUMCLS, NUMROW, NUMCOL
        INTEGER RMCAP(NUMROW, NUMCOL)
C
C LOCAL VARIABLES
        CHARACTER * 6 CLASID
        INTEGER CLASS, CLSIZE, ROOMNO
        LOGICAL FOUND
        INTEGER FLOOR, ROOM
C
C READ AND PROCESS ALL ROOM REQUESTS
        DO 10 CLASS = 1, NUMCLS
           READ*, CLASID, CLSIZE
           CALL ASSIGN(RMCAP, NUMROW, NUMCOL, CLSIZE,
     Z               FLOOR, ROOM, FOUND)
           IF (FOUND) THEN
C             COMPUTE ROOM NUMBER AND RESERVE ROOM
              ROOMNO = FLOOR * 100 + ROOM
              PRINT '(1X,A6,3I10)', CLASID, CLSIZE, ROOMNO,
     Z            RMCAP(FLOOR, ROOM)
              RMCAP(FLOOR, ROOM) = -(RMCAP(FLOOR, ROOM) - CLSIZE)
           ELSE
C             PRINT MESSAGE
              PRINT '(A,A6)', ' NO ROOM ASSIGNED FOR CLASS ',
     Z            CLASID
           ENDIF
     10 CONTINUE
C
        RETURN
        END
```

**Fig. 10.16**   PROCES subroutine for Problem 10B.

COL. The data table for ASSIGN follows; the flow diagram is shown in Fig. 10.17.

**Data Table for Subroutine ASSIGN**

| *Input arguments* | *Output arguments* |
|---|---|
| RMCAP: Represents a two-dimensional integer array to be searched (size determined by arguments NUMROW and NUMCOL) | FOUND: Represents a flag used to indicate whether or not a room has been found. FOUND is set to true if a room is found and false otherwise (logical) |

NUMROW: Represents the number of rows in RMCAP (integer)

NUMCOL: Represents the number of columns in RMCAP (integer)

CLSIZE: Represents the size of the class to be assigned to a room (integer)

FLOOR, ROOM: Represent the indices of the entry in RMCAP corresponding to the room in which the class will be placed (integer)

*Program variables*

ROW: Outer loop control variable and row subscript (integer)

COL: Inner loop control variable and column subscript (integer)

The flow diagrams shown in Fig. 10.17 use the following criteria to locate the room with smallest capacity that is larger than CLSIZE.

- If a room is found with a capacity equal to CLSIZE, this room is chosen as the best-fit room, and the search is complete (Step 2.2.2).

- When the first room with capacity larger than CLSIZE is found, it is chosen to be the best-fit room (Step 2.2.3). If, subsequently, a room of sufficient capacity is located that is smaller than the current best-fit room, the new room becomes the best-fit room (Step 2.2.4).

After all rooms have been examined, the row and column indices of the best-fit room are returned to subroutine PROCES. It will be impossible for this room to be reassigned to another class as its corresponding element of array RMCAP will be negative.

We will implement Steps 2.2.2, 2.2.3, and 2.2.4 using nested decision structures. The program for subroutine ASSIGN is shown in Fig. 10.18a. A sample run of the program is shown in Fig. 10.18b.

**Exercise 10.4:**   Rewrite the data entry statement for RMCAP (main program, Fig. 10.12) to initialize RMCAP on a row-by-row basis.

**Exercise 10.5:**
   a) Write the subroutine TOTAL to compute the total number of seats left over for the entire school building. Remember that for rooms that have been assigned to a class, the negation of the number of remaining seats is stored in RMCAP rather than the number itself.
   b) Complete the program system chart for Problem 10B by writing the subroutine PRTCAP. Your subroutine should produce a nicely labelled, two-dimensional table with each line representing the capacities of the rooms on one floor of the school building. A data table and a flow diagram should be provided.

**Exercise 10.6:**   The algorithm used in subroutine ASSIGN is called a *best-fit* algorithm, because the room having the capacity that was closest to class size was assigned to each class. Another algorithm that might have been used is called a *first-fit* algorithm. In this algorithm, the first room having a capacity greater than or equal to class size is assigned

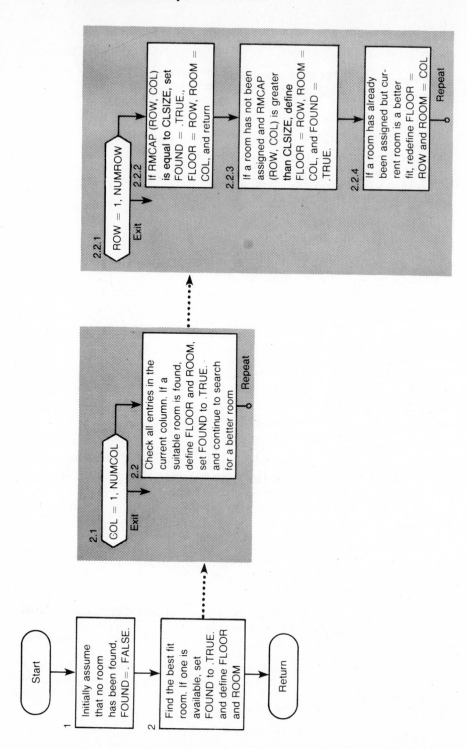

**Fig. 10.17**  Flow diagrams of ASSIGN subroutine for room scheduling problem (10B).

```
      SUBROUTINE ASSIGN (RMCAP, NUMROW, NUMCOL, CLSIZE, FLOOR,
     Z ROOM, FOUND)
C
C USE BEST FIT ALGORITHM TO FIND SMALLEST ELEMENT IN RMCAP
C THAT IS GREATER THAN OR EQUAL TO CLSIZE
C
C ARGUMENT DEFINITIONS --
C   INPUT ARGUMENTS
C     RMCAP - TWO DIMENSIONAL INTEGER ARRAY TO BE SEARCHED
C     NUMROW - NUMBER OF ROWS IN RMCAP
C     NUMCOL - NUMBER OF COLUMNS IN RMCAP
C     CLSIZE - SIZE OF CLASS TO BE ASSIGNED TO A ROOM
C   OUTPUT ARGUMENTS
C     FOUND - TRUE IF A ROOM IS FOUND, FALSE OTHERWISE
C     FLOOR - INDICATES ROW OF RMCAP ENTRY WITH BEST FIT
C     ROOM - INDICATES COLUMN OF RMCAP ENTRY WITH BEST FIT
C
      INTEGER NUMROW, NUMCOL, CLSIZE
      INTEGER RMCAP(NUMROW, NUMCOL)
      LOGICAL FOUND
      INTEGER FLOOR, ROOM
C
C LOCAL VARIABLES
      INTEGER ROW, COL
C
C ASSUME NO ROOM FOUND AT START
      FOUND = .FALSE.
C OUTER LOOP, CHECK COLUMN ENTRIES
      DO 30 COL = 1, NUMCOL
C         INNER LOOP, CHECK ROWS OF EACH COLUMN
          DO 20 ROW = 1, NUMROW
             IF (RMCAP(ROW,COL) .GE. CLSIZE) THEN
                IF (RMCAP(ROW,COL) .EQ. CLSIZE) THEN
C                  SIZE MATCH-- MAKE ASSIGNMENT OF BEST FIT ROOM
                   FOUND = .TRUE.
                   FLOOR = ROW
                   ROOM = COL
                   RETURN
                ELSE IF (.NOT. FOUND) THEN
C                  FIRST AVAILABLE ROOM FOUND
                   FOUND = .TRUE.
                   FLOOR = ROW
                   ROOM = COL
                ELSE IF (RMCAP(ROW,COL) .LT. RMCAP(FLOOR,ROOM)) THEN
C                  SUBSEQUENT AVAILABLE ROOM-- BETTER FIT
                   FLOOR = ROW
                   ROOM = COL
                ENDIF
             ENDIF
   20     CONTINUE
   30 CONTINUE
C
      RETURN
      END
```

**Fig. 10.18a**   ASSIGN subroutine for Problem 10B.

```
        ROOM CAPACITY TABLE, INITIAL CONTENTS
                      ROOM NUMBER
   FLOOR    01    02    03    04    05
      1     30    30    15    30    40
      2     25    30    25    10   110
      3     62    30    40    40    30

   ROOM ASSIGNMENT TABLE
   CLASS ID      SIZE      ROOM      CAPACITY
   CL01           38        303         40
   CL02           41        301         62
   CL03            6        204         10
   CL04           26        101         30
   CL05           28        102         30
   CL06           21        201         25
   CL07           25        203         25
   CL08           97        205        110
   CL09           12        103         15
   CL10           36        304         40
   CL11           28        202         30
   CL12           27        302         30
   CL13           29        104         30
   CL14           30        305         30
   CL15           18        105         40

   REMAINING SEATS IN EACH ROOM
                      ROOM NUMBER
   FLOOR    01    02    03    04    05
      1      4     2     3     1    22
      2      4     2     0     4    13
      3     21     3     2     4     0

   THE NUMBER OF REMAINING SEATS IS   85
```

**Fig. 10.18b**    Sample run for Problem 10B.

to the class (no further searching for a room is carried out). Modify the flow diagram (Fig. 10.17) and program (Fig. 10.18) to reflect the first-fit algorithm. (You will see that this algorithm is simpler than best-fit.) Apply both algorithms using the room capacities shown earlier and the following 15 class sizes: 38, 41, 6, 26, 28, 21, 25, 97, 12, 36, 28, 27, 29, 30, 18. Exactly what is wrong with the first-fit algorithm?

**Exercise 10.7:**    *Continuation of Exercise 10.6.* Modify the first-fit algorithm for your subroutine ASSIGN to include steps to first sort the class size data. Write a subroutine to perform the sort, and don't forget to move class ID's as you arrange the corresponding sizes. Does this modification overcome the problem encountered when using the first-fit algorithm without the sort? Why?

### 10.6.2    Introduction to Computer Art: Drawing Block Letters

Many of you have seen examples of computer art or calendars "drawn" by the computer. Normally the picture consists of lines of numbers or symbols printed so as to depict a pattern. The pattern is composed of different layers of

**Fig. 10.19**  Computer art.

shading and the degree of shading is determined by the density of symbols printed in a given area. (See Fig. 10.19). Since the line printer prints one line at a
time, the "computer artist" must organize the picture as a sequence of print lines
of constant width.

**Example 10.8:**    The program in Fig. 10.20 enters a picture into an array
(PICTUR) and then draws N copies of the picture. The number of lines in the
picture is given by the variable **HEIGHT** (**HEIGHT** $\leq$ 100); the width is 60
characters.

```
C COMPUTER ART PROGRAM
C
      INTEGER N, HEIGHT, NCOPY, NLINE
      CHARACTER * 60 PICTUR(100)
C
C ENTER DATA
      READ*, N, HEIGHT
      READ '(A)', (PICTUR(NLINE), NLINE = 1, HEIGHT)
C
C DRAW N COPIES
      DO 10 NCOPY = 1, N
C         DRAW ONE LINE OF PICTURE AT A TIME
          DO 20 NLINE = 1, HEIGHT
              PRINT*, PICTUR(NLINE)
   20     CONTINUE
C         SEPARATE EACH COPY WITH BLANK LINES
          PRINT 30
   30     FORMAT (//////)
   10 CONTINUE
C
      STOP
      END
```

**Fig. 10.20**  Program to draw a picture.

In the next problem, we will write a program which draws large block letters on the line printer. You may have already seen examples of the output of
such a program in signs or announcements printed by the computer. Block letters are also often used in identifying the name of the owner of a program listing.

**Problem 10C:**  Develop a program that prints a sequence of letters, provided as
input data, in large block letters across a page. Each letter should be printed as a
6 × 5 grid pattern of X's and blanks (6 rows and 5 columns). We will skip two
print columns between letters. If your line printer provides 120 print columns of
output, a maximum of 17 letters can be displayed (17 × 7 = 119).

For example, given the input string 'DRAW ME' the program should produce:

```
XXX       XXXX         X        X    X       X X     XXXXX
X   X     X    X     X   X      X    X      X X X     X
X     X   XXXX       X     X    X    X      X   X     XXXX
X     X   X  X       XXXXX      X X X       X     X   X
X   X     X    X     X     X    X X X       X     X   X
XXX       X      X   X     X     X X        X     X   XXXXX
```

**Discussion:** We will use a two-dimensional character array to store all of the block letters including a space (blank). Each line of a letter will be stored as a character string of width five in the array BLOCK declared as

```
CHARACTER * 5 BLOCK(6, 0:26)
```

In referencing BLOCK, the second subscript denotes a particular letter of the alphabet (1 for A, 2 for B, 3 for C, . . . 26 for Z, and 0 for the blank), and the first subscript denotes a line of the block letter. The first subscript has an upper bound of six since there are six lines in each block letter.

    The block diagram patterns for each letter will be defined through the use of a rather lengthy list of 27 data statements. Some examples of these statements follow; the storage arrangement that they produce is shown in Fig. 10.21. An analysis of this figure should provide a clear picture of the way in which the information in BLOCK will be arranged.

**Data Statements:**

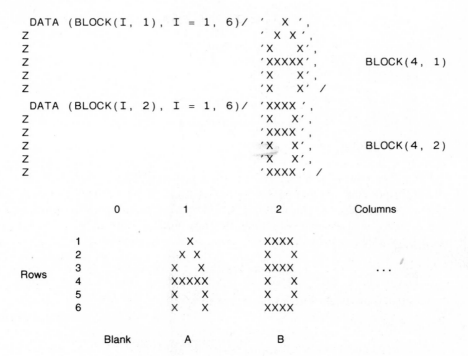

Fig. 10.21    The array BLOCK.

    Now that we have an understanding of the main data structure to be used in solving the problem, we can begin writing a level one algorithm for producing the desired block letter output. The data table is next; the level one flow diagram for this algorithm is shown in Fig. 10.22a.

**Data Table for Block Letter Program (Problem 10C)**

| *Input variables* | *Program variables* | *Output variables* |
|---|---|---|
| STRING: Contains the string of characters to be printed in block form (character * 1, size 17) | BLOCK: Array of block letters (character * 5, size $6 \times 0{:}26$) | Selected elements from the array BLOCK |

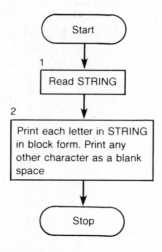

**Fig. 10.22a**   Level one flow diagram for block letter program (Problem 10C).

In the refinement of Step 2, the library function INDEX will be used to identify each letter in STRING by searching the character string ALFBET('AB . . . Z'). The relative position of each input character in ALFBET will determine the corresponding column of the array BLOCK to be printed. For each input character that is not a letter, INDEX will return a value of zero, and the block character for a blank will be printed. Rather than search for all input characters each time a line is printed, we will perform the search first and save the search results in array XLIST. The new data table entries for the Step 2 refinement follow; the refinement is shown in Fig. 10.22b.

**Additional Data Table Entries for Problem 10C**

*Program parameters*

ALFBET = 'AB . . . Z', the alphabet string (character * 26)

*Program variables*

XLIST: The array of indices corresponding to the characters in STRING (integer, size 17)

NEXT: Selects the current
character (integer)

ROW: The row of array
BLOCK being printed
(integer)

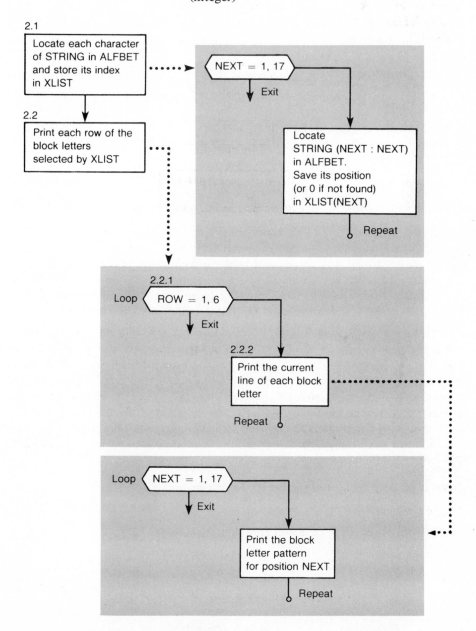

**Fig. 10.22b**   Flow diagram for refinement of Step 2 (Fig. 10.22a).

If the input string is 'DRAW ME', Step 2.1 would define the array **XLIST** as follows:

| XLIST(1) | XLIST(2) | XLIST(3) | XLIST(4) | XLIST(5) | XLIST(6) | XLIST(7) | XLIST(8) | ... | XLIST(17) |
|---|---|---|---|---|---|---|---|---|---|
| 4 | 18 | 1 | 23 | 0 | 13 | 5 | 0 | | 0 |
| D | R | A | W | | M | E | | | |

Step 2.2 would then print the strings in rows 1 through 6 of columns 4, 18, 1, 23, 0, 13, and 5 of the array **BLOCK** (followed by ten copies of the entry in column 0 of BLOCK (the blank)).

For this example, the array elements are printed as indicated below:

position 1      position 2      position 3 . . . position 17

Line 1      BLOCK(1,4), BLOCK(1,18), BLOCK(1,1), . . .BLOCK(1,0)
Line 2      BLOCK(2,4), BLOCK(2,18), BLOCK(2,1), . . .BLOCK(2,0)

Line 6      BLOCK(6,4), BLOCK(6,18), BLOCK(6,1), . . .BLOCK(6,0)

In the program (Fig. 10.23), note that loop 2.2.2 is implemented as an implied DO loop (nested within an explicit DO loop). This is to ensure that all of the character strings that correspond to a particular row of the block-letter pattern are displayed on the same line of computer printout. If an explicit DO were used instead, these character strings would be printed on separate lines. Format 29 specifies single spacing between lines and two blank print columns between letters.

```
C PRINT THE BLOCK LETTER PATTERN FOR AN INPUT STRING
C
C PROGRAM PARAMETERS
      CHARACTER * 26 ALFBET
      PARAMETER (ALFBET = 'ABCDEFGHIJKLMNOPQRSTUVWXYZ')
C
C VARIABLES
      INTEGER ROW, NEXT, XLIST(17)
      CHARACTER * 5 BLOCK(6, 0:26)
      CHARACTER * 17 STRING
      CHARACTER * 1 LETTER
C INITIALIZE ARRAY BLOCK
      DATA (BLOCK(I, 0), I = 1, 6)/ '     ',
     X                             '     ',
     X                             '     ',
     X                             '     ',
     X                             '     ',
     X                             '    '/
```

*(Continued)*

```
       DATA (BLOCK(I, 1), I = 1, 6)/ '  X  ',
      X                               ' X X ',
      X                               'X   X',
      X                               'XXXXX',
      X                               'X   X',
      X                               'X   X'/
       DATA (BLOCK(I, 2), I = 1, 6)/ 'XXXX ',
      X                               'X   X',
      X                               'XXXX ',
      X                               'X   X',
      X                               'X   X',
      X                               'XXXX '/
                      .
                      .
                      .
       DATA (BLOCK(I,26), I = 1, 6)/ 'XXXXX',
      X                               '    X',
      X                               '   X ',
      X                               '  X  ',
      X                               ' X   ',
      X                               'XXXXX'/
C
C READ DATA STRING
       READ '(A17)', STRING
C
C FIND THE COLUMN INDICES IN THE BLOCK CORRESPONDING TO EACH
C CHARACTER IN THE STRING
       DO 10 NEXT = 1, 17
           XLIST(NEXT) = INDEX(ALFBET, STRING(NEXT : NEXT))
    10 CONTINUE
C
C PRINT EACH ROW OF THE BLOCK LETTERS SELECTED BY XLIST
       DO 20 ROW = 1, 6
C          PRINT 17 LETTERS ACROSS A LINE.  THE COLUMN SUBSCRIPT
C          OF EACH ELEMENT OF BLOCK PRINTED IS FOUND IN XLIST
C          THE ROW SUBSCRIPT IS THE NUMBER OF THE LINE BEING PRINTED.
           PRINT 29, (BLOCK(ROW, XLIST(NEXT)), NEXT = 1, 17)
    20 CONTINUE
    29 FORMAT (1X, 17(A5, 2X))
C
       STOP
       END
```

**Fig. 10.23**   Program for Problem 10C.

**Exercise 10.8:**   Describe how you would modify the program to eliminate printing trailing blanks.

## 10.7   COMMON PROGRAMMING ERRORS

The errors encountered using multidimensional arrays are similar to those encountered in processing one-dimensional arrays. The most frequent errors are likely to be subscript range errors. These errors may be more common now because multiple subscripts are involved in an array reference, introducing added complexity and confusion. Your compiler should print a diagnostic message if a subscript range error occurs during execution. You should verify that the subscript boundary values are correct particularly for complicated subscript expressions.

Other kinds of errors arise because of the complex nesting of DO loops and implied DO loops when they are used to manipulate multidimensional arrays. Care must be taken to ensure that the subscript order is consistent with the nesting structure of the loops. Inconsistent usage will cause a program error.

An additional source of error involves the use of subprogram arguments in dummy array declarations. If the range of a subscript is passed through the argument list, care must be taken to ensure that the value passed is correct. Otherwise, the address computation performed within the subroutine will cause the wrong array elements to be manipulated and out-of-range errors may occur.

## 10.8   SUMMARY

In this chapter, we have introduced a more general form of the array. This form is useful in representing data that are most naturally thought of in terms of multidimensional structures. The multidimensional array is convenient for representing rectangular tables of information, matrices, game-board patterns, $n$-dimensional spaces, and tables involving data with multiple category decompositions (such as that found in the university enrollment example, Example 10.3).

We have seen examples of the manipulation of individual array elements through the use of nests of DO loops. The correspondence between the loop control variables and the array subscripts determines the order in which the array elements are processed.

Reading, printing and initializing multidimensional arrays were also described in this chapter. Specifically, we saw that an unsubscripted reference to a two-dimensional array in a READ or PRINT statement causes data to be manipulated on a column-by-column basis. We have written nested loops that can be used to read and write data on a row-by-row basis instead.

## PROGRAMMING PROBLEMS

**10D**   Write a program that reads in a tic-tac-toe board and determines the best move for player X. Use the following strategy: Consider all squares that are empty and evaluate potential moves into them. If the move fills the third square in a row, column, or diagonal that already has two X's, add 50 to the score; if it fills the third square in a row, column, or diagonal with two O's, add 25 to the score; for each row, column, or diagonal containing this move that will have two X's and one blank, add 10 to the score; add 8 for each row, column, or diagonal through this move that will have one O, one X, and one blank; add four for each row, column, or diagonal which will have one X and the rest blanks. Select the move that scores the highest.

The possible moves for the board below are numbered. Their scores are shown to the right of the board. Move 5 is selected.

|   | O | X |
|---|---|---|
| 2 | X | 3 |
| O | 4 | 5 |

(with "1" to the left of the first row)

$1-10 + 8 = 18$
$2-10 + 8 = 18$
$3-10 + 10 = 20$
$4-8$
⑤$-10 + 10 + 8 = 28$

**10E** Each card of a poker hand will be represented by a pair of integers: the first integer represents the suit; the second integer represents the value of the card. For example, 4, 10 would be the 10 of spades, 3, 11 the jack of hearts, 2, 12 the queen of diamonds, 1, 13 the king of clubs, 4, 14 the ace of spades. Read five cards in and represent them in a $4 \times 14$ array. A mark should be placed in the five array elements with row and column indices corresponding to the cards entered. Evaluate the poker hand. Provide subroutines to determine whether the hand is a flush (all one suit), a straight (five consecutive cards of different suits), a straight flush (five consecutive cards of one suit), 4 of a kind, a full house (3 of one kind, 2 of another), 3 of a kind, 2 pair, or 1 pair.

**10F** Represent the cards of a bridge hand by a pair of integers, as described in Problem 10E. Read the thirteen cards of a bridge hand into a $4 \times 14$ array. Compute the number of points in the hand. Score 4 for each ace, 3 for a king, 2 for a queen, 1 for a jack. Also, add 3 points for any suit not represented, 2 for any suit with only one card that is not a face card (jack or higher), 1 for any suit with only two cards, neither of which is a face card.

**10G** *Continuation of Problem 10B.* If, in the room scheduling problem (10B), we removed the restriction of a single building, and wished to write the program to accommodate an entire campus of buildings, each with varying numbers of floors and varying numbers of available rooms on each floor, the choice of a two-dimensional array for storing room capacities may prove inconvenient. Instead, we would have to use two *parallel arrays*, RMID and RMSIZE to store the identification of each room (building and number) and its size. Write a program, with appropriate subroutines, to solve the room scheduling problem using the 15 class sizes given in Exercise 10.6, and the campus room table shown below.

|  Room ID |  |  |
|---|---|---|
| *Building* | *Number* | *Room size* |
| HUMA | 1003 | 30 |
| MATH | 11 | 25 |
| MUSI | 2 | 62 |
| LANG | 701 | 30 |
| MATH | 12 | 30 |
| ART | 2 | 30 |
| EDUC | 61 | 15 |
| HUMA | 1005 | 25 |
| ART | 1 | 40 |
| ENG | 101 | 30 |
| MATH | 3 | 10 |
| EDUC | 63 | 40 |
| LANG | 702 | 40 |
| MUSI | 5 | 110 |
| HUMA | 1002 | 30 |

**10H**  Implement the block letter program using only the letters A through E as characters that may be printed.

**10I**  Instead of block letters, store a collection of stick figures in an array. Assign each stick figure a single-digit identification number. Your input data cards should indicate the stick figure to be drawn and its relative placement on the output page.

**Example:** For the three data cards below,

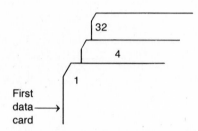

Figure 1 will be drawn above and to the left of Fig. 4. Figures 3 and 2 will be placed next to each other at the bottom of the drawing. Their horizontal displacement will be such that they will be drawn between Figs. 1 and 4.

**10J**  Write a set of subroutines to manipulate a pair of matrices. You should provide subroutines for addition, subtraction, and multiplication. Each subroutine should validate its input arguments (i.e., check all matrix dimensions) before performing the required data manipulation.

**10K**  The results from the mayor's race have been reported by each precinct as follows, one input card per precinct:

| Precinct | Candidate A | Candidate B | Candidate C | Candidate D |
|---|---|---|---|---|
| 1 | 192 | 48 | 206 | 37 |
| 2 | 147 | 90 | 312 | 21 |
| 3 | 186 | 12 | 121 | 38 |
| 4 | 114 | 21 | 408 | 39 |
| 5 | 267 | 13 | 382 | 29 |

Write a program to do the following:
a) Print out the table with appropriate headings for the rows and columns.
b) Compute and print the total number of votes received by each candidate and the percent of the total votes cast.
c) If any one candidate received over 50% of the votes, the program should print a message declaring that candidate the winner.
d) If no candidate received over 50% of the votes, the program should print a message declaring a run-off between the two candidates receiving the largest number of votes; the two candidates should be identified by their letter names.
e) Run the program once with the above data and once with candidate C receiving only 108 votes in precinct 4.

The card format is as follows:

> Card column 1 : precinct number
> Card columns 3–5: candidate A's votes
> Card columns 7–9: candidate B's votes
> Card columns 11–13: candidate C's votes
> Card columns 15–17: candidate D's votes

**10L**    The game of Life, invented by John H. Conway, is supposed to model the genetic laws for birth, survival, and death. (See *Scientific American*, October 1970, p. 120.) We will play it on a board consisting of 25 squares in the horizontal and vertical directions. Each square can be empty or contain an X indicating the presence of an organism. Every square (except the border squares) has eight neighbors. The small square shown in the segment of the board drawn below connects the neighbors of the organism in row 3, column 3.

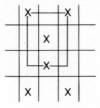

Read in an initial configuration of organisms. Print the original game array, calculate the next generation of organisms in a new array, copy the new array into the original game array, and repeat the cycle for as many generations as you wish. Provide a program system chart. *Hint:* Assume that the borders of the game array are infertile regions where organisms can neither survive nor be born; you will not have to process the border squares.

# GLOSSARY
# OF FORTRAN STATEMENTS
# AND STRUCTURES

| Statement | Examples | Page |
|---|---|---|

---

### 1. *Declarations*

| Statement | Examples | Page |
|---|---|---|
| Type declarations | INTEGER K, NMBR, COUNTR, LIST(50), N | 16, 183 |
| | REAL GROSS, NET, X(25,25), W(-5:5) | 16, 183, 222 |
| | LOGICAL FLAG, FOUND, SWITCH | 133 |
| | CHARACTER*26 ALFBET | 134 |
| | CHARACTER TICTAC(3,3) | 437 |
| Common declaration | COMMON BETA(25,50), A | 324 |
| | COMMON /TXTBLK/ TEXT(240) | |
| Parameter declaration | PARAMETER (MIN = 100.00, TAX = 25.00) | 75 |
| | PARAMETER (TITLE = 'SOLVING PROBLEMS') | |
| Data declaration | DATA K,GROSS /5,-2.5/, FLAG /.TRUE./ | 212 |
| | DATA X /625*0/ | |
| | DATA (LIST(I),I=1,3) /10,20,30/ | |
| Save statement | SAVE ERRCNT, ERRFLG, /INBLK/ | 326 |

### 2. *Data Manipulation*

| Statement | Examples | Page |
|---|---|---|
| Arithmetic assignment | NMBR = LIST(I) + 6 | 136 |
| | GROSS = (5.0 * X) ** 2 | |
| | CSUM = CSUM + ENRANK(1, 3, K) | |
| Logical Assignment | FLAG = .TRUE. | 391 |
| | SWITCH = FOUND .AND. (K .LT. NMBR) | |

| | | |
|---|---|---|
| Character assignment | ALFBET = 'ABCDEFGHIJKLMNOPQRSTUVWXYZ' | 144, 404 |
| | INTSTR = ALFBET(9 : 14) | |
| | RLSTR = ALFBET( : 8) // ALFBET(15 : ) | |

### 3. *Execution Control*

| | | |
|---|---|---|
| Terminate execution | STOP | 22 |
| Unconditional transfer | GO TO 30  (label between 1 and 99999) | 94 |
| Logical IF | IF (GROSS .LT. 100.0) TAX = 0.0 | 88 |

### 4. *Input and Output*

| | | |
|---|---|---|
| Input | READ*, HOURS, RATE | 20 |
| | READ*, N, (X(I), I = 1, N) | 196 |
| | READ 26, ALPHA | 352 |
| | 26  FORMAT (8F10.3/) | |
| | READ 32, N, (LIST(K), K = 1, N) | 359 |
| | 32  FORMAT (I3/(10I8/)) | |
| Output | PRINT*, 'THE NUMBER OF ITEMS IS', N | 21 |
| | PRINT*, '      NAME      SCORE' | 100 |
| | PRINT '(A, 3X, F7.2, 2X, I2)', | 167 |
| | Z   NAME, GROSS, DEPEND | |
| | PRINT*, (X(I), I = 1, N) | 196 |
| | PRINT 401, ALFBET | 346 |
| | 401  FORMAT (' ALPHABET IS ', A26) | |
| | PRINT 402, COUNTR | 347 |
| | 402  FORMAT ('1', 70X, 'PAGE ', I4) | |

### 5. *Control Structures*

Decision structures

| | | |
|---|---|---|
| Single-alternative | IF (condition) THEN | 84 |

```
        --- ⎤
        ---  ⎬  executed if
        ---  ⎭  condition true
        ---
        ---
      ENDIF
```

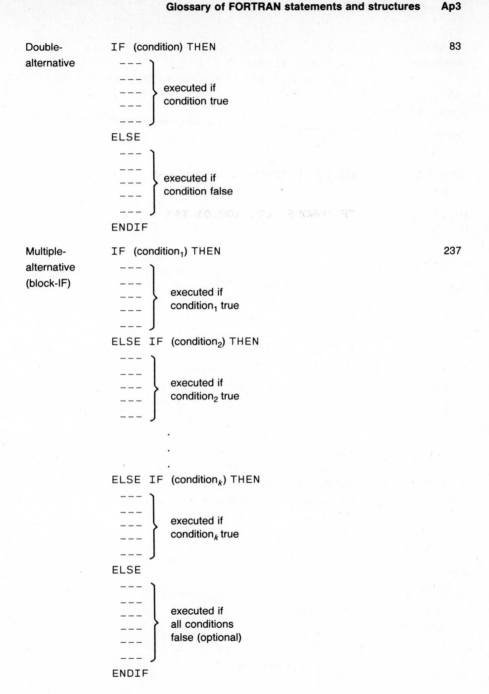

Double-
alternative

```
IF (condition) THEN                                              83
   ---  ⎫
   ---  ⎪
   ---  ⎬  executed if
   ---  ⎪  condition true
   ---  ⎭
ELSE
   ---  ⎫
   ---  ⎪
   ---  ⎬  executed if
   ---  ⎪  condition false
   ---  ⎭
ENDIF
```

Multiple-
alternative
(block-IF)

```
IF (condition₁) THEN                                           237
   ---  ⎫
   ---  ⎪
   ---  ⎬  executed if
   ---  ⎪  condition₁ true
   ---  ⎭
ELSE IF (condition₂) THEN
   ---  ⎫
   ---  ⎪
   ---  ⎬  executed if
   ---  ⎪  condition₂ true
   ---  ⎭

          .
          .
          .

ELSE IF (conditionₖ) THEN
   ---  ⎫
   ---  ⎪
   ---  ⎬  executed if
   ---  ⎪  conditionₖ true
   ---  ⎭
ELSE
   ---  ⎫
   ---  ⎪
   ---  ⎬  executed if
   ---  ⎪  all conditions
   ---  ⎭  false (optional)
   ---
ENDIF
```

Repetition
structures

| | | |
|---|---|---|
| WHILE | (Non-Standard) | 92 |
| loop | `WHILE (condition) DO` | |

```
      - - -  ⎫
      - - -  ⎪
      - - -  ⎬  loop body
      - - -  ⎪  (executed while
      - - -  ⎭  condition is true)
ENDWHILE
```

| | | |
|---|---|---|
| WHILE | (Standard FORTRAN Implementation) | 95 |
| loop | `30 IF (X .LT. Y) GO TO 39` | |

```
      - - -  ⎫
      - - -  ⎪
      - - -  ⎬  loop body (executed while
      - - -  ⎭  condition is false)
   GO TO 30
39 CONTINUE
```

| | | |
|---|---|---|
| DO loop | `DO sn lcv = iv, fv, sv` | 107, 245 |

```
      - - -  ⎫
      - - -  ⎪  (lcv — integer or real
      - - -  ⎬  variable; loop
      - - -  ⎪  parameters — integer or real valued
      - - -  ⎭  expressions)
sn CONTINUE
```

## 6. *Subprogram Features*

| | | |
|---|---|---|
| Subprogram definition | `SUBROUTINE SRCHAR` | 300 |
| | `SUBROUTINE EXCH (NMBR,COUNTR)` | |
| | `LOGICAL FUNCTION SAME (A,B,C)` | 281 |
| | `INTEGER FUNCTION SEEK (LIST)` | |
| Statement function | `F(X) = X ** 2 + 1.0` | 286 |
| Subprogram return | `RETURN` | 281 |
| Subprogram call | `CALL SRCHAR` | 301 |
| | `CALL EXCH(LIST(I), LIST(J))` | |
| | `Y = SQRT(X)` | 158, 284 |
| Subprogram end | `END` | 22, 281 |

| | | |
|---|---|---|
| *7. Formats* | 10 FORMAT (I3, 3X, F4.1) | 346 |
| | 17 FORMAT ('1LIST OF ITEMS...'/(10F12.3)) | 349 |
| | PRINT '(A, I3)', ' NAME', COUNT | 167 |

*8. File Control*

| | | |
|---|---|---|
| Open | OPEN (UNIT = 1, FILE = 'EZRA', STATUS = 'OLD',<br>Z FORM = 'UNFORMATTED', ACCESS = 'SEQUENTIAL') | 365 |
| Endfile | ENDFILE (UNIT = 3) | 365 |
| Close | CLOSE (UNIT = 3) | 366 |
| General<br>Read/Write | READ (UNIT = 1, FMT = 10, END = 11) N, A<br>WRITE (UNIT = 1, REC = I/2) N, A | 367 |
| Backspace | BACKSPACE (UNIT = 1) | 369 |
| Rewind | REWIND (UNIT = 1) | 370 |

# Programs and Subprograms

# Program Style Displays

# Reference Tables and Displays

# ANSWERS TO
# SELECTED EXERCISES

*Chapter 1*

1.1
| cell number | contents |
|---|---|
| 0 | −27.2 |
| 2 | MINE |
| 997 | 0.05 |

1.2
| contents | cell address |
|---|---|
| 12.5 | 004 |
| −26 | 003 |
| 998.0 | 998 |

1.5  The statement that prints the hours and rate may be placed anywhere between the statement to read hours and rate and the STOP. No other statement can be moved because of the dependencies among statements. For example, the computation of NET can only be performed after the computation of GROSS, and printing GROSS and NET only makes sense after these two values have been defined.

1.7  Values printed:

```
40.0  16.25  650.00  117.00  533.00
```

1.8

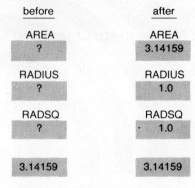

before | after

AREA
?

AREA
3.14159

RADIUS
?

RADIUS
1.0

RADSQ
?

RADSQ
1.0

3.14159

3.14159

Area = 12.56636

1.14     insert
```
PRINT*, 'ENTER DISTANCE (IN MILES) AND SPEED (IN MILES',
Z 'PER HOUR)'
```
before the first **READ** statement and remove the two **PRINT** statements that follow this **READ**.

insert
```
PRINT*, 'ENTER MILES PER GALLON AND COST PER GALLON',
Z '(IN DOLLARS)'
```
before the second **READ** statement and remove the two **PRINT** statements that follow this **READ**.

1.15     The value of X cannot be determined unless the value of COUNT is given.

*Chapter 2*

2.1
```
C READ AND PRINT DATA
      PRINT*, 'ENTER FIRST NUMBER'
      READ*, NUM1
      PRINT*, 'ENTER SECOND NUMBER'
      READ*, NUM2
```

2.2     *Data Table:*

| *Input variables* | *Output variables* |
|---|---|
| NUM1: First number to be used in computation | SUM: Sum of four numbers |
| NUM2: Second number to be used in computation | AVE: Average of four numbers |
| NUM3: Third number to be used in computation | |
| NUM4: Fourth number to be used in computation | |

*Algorithm:*

Step 1:   Read data items into variables NUM1, NUM2, NUM3, NUM4 and echo print values if using batch.

Step 2:   Compute the sum of the data items NUM1, NUM2, NUM3, NUM4, and store the result in the variable SUM.

Step 3:   Compute the average of the data items NUM1, NUM2, NUM3, NUM4 and store the result in the variable AVE.

Step 4:   Print the values in the variables SUM and AVE.

Step 5:   Stop.

2.4   a)

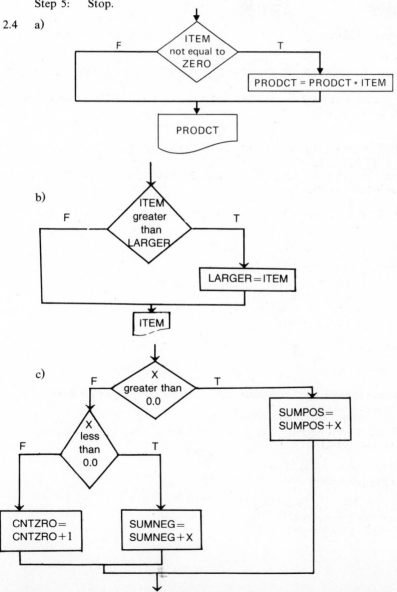

2.5        HOURS = 37.5
           RATE = 3.75
           GROSS = 37.5 * 3.75 = <u>140.63</u>
           NET = 140.63 − 25.00 = <u>115.63</u>

           HOURS = 20
           RATE = 4.00
           GROSS = 20 * 4.00 = <u>80.00</u>
           NET = <u>80.00</u>

2.8

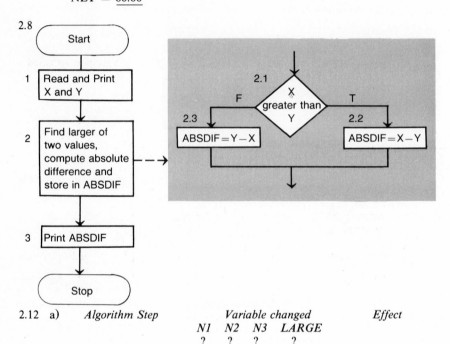

2.12  a)    *Algorithm Step*        *Variable changed*          *Effect*

| | N1 | N2 | N3 | LARGE | |
|---|---|---|---|---|---|
| | ? | ? | ? | ? | |
| 1 | 5 | 20 | 15 | | 5 > 20 is false |
| 2.1 | | | | | execute 2.3 |
| 2.3 | | | | 20 | |
| 2.4 | | | | | 15 > 20 is false |
| 3 | | | | | Print 20 |

The trace followed PATH 4: Step 2.1-F, Step 2.4 F.

b)  PATH 1: Step 2.1-T, Step 2.4-T

| N1 | N2 | N3 | LARGE |
|---|---|---|---|
| 20 | 10 | 30 | 30 |

The algorithm would work for these data.

c)  N1 = 16, N2 = −20, N3 = 0, LARGE = 16.
This would follow PATH 2: Step 2.1-T, Step 2.4-F
Yes as all paths have been tested.

2.17    C PROGRAM PARAMETERS
              INTEGER NRITMS
              PARAMETER (NRITMS = 2000)

The program otherwise remains the same, except that the parameter NRITMS should be used everywhere the constant 2000 appears.

2.18    Using MINSAL and TAX (instead of 100.00 and 25.00) makes the computations clearer. Also, if those values change, only the PARAMETER statement must be changed to reuse the program.

*Chapter 3*

3.1    Trace for X = 7.2, Y = 3.5

| Program Trace *FORTRAN statements* | Variables Affected | | |
|---|---|---|---|
| | *X* | *Y* | *TEMP* |
| READ*, X | 7.2 | | |
| READ*, Y | | 3.5 | |
| TEMP = X | | | 7.2 |
| X = Y | 3.5 | | |
| Y = TEMP | | 7.2 | |

Trace for X = Y = 6.2

| | *X* | *Y* | *TEMP* |
|---|---|---|---|
| READ*, X | 6.2 | | |
| READ*, Y | | 6.2 | |
| TEMP = X | | | 6.2 |
| X = Y | 6.2 | | |
| Y = TEMP | | 6.2 | |

3.3    a)

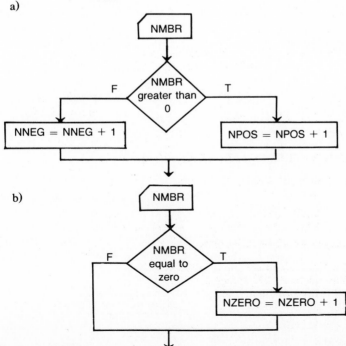

b)

c)

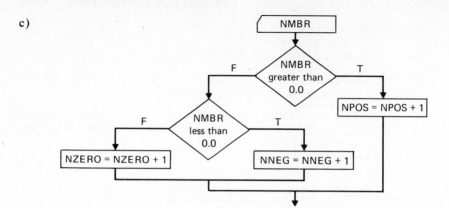

3.5   i)

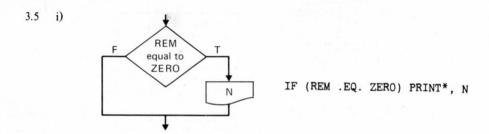

```
IF (REM .EQ. ZERO) PRINT*, N
```

ii)

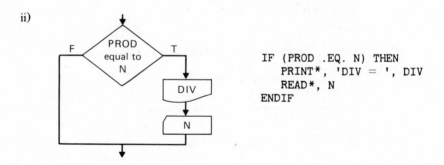

```
IF (PROD .EQ. N) THEN
    PRINT*, 'DIV = ', DIV
    READ*, N
ENDIF
```

iii)

```
IF (NBRLTE .GT. 25.0) THEN
    GALREQ = MILES/14.0
ELSE
    GALREQ = MILES/22.5
ENDIF
```

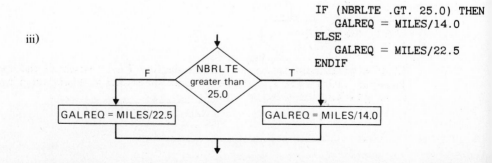

3.6

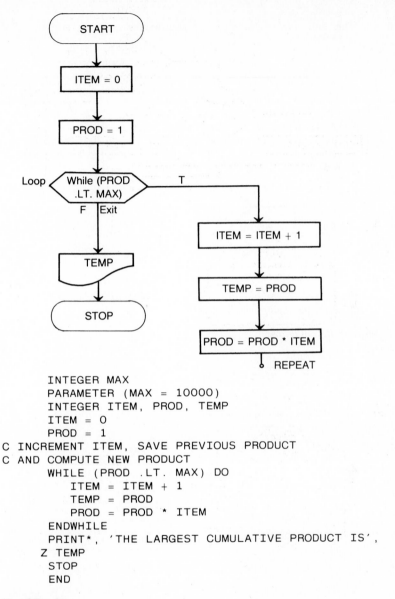

```
        INTEGER MAX
        PARAMETER (MAX = 10000)
        INTEGER ITEM, PROD, TEMP
        ITEM = 0
        PROD = 1
C INCREMENT ITEM, SAVE PREVIOUS PRODUCT
C AND COMPUTE NEW PRODUCT
        WHILE (PROD .LT. MAX) DO
            ITEM = ITEM + 1
            TEMP = PROD
            PROD = PROD * ITEM
        ENDWHILE
        PRINT*, 'THE LARGEST CUMULATIVE PRODUCT IS',
      Z TEMP
        STOP
        END
```

3.7   a)   X .GE. Y
       b)   COUNTR .NE. NRITMS
       c)   ITEM .EQ. SNTVAL
       d)   2.0 .LE. 3.0

3.8      **LARGE** would be initialized to $-1$ (via the first **READ** statement and the
         subsequent assignment). The loop body would not execute at all, and the print-
         ed value for **LARGE** would be $-1.0$.

3.10     The data table should have additional output variables, SMALL and RANGE.

> SMALL: The smallest of all
>     scores processed at any point.
> RANGE: The range of data,
>     LARGE–SMALL.

SMALL should be initialized to the first value read in as is the case with LARGE. Then the following test should be added to the loop body portion of the flow diagram. The change should be reflected in the level one flow diagram as well.

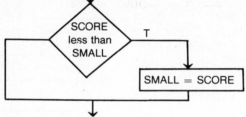

The RANGE computation and print should be inserted into the level one flow diagram just before the Stop.

3.11  *Refined flow diagram:*

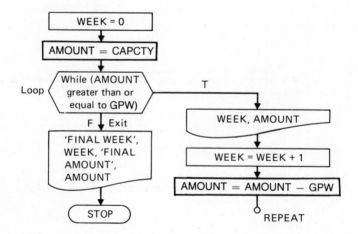

```
            INTEGER CAPCTY, GPW
            PARAMETER (CAPCTY = 10000, GPW = 183)
            INTEGER WEEK, AMOUNT
      C INITIALIZE PROGRAM VARIABLES
            WEEK = 0
            AMOUNT = CAPCTY
            PRINT*, 'WEEK AMOUNT'
      C DEDUCT USAGE FOR A WEEK IF ENOUGH WATER IS LEFT
            WHILE (AMOUNT .GE. GPW) DO
                PRINT*, WEEK, AMOUNT
                WEEK = WEEK + 1
                AMOUNT = AMOUNT - GPW
            ENDWHILE
            PRINT*, 'FINAL WEEK ', WEEK,
          Z '  FINAL AMOUNT = ', AMOUNT
            STOP
            END
```

3.12
```
      VALUE = STVAL
      YEAR = 1
      WHILE (YEAR .LE. CTTERM) DO
          INTRST = VALUE * RATE
          VALUE = VALUE + INTRST
          PRINT*, YEAR, INTRST, VALUE
          YEAR = YEAR + 1
      ENDWHILE
```

*Chapter 4*

4.1    a)  Type real
       b)  Illegal (could be type character, but apostrophes are missing)
       c)  Illegal (only .TRUE. or .FALSE. may be delimited with periods as logical constants)
       d)  Type character
       e)  Type integer
       f)  Type real
       g)  Illegal (needs an E in front of + to be real)
       h)  Type character
       i)  Illegal (could be type real without the $)
       j)  Illegal (could be type character but the apostrophes are missing)

4.4    $(X/Y) * Z = (6.0/3.0) * 4.0 = 2.0*4.0 = 8.0$
       $X/ (Y * Z) = 6.0/(3.0 * 4.0) = 6.0/12.0 = 0.5$
       The compiler computes $X/Y * Z$ as $(X/Y) * Z$.

4.5    One possible FORTRAN equivalent is:

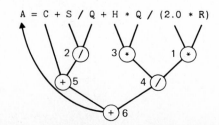

4.6   a)  $\dfrac{w + x}{y + z}$                           b)   $gh - fw$

c)  $ab^2$                                    d)   $(b^2 - 4\,ac)^{1/2}$
e)  $(x^2 - y^2)^{1/2}$                    f)   $(x^2 + r/365)^w$

g)  $p_2 - \dfrac{p_1}{t_2} - t_1$

4.9   a) X = 4.0 * A * C              b) A = A * C
c) I = 2 * (- J)                  d) K = 3 * (I + J)
e) X = A / B * C                  f) I = J * 3

4.10  a)  X * (Y + W) * Z
b)  X ** 2 + Y ** 2
c)  (X * Y) ** 2
d)  (A * X + B * G) / (A * W + B * Y)
e)  3.0 * (X * Y + W * Z) - 12.0
f)  (3.2 - 7.5 * Y) ** 2
g)  3.14 E + 6 - .013 E-4 * (X + Y)
h)  (18.0 * X - 12.5 E + 3
(X ** (N - 3) + Y ** (N - 2))

4.13      (the **AND** example)
```
      IF (SCORE .GE. 0) THEN
         IF (SCORE .LE. 100) THEN
C            PROCESS A VALID SCORE

               .  .  .

         ELSE
            PRINT*, SCORE, 'OUT OF RANGE AND IS IGNORED.'
         ENDIF
      ELSE
         PRINT*, SCORE, 'OUT OF RANGE AND IS IGNORED.'
      ENDIF
```

(the **OR** example)
```
      IF (SCORE .LT. 0 ) THEN
         PRINT*, SCORE, 'OUT OF RANGE AND IS IGNORED.'
      ELSE
         IF (SCORE .GT. 100) THEN
            PRINT*, SCORE, 'OUT OF RANGE AND IS IGNORED.'
         ELSE
C            PROCESS A VALID SCORE

               .  .  .

         ENDIF
      ENDIF
```

4.15

```
C DEMONSTRATION OF SIN AND COS
C
      REAL PI
      PARAMETER (PI = 3.14159)
      REAL X
      INTEGER I
C
C     PRINT SIN AND COS TABLE
      PRINT*, ' X (IN DEGREES)  SIN(X)  COS(X) '
      DO 30 I = 0, 180, 15
         X = REAL(I)
         PRINT*, X, SIN(X *PI / 180.0), COS(X * PI /
   Z     180.0)
   30 CONTINUE
C
      STOP
      END
```

4.17

```
      LOGICAL BADAIM                                      ⇐ (Add)
         .
         .
         .
      RADIAN = THETA * (PI / 180.0)
      BADAIM = .TRUE.                                     ⇐ (Add)
      WHILE (BADAIM) DO                                   ⇐ (Add)
         T = X / (V * COS(RADIAN))
         H = V * T * SIN(RADIAN) - G/2.0 * T ** 2
         IF (H .LT. 100.0) THEN
            PRINT*, 'ARROW TOO LOW'
            IF (H .LT. 0.0) PRINT*,'ARROW DID NOT REACH TOWER'
            V = V + 10.0                                  ⇐ (Add)
         ELSE
            IF (H .GT. 110.0) THEN
               PRINT*, 'ARROW TOO HIGH'
               V = V - 8.0                                ⇐ (Add)
            ELSE
               PRINT*, 'GOOD SHOT PRINCE'
               BADAIM = .FALSE.                           ⇐ (Add)
            ENDIF
         ENDIF
      ENDWHILE
      STOP
      END
```

4.20  a)  1.666 . . . (floating point) REAL (COLOR)
      b)  .666 . . . (floating point) REAL (COLOR)
      c)  0        (fixed point)
      d)  −3.0     (floating point) REAL (COLOR+STRAW)
      e)  11       (fixed point)    REAL (RED), apply INT to entire expression
      f)  1        (fixed point)    REAL (STRAW) and REAL (COLOR), apply
                                    INT to entire expression

4.21    Use a DO loop and the function EXP and ALOG. Remember both EXP and
        ALOG require real arguments.

4.24  a.  correct
    b.  incorrect (4I should be written as I4)
    c.  incorrect (missing comma between the descriptors
             F16.3 and 5X)
    d.  correct
    e.  correct

4.25  a) and b)   1234 □□□□ 555.4567

    c)  K □ = □ 1234

    ALPHA □ = □□□ 555.46

    d)  K □ = □□ 1234 □□□□□□□□□□ ALPHA □ = □□□□ 555.457

*Chapter 5*

5.1   CHARACTER * 8 TEAM (6)

### Array TEAM

| TEAM (1) | TEAM (2) | TEAM(3) | TEAM(4) | TEAM(5) | TEAM(6) |
|---|---|---|---|---|---|
| ORIOLES□ | RED□ SOX□ | YANKEES□ | ANGELS□□ | ROYALS□□ | TWINS□□□ |

INTEGER RUNS(6), HITS(6), RBI(6)

### Array RUNS

| RUNS(1) | RUNS(2) | RUNS(3) | RUNS(4) | RUNS(5) | RUNS(6) |
|---|---|---|---|---|---|
| 527 | 612 | 710 | 693 | 716 | 425 |

### Array HITS

| HITS(1) | HITS(2) | HITS(3) | HITS(4) | HITS(5) | HITS(6) |
|---|---|---|---|---|---|
| 2095 | 2092 | 2695 | 1942 | 2888 | 1816 |

### Array RBI

| RBI(1) | RBI(2) | RBI(3) | RBI(4) | RBI(5) | RBI(6) |
|---|---|---|---|---|---|
| 461 | 563 | 681 | 647 | 655 | 399 |

```
REAL BA(6)
```

| BA(1) | BA(2) | BA(3) | BA(4) | BA(5) | BA(6) |
|-------|-------|-------|-------|-------|-------|
| .263 | .271 | .262 | .233 | .278 | .255 |

```
LOGICAL EASTDV(6)
```

| EASTDV(1) | EASTDV(2) | EASTDV(3) | EASTDV(4) | EASTDV(5) | EASTDV(6) |
|-----------|-----------|-----------|-----------|-----------|-----------|
| .TRUE. | .TRUE. | .TRUE. | .FALSE. | .FALSE. | .FALSE. |

5.2    for ISUB=4:
X(ISUB) refers to the 4th element of array X.
X(4) refers to the 4th element of array X.
X(2*ISUB) refers to the 8th element of array X.
X(5*ISUB—6) refers to the 14th element of array X.
   (This subscript is out of the legal range of subscript values of 1 through
   10) for the array X.)

5.6
```
INTEGER PRIME (10)
READ*, PRIME
```

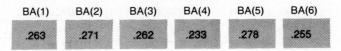
```
2 3 5 7 11 13 17 19 23 29
```
```
PRINT*, 'PRIME ARRAY = ', PRIME
```

5.7
```
    PRINT*,'        N              PRIME (N)'
    DO 20  N = 1, 10
        PRINT*, N, PRIME(N)
 20 CONTINUE
```

5.8  a)
```
      DO 40 I = 1, 8
          PRINT*, PRIME(I)
  40 CONTINUE
```
b)
```
      DO 40 I = 3, 8
          PRINT*, PRIME(I)
  40 CONTINUE
```
c)
```
      DO 40  I = 7, 10
          PRINT*, PRIME(I)
  40 CONTINUE
```
d)
```
      DO 40 I = 1, K
          PRINT*, PRIME(I)
  40 CONTINUE
```

5.9
```
      DO 40 I = 1, 7
          READ*, PRIME(I)
  40 CONTINUE
```
The data must be punched one item per card. The card in Exercise 5.6 won't
work: only the first item would be read from this card—the last 9 would be
lost.

5.12
```
IF (N .LT. 1 .OR. N. GT. 10) THEN
    PRINT*, 'N IS OUT OF RANGE OF THE SIZE OF SCORES'
    PRINT*, 'EXECUTION IS TERMINATED'
    STOP
ENDIF
```
N serves as the end value for the implied loops used to read and print the SCORES. The loop-control variable (I) for these loops is used as the subscript for the array SCORES. Thus, the value of COUNT must be within the range of SCORES so as to ensure that the loop-control variable will represent legal subscript values. The value of N must also be compatible with the rules for DO loop structures; i.e. greater than or equal to the initial value 1.

5.16    "Key not found" would be printed after each failure to match the KEY to an element of the array A.

5.17

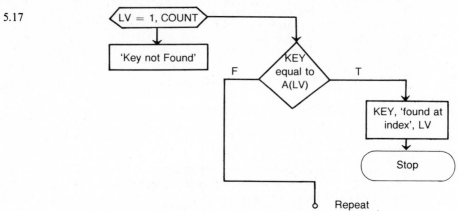

No, the message 'KEY NOT FOUND' would only be printed if the key is not in the array. Otherwise index would be printed and the program stopped.

5.22
```
TABIND = INT(250.0/500.0)+1 = INT(0.5)+1 = 1
TABIND = INT(1275.0/500.0)+1 = INT(2.55)+1 = 3
TABIND = INT(2750.0/2000.0)+4 = INT(1.38)+4 = 5
TABIND = INT(4000.0/2000.0)+4 = INT(2.0)+4 = 6
TABIND = INT(11700.0/2000.0)+4 = INT(5.85)+4 = 9
TABIND = INT(23000.0/2000.0)+4 = INT(11.5)+4 = 15
    'Salary out of range—ignored'
```

5.23    *Search Algorithm*

```
INTEGER CATCNT, SNTVAL
PARAMETER (CATCNT = 5, SNTVAL = -99)
INTEGER LOWGRD(CATCNT), SCORE, I, INDEX
CHARACTER * 1 LETTER(CATCNT)
DATA LOWGRD /90, 80, 70, 60, 0/
DATA LETTER / 'A', 'B', 'C', 'D', 'F'/
PRINT*, 'LIST OF SCORES AND GRADES ...'
READ*, SCORE
WHILE  (SCORE .NE. SNTVAL) DO
    IF (SCORE .LT. 0 .OR. SCORE .GT. 100) THEN
        PRINT*, 'ILLEGAL SCORE ', SCORE, 'IGNORED'
    ELSE
        DO 10 I = 1, CATCNT
            IF (SCORE .GT. LOWGRD(I)) THEN
                INDEX = I
                GO TO 20
            ENDIF
10      CONTINUE
20      PRINT*, SCORE, LETTER(INDEX)
    ENDIF
    READ*, SCORE
ENDWHILE
```

*Direct Computation Algorithm*

```
INTEGER CATCNT, SNTVAL
PARAMETER (CATCNT = 5, SNTVAL = -99)
INTEGER LOWGRD(CATCNT), SCORE, INDEX, I
CHARACTER * 1 LETTER(CATCNT)
DATA LOWGRD /90, 80, 70, 60, 0/
DATA LETTER / 'A', 'B', 'C', 'D', 'F'/
PRINT*, 'LIST OF SCORES AND GRADES ...'
READ*, SCORE
WHILE (SCORE .NE. SNTVAL) DO
    IF (SCORE .LT. 0 .OR. SCORE .GT. 100) THEN
        PRINT*, 'ILLEGAL SCORE ', SCORE, ' IGNORED'
    ELSE
        INDEX = 10 - SCORE/10
        IF (INDEX .GT. 5) INDEX = 5
        IF (INDEX .EQ. 0) INDEX = 1
        PRINT*, SCORE, LETTER(INDEX)
    ENDIF
    READ*, SCORE
ENDWHILE
```

5.25  a)  Legal      Size 6      Subscript range 1–6
      b)  Legal      Size 21     Subscript range −10, −9 . . . 0 . . . 9, 10
      c)  Illegal     lowbound exceeds highbound
      d)  Illegal     lowbound exceeds highbound
      e)  Legal      Size 10     Subscript range 0, 1, . . . 9
      f)  Legal      Size 11     Subscript range 10, 11 . . ., 20

5.26  a)  for I=2
          AGE(2\*I−1)                  Legal: AGE(2\*2−1) = AGE(3)
          AGE(4\*I)                    Illegal: AGE(8) out of range
      b)  for I=5
          ORD(2\*I−1)                  Legal: ORD(2\*5−1) = ORD(9)
          ORD(2\*I+10)                 Illegal: ORD(2\*5+10) = ORD(20)
      c)  for K=3
          DIGIT(3\*K−2)                Legal: DIGIT(3\*3−2) = DIGIT(7)
          DIGIT(4\*K−2)                Illegal: DIGIT(4\*3−2) = DIGIT(10)
      d)  for INDEX=2
          FLAG(5\*INDEX)               Legal: FLAG(5\*2) = FLAG(10)
          FLAG(10\*INDEX)              Legal: FLAG(10\*2) = FLAG(20)

*Chapter 6*

6.3

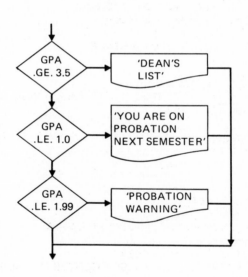

```
IF (GPA .GE. 3.5) THEN
    PRINT*, 'DEAN''S LIST'
ELSE IF (GPA .LE. 1.0) THEN
    PRINT*, 'YOU ARE ON PROBATION NEXT SEMESTER'
ELSE IF (GPA .LE. 1.99) THEN
    PRINT*, 'PROBATION WARNING'
ENDIF
```

6.4     Problem 4A
```
IF (VOTER .LT. 'J') THEN
    CLERK = 'ABRAHAM'
ELSE IF (VOTER .LT. 'S') THEN
    CLERK = 'MARTIN'
ELSE
    CLERK = 'JOHN'
ENDIF
```

*Example 4.19:*
```
IF (H .LT. 0.0) THEN
    PRINT*, 'ARROW DID NOT REACH THE TOWER'
ELSE IF (H .LT. 100.0) THEN
    PRINT*, 'ARROW TOO LOW'
ELSE IF (H .GT. 110.0) THEN
    PRINT*, 'ARROW TOO HIGH'
ELSE
    PRINT*, 'GOOD SHOT PRINCE'
ENDIF
```

6.8
```
IF (N .GE. MAXSIZ) THEN
    PRINT*, N, ' IS TOO BIG - INSERTION OF ', IDNUM,
   Z ' IGNORED'
ELSE
C       SHIFT EACH VALUE ONE ELEMENT TO THE RIGHT
                          .
                          .
                          .
```

6.9   a)
```
        READ*, NDROP
        IX = 1
        WHILE (NDROP .NE. CLIST(IX)) DO
          IX = IX + 1
          IF (IX .GT. N) THEN
              PRINT*, 'ID TO BE DELETED ', NDROP, ' NOT FOUND.'
              STOP
          ENDIF
        ENDWHILE
        WHERE = IX
```

      b)
```
        READ*, IDNUM
        IX = 1
        WHILE (CLIST(IX) .LE. IDNUM) DO
          IX = IX + 1
          IF (IX .GT. N) GO TO 20
        ENDWHILE
   20 CONTINUE
        WHERE = IX
```

6.10    The decision structure and **DO** loop overlap.

6.11
```
OUTER    1
INNER J 1  1
INNER J 1  3
INNER K 1  2
INNER K 1  4
OUTER    2
INNER J 2  1
INNER J 2  3
INNER K 2  2
INNER K 2  4
```

6.13    Change the DO loop limit value from COUNT-1 to COUNT-NLOOP, where NLOOP represents the number of passes performed so far. NLOOP should be initialized to zero and incremented by one before each repetition of the outer loop.

6.14    Change the exchange condition to
$$(M(INDEX) .LT. M(INDEX + 1))$$

6.15

```
INTEGER MEDIAN, MEDIX



PRINT*, (M(I), I=1, COUNT)
MEDIX = COUNT / 2 + 1
IF (MOD(COUNT, 2) .EQ. O) THEN
    MEDIAN = (M(MEDIX)+M(MEDIX-1)) / 2
ELSE
    MEDIAN = M(MEDIX)
ENDIF
PRINT*, 'MEDIAN = ', MEDIAN
STOP
```
Inserted statements for computing median

*Chapter 7*

7.1    `MAX = MAXO(NUM1, NUM2, NUM3, NUM4)`

7.2
```
INTEGER FUNCTION SQR (N)
INTEGER N
SQR = N * N
RETURN
END
```

7.4
```
LOGICAL FUNCTION SAME (X, Y, Z)
REAL X, Y, Z
IF (X .EQ. Y .AND. Y .EQ. Z) THEN
    SAME = .TRUE.
ELSE
    SAME = .FALSE.
ENDIF
RETURN
END
```

7.8

```
          REAL FUNCTION SUM (X, N)
C
C FIND THE SUM OF N REAL DATA ITEMS
C
C ARGUMENT DEFINITIONS (ALL INPUT) --
C   X - ARRAY OF DATA TO BE SUMMED
C   N - NUMBER OF ITEMS IN X
C
          INTEGER N
          REAL X(N)
C
C LOCAL VARIABLES
          INTEGER I
          REAL TEMSUM
C
C ACCUMULATE SUM
          TEMSUM = 0.0
          DO 10 I = 1, N
              TEMSUM = TEMSUM + X(I)
      10 CONTINUE
C
C RETURN RESULT
          SUM = TEMSUM
C
          RETURN
          END
```

7.11

```
          REAL FUNCTION SMALL (A, NRITMS)
C
C DETERMINE THE SMALLEST ITEM IN AN ARRAY OF REAL DATA
C ITEMS
C
C ARGUMENT DEFINITIONS (ALL INPUT) --
C   A - ARRAY CONTAINING THE DATA TO BE PROCESSED
C   NRITMS - NUMBER OF TIMES IN THE ARRAY
C
          INTEGER NRITMS
          REAL A(NRITMS)
C
C LOCAL VARIABLES
          REAL CURSML
          INTEGER I
C
C INITIALIZE CURRENT SMALLEST ITEM
          CURSML = A(1)
C LOOK FOR AS ITEM THAT IS SMALLER THAN CURSML
C REDEFINE CURSML WHEN A NEW SMALLEST ITEM IS FOUND
          DO 40 I = 1, NRITMS
              IF (A(I) .LT. CURSML) CURSML = A(I)
      40 CONTINUE
C
C RETURN VALUE OF THE SMALLEST ITEM WHEN THE SEARCH IS
C COMPLETE
          SMALL = CURSML
C
          RETURN
          END
```

7.12  a) 8    b) 5    c) 3    d) 1

7.13  a)  The 2nd actual argument is an array name.
 b)  ONOFF is not declared in the main program.
 c)  LIST is not declared in the main program.
 d)  Y(16) is out-of-range for array Y.
 e)  X(15) is a single array element; the first actual argument must be an array name without a subscript
 f)  There are only three actual arguments.

7.14  a)  illegal - the array size N should be the first argument
 b)  illegal - array Z is used as an input argument; however, Z is undefined
 c)  illegal - X(*) can only be used inside the subroutine
 d)  illegal - A, B, and C are not declared as arrays in the main program
 e)  Z(1) = 22.5
     Z(2) = −2.5
     Z(3) = −2.9
 f)  X(1) = 22.5
     X(2) = −2.5
     X(3) = −2.9

7.18
```
        REAL FUNCTION FNDAVE (X, N)
C
C FIND THE AVERAGE OF AN ARRAY OF REAL DATA ITEMS
C
C ARGUMENT DEFINITIONS (ALL INPUT) --
C   X - ARRAY OF DATA TO BE AVERAGED
C   N - NUMBER OF ELEMENTS IN X
C
        INTEGER N
        REAL X(N)
C
C LOCAL VARIABLES
        INTEGER I
        REAL SUM
C
        SUM=0.0
        DO 20 I = 1, N
           SUM = SUM + X(I)
   20 CONTINUE
        FNDAVE = SUM / REAL(N)
C
        RETURN
        END
```

7.19    This is done in function FNDMED—not the main program.

7.20  a)  RANGE = LARGE(TABLE, XCOUNT) − SMALL(TABLE, XCOUNT)
     MEAN = FNDAVE(TABLE, XCOUNT)
     MEDIAN = FNDMED(TABLE, XCOUNT)
 b)  No changes are needed in the subprograms.

7.25  a)  Y = 16.0
 b)  A = −2.0
 c)  X(3) = 1.0
 d)  X(5) = 10.2,  assuming X(3) has been changed to 1.0

7.28  a)  1.0   2.0   4.0
      b)  7, 5, 3, 1, 9
          There is no relationship. Because the array NEXT is an argument in the call to
          DEFINE.

*Chapter 8*

8.1       Disregarding carriage control, the three lines printed would be
          (A, A)        HI
          HI            (A, A)
          HI            (A, A)        (A, A)
          On a high speed line printer, the first character of each line above would be
          used for carriage control. The character "(" is an illegal carriage control charac-
          ter.

8.2   b) and d)

8.4   a)      PRINT '('' NAME'', 10X, ''SALARY'', 10X,
             Z ''DEPENDENTS'')'
              PRINT 30
          30 FORMAT ('NAME', 10X, 'SALARY', 10X, 'DEPENDENTS')

      b)      PRINT '('' EMPLOYEE = '', A)', NAME
              PRINT 40, NAME
          40 FORMAT (' EMPLOYEE = ', A)

      c)      PRINT '('' SALARY = '', F6.0, 3X,
             Z          '' DEPENDENTS = '', I2)',
             Z          GROSS, DEPEND
            PRINT 50, GROSS, DEPEND
          50 FORMAT (' SALARY = ', F6.0, 3X, ' DEPENDENTS =', I2)

8.5   a)  REGGIE      PLAYS FOR THE YANKEES

      b)  LAST YEAR HE HIT 41 HOME RUNS AND BATTED .300

      c)  NAME        TEAM        HOME RUNS AVERAGE
          REGGIE      YANKEES           41        .300

8.6           INTEGER AGE, YEAR, OCCODE
              CHARACTER * 11 SSNO
              CHARACTER * 1  INIT
              CHARACTER * 13 FIRST
              CHARACTER * 20 LAST
              READ 10, SSNO, LAST, FIRST, INIT, AGE, YEAR, OCCODE
          10 FORMAT(A11, A20, A13, A1, 3X, I2, 4X, I5, 4X, I3)

552–63–0179BROWN          JERRY        L   38      23       12
1 2 3 4 5 6 7 8 9 10 11 12 13 14 15 16   •••   32   •••   45 46 47 48 49 50   •••   58 59 60 61 62 63 64 65 66   •••

8.9     30 FORMAT ( F7.1 / I5, 2X, A / F6.1, I3)

8.10
```
/16312DAWN-321     49
 1 2 3 4 5 6 7 8 9 10 11 12 13 14 15 16 17 18 19
```

8.11     There would be 12 lines of 10 values each. Each value would be right justified in a field of width 12 with 3 decimal places printed.

8.15  a)     PRINT 10, X
        10 FORMAT (5(1X, 4F8.2/)) or FORMAT (1X, 4F8.2)
     b)     PRINT 20, N, (X(I), I = 1, N))
        20 FORMAT (1X, 'N = ', I2/5 (1X, 4F8.2/))
     c)     PRINT 30, QUEUE
        30 FORMAT (170(1X, 6E15.6/)) or FORMAT (1X, 6E15.6)
     d)     DO 40 I = 1, 120
              PRINT 50, ROOM(I), TEMP(I)
        40 CONTINUE
        50 FORMAT (1X, I3, 5X, F6.1)

8.16     OPEN (UNIT = 3, FILE = 'TEXT', ACCESS = 'SEQUENTIAL',
        Z     FORM = 'UNFORMATTED', STATUS = 'OLD')

8.20     An additional alternative should be added to the block-IF structure shown in Fig. 8.6.
        ELSE IF (OLDATA(1) .EQ. UPDATA(1)) THEN
            WRITE (UNIT = 3) UPDATA
            READ (UNIT = 1, END = 40) OLDATA
            READ (UNIT = 2, END = 40) UPDATA

8.22     Modify Fig. 8.3a as follows. Use the OPEN statement
        OPEN (UNIT = 1, FILE = 'DIRINV', ACCESS = 'DIRECT',
        Z     FORM = 'FORMATTED', STATUS = 'NEW', RECL = 80)
        Replace the file WRITE statement with
        IF (STOCK .GT.0 .AND. STOCK .LE. MAXSTK) THEN
            WRITE (UNIT = 1, FMT = 21, REC = STOCK)
        Z     STOCK, AUTHOR, TITLE, PRICE, QUANT
        ELSE
            PRINT*, 'RECORD ', STOCK, ' IS OUT OF RANGE'
        ENDIF
        Format 21 and parameter MAXSTK are assumed to be defined as in Fig. 8.8.

*Chapter 9*

9.1  a)  Logical operator .AND. must have logical expressions as operands.
         (I .LT. 1) .AND. (I .LT. 2) .AND. (I .LT. 3)
    b)  Real variable (Z) cannot be used as operand of logical operator .OR.
         (X .EQ. Y) .OR. (X .EQ. Z)
    c)  Real variable (X) cannot be used as operand of logical operator .OR.
         (X .LT. Z) .OR. (Y .LT. Z)
    d)  Logical expressions (FLAG2, .TRUE.) cannot be used as operands of relational operator .EQ.
         FLAG1 .OR. FLAG2
    e)  Logical expressions (FLAG1, FLAG2) cannot be used as operands of relational operator .EQ.
         (FLAG1 .EQV. FLAG2) .OR. (X .NE. Y)

9.2    True; false

9.4    START must be added to the argument list, the argument definition list (input) and the argument declarations (integer). A check should be made to ensure that START lies between 1 and SIZE. The statement
                    NEXT = 1
       should be changed to read
                    NEXT = START

9.6  1)  042
     2)  , JOHN QUINCY
     3)  ADAMS,
     4)  ADAMS, JOHN QUINCY

9.7    Change the first **PRINT** statement to
       PRINT*, SENTNC(FIRST+1 : NEXT-1) // SENTNC(FIRST : Z FIRST) // 'AY'
       change the second **PRINT** statement to
       PRINT*, SENTNC(FIRST+1 : LENGTH-1) // SENTNC(FIRST : Z FIRST) // 'AY'

9.10 1)  **JOHN ADAMS**
     2)  **ADAMS J.Q.**

9.11 a) 'ARE STRUCTURED PROGRAMS' is assigned to QUOTE
b) 'RAMS ARE STRUCTURED PROG' is assigned to QUOTE2
c) above string is assigned to QUOTE
d) 'HIPS' is assigned to HIPPO
e) 'LARGE' is assigned to BIGGER
f) The blank string is assigned to BIGGER
g) The string 'ST' is inserted in positions 6 and 7 of BIGGER

9.13 It is possible for a function reference to have commas in the argument list. Some array references may have commas as well.

9.14 It would be illegal to assign TEXT a new value by rearranging the substrings of the current TEXT.

9.15 Add an additional alternative to Step 2.2.4 as shown below
```
ELSE IF (NEWLEN .EQ. OLDLEN) THEN
    COPY = TEXT
    COPY(POSOLD : POSOLD+OLDLEN-1) = NEW
```
Alternatively, simply replace OLD by NEW directly in TEXT
```
TEXT(POSOLD : POSOLD+OLDLEN-1) = NEW
```

9.16 The main program segment is provided next. GETLEN is the function shown in Fig. 9.9 that determines the actual length of a character variable excluding blank padding. ENTER is a subroutine that reads a text string up to MAXLEN characters that is terminated by a $.
```
          INTEGER MAXLEN
          PARAMETER (MAXLEN = 1000)
          INTEGER CURLEN, OLDLEN, NEWLEN, GETLEN
          CHARACTER * (MAXLEN) TEXT
          CHARACTER * 80 OLD, NEW
C
C READ TEXT AND SET CURLEN
          CALL ENTER(TEXT, MAXLEN, CURLEN)
C
C PROCESS EACH EDIT REQUEST
          READ '(A)', OLD
          WHILE (OLD .NE. '***') DO
              OLDLEN = GETLEN(OLD)
              READ '(A)', NEW
              NEWLEN = GETLEN(NEW)
              CALL REPLAC(TEXT, MAXLEN, CURLEN, OLD, OLDLEN,
     Z                  NEW, NEWLEN)
              READ '(A)', OLD
          ENDWHILE
```
The data cards should begin with TEXT (terminated by a $)
There should be a pair of data cards for each edit request. Each card should contain a string starting in column 1 as shown next.
a) FRAC
STRUC
b) IN
ON
c) BOOK
TEXT
d) AMI
AMMI
e) GRR
GR

*Chapter 10*

10.1  a)
```
      JSUM = 0
      DO 10 I = 1, 50
          JSUM = JSUM + ENRANK (I, 3, 3)
   10 CONTINUE
```

b)
```
      SSUM = 0
      DO 10 J = 1, 5
          SSUM = SSUM + ENRANK (25, J, 2)
   10 CONTINUE
```

c)
```
      CLTOT = 0
      DO 10 I = 1, 50
          CRSSUM = 0
          DO 20 K = 1, 4
             CRSSUM = CRSSUM + ENRANK (I, 1, K)
   20     CONTINUE
          PRINT*, 'NO. OF STUDENTS IN COURSE', I,
   Z         'AT CAMPUS 1 = ', CRSSUM
          CLTOT = CLTOT + CRSSUM
   10 CONTINUE
      PRINT*, 'TOTAL NUMBER OF STUDENTS AT CAMPUS 1 = ',
   Z CLTOT
```

d)
```
      TOTSUM = 0
      DO 10 J = 1, 5
          CAMSUM = 0
          DO 20 I = 1, 50
             DO 30 K = 3, 4
                 CAMSUM = CAMSUM + ENRANK (I, J, K)
   30        CONTINUE
   20     CONTINUE
          PRINT*, 'NO. OF UPPER-CLASS PERSONS AT CAMPUS',
   Z         J, '=', CAMSUM
          TOTSUM = TOTSUM + CAMSUM
   10 CONTINUE
      PRINT*, 'TOTAL NO. OF UPPER-CLASSPERSONS AT ALL',
   Z ' CAMPUSES = ', TOTSUM
```

10.2
```
      LOGICAL FUNCTION SAME (VAR1, VAR2, VAR3)
      CHARACTER * 1 VAR1, VAR3, VAR2
      INTEGER BLANK
      PARAMETER (BLANK = ' ')
      SAME = .FALSE.
      IF (VAR1 .EQ. VAR2 .AND. VAR2 .EQ. VAR3 .AND. VAR3
   Z    .NE. BLANK) SAME = .TRUE.
      RETURN
      END
```

10.3 a) Prints only elements on the diagonal starting with TABLE(1,1).

     b) Row number determines the number of row elements printed; e.g., from row 1 only element TABLE(1,1) is printed; for row 2 only elements TABLE(2,1) and TABLE(2,2) are printed; for row M all elements are printed: TABLE(M,1), TABLE (M,2),..., TABLE(M,M).

     c)

| Row 1 | All elements |
|-------|--------------|
| Row 2 | TABLE(2,2),..., TABLE(2,M) |
| Row 3 | TABLE(3,3),..., TABLE(3,M) |
| . | . |
| . | . |
| . | . |
| **Row M** | **TABLE(M,M)** |

10.4
```
      DO 10 I = 1, NUMROW
          READ*, (RMCAP(I,J), J = 1, NUMCOL)
   10 CONTINUE
```

10.5 a)
```
            SUBROUTINE TOTAL (RMCAP, NUMROW, NUMCOL, SUM)
C
C COMPUTE THE TOTAL NUMBER OF REMAINING SEATS
C
C ARGUMENT DEFINITIONS --
C    INPUT ARGUMENTS
C       RMCAP - ROOM CAPACITY ARRAY
C       NUMROW - NUMBER OF ROWS IN THE ARRAY
C       NUMCOL - NUMBER OF COLUMNS IN THE ARRAY
C
C    OUTPUT ARGUMENTS
C       SUM - TOTAL NUMBER OF AVAILABLE SEATS LEFT
C
        INTEGER NUMROW, NUMCOL, RMCAP(NUMROW,NUMCOL),
       Z SUM
C
C    COMPUTE SUM
        SUM = 0
        DO 20 I = 1, NUMROW
           DO 10 J = 1, NUMCOL
              SUM = SUM + IABS(RMCAP(I,J))
   10      CONTINUE
   20 CONTINUE
C
        RETURN
        END
```

b)
```
          SUBROUTINE PRTCAP (RMCAP, NUMROW, NUMCOL)
C
C PRINT THE CONTENTS OF THE ROOM CAPACITY ARRAY
C
C ARGUMENT DEFINITIONS --
C   INPUT ARGUMENTS
C     RMCAP - ROOM CAPACITY ARRAY
C     NUMROW - NUMBER OF ROWS IN THE ARRAY
C     NUMCOL - NUMBER OF COLUMNS IN THE ARRAY
C
          INTEGER NUMROW, NUMCOL
          INTEGER RMCAP(NUMROW, NUMCOL)
C
C LOCAL VARIABLES
          INTEGER ROW, COL
C
C PRINT THE ROOM CAPACITY ARRAY
          PRINT '(A)', '0      ROOM NUMBER'
          PRINT '(A)', ' FLOOR  01  02  03  04  05'
          DO 10 ROW = 1, NUMROW
              PRINT '(A,615)', '0', ROW,
     Z          (IABS(RMCAP(ROW,COL)), COL = 1, NUMCOL)
     10   CONTINUE
C
          RETURN
          END
```

10.8    Use function GETLEN to find the actual length, ACTLEN, of XLIST. The implied DO loop in Fig 10.23 should use ACTLEN as its limit parameter instead of 17.

# INDEX

# ABOUT THE AUTHORS

Frank L. Friedman is an Associate Professor of Computer and Information Sciences at Temple University. Formerly, he was an instructor in mathematics at Goucher College, Towson, Maryland. Dr. Friedman did his undergraduate work at Antioch College and received Master's degrees from Johns Hopkins University and Purdue University. He was awarded the Ph.D. in Computer Sciences from Purdue University in 1974.

Elliot B. Koffman is a Professor of Computer and Information Sciences at Temple University, Philadelphia. He has also been an Associate Professor in the Electrical Engineering and Computer Science Department at the University of Connecticut. Dr. Koffman received his Bachelor's and Master's degrees from the Massachusetts Institute of Technology and earned his Ph.D. from Case Institute of Technology in 1967.

The authors also collaborated on a textbook titled *Problem Solving and Structured Programming in BASIC.*